BACTERIOCINS OF LACTIC ACID BACTERIA

FOOD SCIENCE AND TECHNOLOGY

International Series

SERIES EDITOR

Steve L. Taylor
University of Nebraska

ADVISORY BOARD

John E. Kinsella
University of California, Davis

Douglas Archer
FDA Washington, D.C.

Jesse F. Gregory, III
University of Florida

Susan K. Harlander
University of Minnesota

Daryl B. Lund
Rutgers The State University of New Jersey

Barbara O. Schneeman
University of California, Davis

Robert Macrae
University of Hull, United Kingdom

A complete list of all the books in this series appears at the end of this volume.

BACTERIOCINS OF
LACTIC ACID BACTERIA

EDITED BY

Dallas G. Hoover
Department of Food Science
University of Delaware
Newark, Delaware

Larry R. Steenson
Raskas Foods, Inc.
St. Louis, Missouri

ACADEMIC PRESS, INC.
A Division of Harcourt Brace & Company
San Diego   New York   Boston   London
Sydney   Tokyo   Toronto

This book is printed on acid-free paper. ∞

Copyright © 1993 by ACADEMIC PRESS, INC.
All Rights Reserved.
No part of this publication may be reproduced or transmitted in any form or by any means, electronic or mechanical, including photocopy, recording, or any information storage and retrieval system, without permission in writing from the publisher.

Academic Press, Inc.
1250 Sixth Avenue, San Diego, California 92101-4311

United Kingdom Edition published by
Academic Press Limited
24-28 Oval Road, London NW1 7DX

Library of Congress Cataloging-in-Publication Data

Bacteriocins of lactic acid bacteria / edited by Dallas G. Hoover.
  Larry R. Steenson.
       p. cm. – (Food science and technology)
    Includes bibliographical references and index.
    ISBN 0-12-355510-8
    1. Bacteriocins. 2. Lactic acid bacteria. I. Hoover, Dallas G.
  II. Steenson, Larry R. III. Series
    QR92.B3B35   1993
    664'.0287–dc20   92-41625
                                                                    CIP

PRINTED IN THE UNITED STATES OF AMERICA
93 94 95 96 97 98   MM   9 8 7 6 5 4 3 2 1

DEDICATION

To Pat, Alison, and Ben, and my mother and father. (DGH)

ACKNOWLEDGMENTS

I wish to thank Mary Ann Bagnatori for assistance in word processing and Reba Lytle and Dennis Collins for additional help and support. (DGH)

Contents

Contributors   xv

Foreword   xvii

Preface   xix

CHAPTER I

Antimicrobial Proteins: Classification, Nomenclature, Diversity, and Relationship to Bacteriocins
THOMAS J. MONTVILLE AND ALAN L. KAISER

- I. Introduction   1
    - A. Current Interest in Bacteriocins   1
    - B. Bacteriocins Defined   2
    - C. Nomenclature   5
    - D. Occurrence of Bacteriocins   6
- II. Colicins   7
    - A. General Features   7
    - B. Mode of Action   8
    - C. Immunity   10
- III. Killer Toxins, Yeast Antimicrobial Proteins   10
- IV. Thionins, Plant Antimicrobial Proteins   11
- V. Defensins, Animal Antimicrobial Proteins   14
- VI. Conclusions   16
    - References   17

CHAPTER 2

Screening Methods for Detecting Bacteriocin Activity
DALLAS G. HOOVER AND SUSAN K. HARLANDER

I. Historical Perspective   23
II. Agar Diffusion Techniques   24
     A. Introduction   24
     B. Plating Methods   27
     C. Media Composition and Conditions of Incubation   29
     D. Adaptations of Agar Diffusion Methods Used with Lactic Acid Bacteria   31
III. Liquid Media   35
IV. Titration of Bacteriocins: Critical Dilution Method   36
V. Survivor Counts   37
     References   37

CHAPTER 3

Biochemical Methods for Purification of Bacteriocins
PETER M. MURIANA AND JOHN B. LUCHANSKY

I. Introduction   41
II. Detection and Assay of Bacteriocin Activity   42
III. Production of Bacteriocins   42
     A. Production on Agar   42
     B. Production in Broth   44
IV. Bacteriocin Purification   48
     A. Biochemical Methods   48
     B. Purified LAB Bacteriocins   50
V. Applications of Purified Bacteriocins   53
     A. Biopreservation Systems   53
     B. Characterization of Physical Properties   55
     C. Immunological Studies   55
     D. Protein Sequence Determinations and "Reverse Genetics"   56
VI. Summary   56
     References   57

CHAPTER 4

Applications and Interactions of Bacteriocins from Lactic Acid Bacteria in Foods and Beverages
MARK A. DAESCHEL

- I. Introduction 63
- II. Using Bacteriocinogenic Lactic Acid Bacteria and Bacteriocins to Control Food-Borne Pathogens 66
    - A. Sensitivity of *Listeria monocytogenes* to LAB Bacteriocins 66
    - B. Sensitivity of *Clostridium botulinum* to LAB Bacteriocins 69
- III. Using Bacteriocinogenic Lactic Acid Bacteria and Bacteriocins to Direct Food and Beverage Fermentations 70
- IV. Factors Affecting the Efficacy of Bacteriocins in Foods and Beverages 76
    - A. Interactions with Food Components 76
    - B. Enhancement of Bacteriocin Activity 80
    - C. Resistance of Target Microorganisms to Bacteriocins 82
- V. Assays for Bacteriocins in Foods 84
- VI. Regulatory (United States) and Safety Considerations 85
    - References 86

CHAPTER 5

The Molecular Biology of Nisin and Its Structural Analogues
J. NORMAN HANSEN

- I. An Historical Perspective of Nisin 93
- II. Significance of Posttranslationally Modified Peptides 95
- III. Lantibiotics Could Be Adapted to Multiple Purposes 97
- IV. A Dilemma Posed by Nisin Resistance 98
- V. The Molecular Biology of Nisin Biosynthesis Is of Unknown Complexity 100
- VI. Cloning of the Genes for the Nisin and Subtilin Precursor Peptides 100
- VII. Evolutionary and Functional Relationships between Nisin and Subtilin Implied by Comparison of Their Structural Genes 101

VIII. Expression of the Genes for Nisin and Subtilin and Characterization of Their Transcripts   103
IX. The Ability to Produce Subtilin Can Be Transferred among Strains of *Bacillus subtilis*   105
X. The Ability to Produce Nisin Can Be Transferred between Strains of *Lactococcus lactis*   108
XI. What Is Known about the Organization of Genes Associated with Nisin Biosynthesis   109
XII. What Is Known about the Organization of Genes Associated with Subtilin Biosynthesis   110
XIII. Strategies and Systems to Express the Structural Genes for Nisin and Other Lantibiotics   112
XIV. Processing of Chimeric Precursor Peptides   114
XV. Production of Natural and Engineered Nisin Analogues in *Bacillus subtilis*   115
XVI. Structural and Functional Analysis of Lantibiotic Analogues: The Dehydro Residues Provide a Window through Which the Chemical State of Nisin and Subtilin Can Be Observed   116
XVII. Conclusions and Future Prospects   118
References   119

CHAPTER 6

Nonnisin Bacteriocins in Lactococci: Biochemistry, Genetics, and Mode of Action

JAN KOK, HELGE HOLO, MARCO J. VAN BELKUM, ALFRED J. HAANDRIKMAN, AND INGOLF F. NES

I. Summary   121
II. Introduction   122
III. Nomenclature   122
IV. Diplococcin   123
   A. Identification and Purification of Diplococcin   123
   B. Effects of Diplococcin on Bacterial Cells   124
   C. Plasmids and Diplococcin Production   124
V. Lactostrepcins   125
   A. Identification and Definition of Lactostrepcin   125

       B. Purification and Mode of Action of Lactostrepcin 5   125
       C. Plasmid Curing and Loss of Las 5 Production   126
VI. Lactococcins   127
       A. Genetics of Lactococcins   127
       B. The Lactococcin A Secretion Machinery   134
       C. Biochemistry of Lactococcin A   137
       D. Mode of Action of Lactococcin A   141
VII. Conclusions and Future Prospects   144
       References   146

CHAPTER 7

Molecular Biology of Bacteriocins Produced by Lactobacillus
T. R. KLAENHAMMER, C. FREMAUX, C. AHN, AND K. MILTON

I. Introduction   151
II. Evidence and Roles   152
III. Classification and Biochemical Characteristics   156
IV. Genetic Organization of Bacteriocin Operons   160
       A. Helveticin J   160
       B. Lactacin F   167
V. Common Processing Sites in Peptide Bacteriocins   173
VI. Perspectives and Conclusions   175
       References   177

CHAPTER 8

Pediocins
BIBEK RAY AND DALLAS G. HOOVER

I. Introduction   181
II. Description of the Genus *Pediococcus*   181
III. Bacteriocin Activity in Pediococci   183
IV. Pediocin AcH   188
       A. Properties   188
       B. Toxicity   191
       C. Mode of Action   192

    D. Influence of Growth Conditions on Production
       of Pediocin AcH   196
    E. Genetics   197
    F. Antibacterial Effectiveness of Pediocin AcH   198
    G. Pediocin AcH in Foods   199
V. Additional Studies of Pediocins in Meat Systems   204
VI. Other Potential Applications for Pediocins   205
       References   206

CHAPTER 9

Bacteriocins from Carnobacterium *and* Leuconostoc
MICHAEL E. STILES

I. Description of the Genera *Carnobacterium* and *Leuconostoc*   211
II. Habitats and Sources of *Carnobacterium* and *Leuconostoc* Species   212
III. Bacteriocins Produced by *Carnobacterium* and *Leuconostoc* Species   213
    A. Bacteriocins Produced by *Carnobacterium* Species   214
    B. Bacteriocins Produced by *Leuconostoc* Species   214
IV. Chemical Characterization of *Leuconostoc* and *Carnobacterium* Bacteriocins   215
V. Potential for Application of Carnobacteriocins or Leucocin A in Meat Preservation   216
       References   217

CHAPTER 10

Bacteriocins from Dairy Propionibacteria and Inducible Bacteriocins of Lactic Acid Bacteria
SUSAN F. BAREFOOT AND DALE A. GRINSTEAD

I. Bacteriocins from Dairy Propionibacteria   219
    A. Introduction to the Propionibacteria   219
    B. Nonbacteriocin Inhibitors Produced by the Propionibacteria   220
    C. Propionibacteria Bacteriocins   220

D. Genetics of Propionibacteria and Their Bacteriocins   223
II. Inducible Bacteriocins of Lactic Acid Bacteria   226
    References   228

CHAPTER 11

Regulatory Aspects of Bacteriocin Use
SUSAN K. HARLANDER

I. Introduction   233
II. Characteristics of Bacteriocins   234
III. Potential Uses of Bacteriocins or Bacteriocin-Producing Organisms in Foods and Pharmaceuticals   235
    A. Food Preservatives   235
    B. Health Care Products   235
    C. Starter Cultures   235
    D. Genetically Engineered Starter Cultures   236
    E. Probiotic Organisms   236
    F. Markers for Food-Grade Cloning Vector Construction   236
IV. Factors Affecting Regulatory Approval of Bacteriocins as Food Ingredients   237
    A. Are Bacteriocins Food Additives or GRAS Ingredients?   237
    B. Data Required by FDA   238
V. Factors Affecting Regulatory Approval of Naturally Occurring Bacteriocin-Producing Strains   239
VI. Factors Affecting Regulatory Approval of Genetically Engineered Bacteriocin-Producing Strains   240
VII. Players in the Regulatory Arena   243
    A. U.S. Food and Drug Administration   243
    B. The International Food Biotechnology Council   244
    C. Individual States   244
    D. The President's Council on Competitiveness   244
    E. Others   245
VIII. Future Challenges   245
    A. Consumer Acceptance   245
    B. Competition in the Marketplace   246
    C. Labeling of Biotechnology-Derived Foods   246
IX. Conclusion   246
    References   246

CHAPTER 12

Future Prospects for Research and Applications of Nisin and Other Bacteriocins
WILLEM M. DE VOS

    I. From Past to Present   249
   II. Research Approaches and Incentives   251
  III. Genetic Engineering   252
  IV. Traditional and Molecular Screening   256
   V. Protein Engineering   258
 VI. Concluding Remarks   261
        References   262

Index   267

Contributors

Numbers in parentheses indicate the pages on which the authors' contributions begin.

Cheol Ahn (151), Department of Applied Biology and Technology, Kangwon National University, Chuncheon, Kangwondo, Korea 200–701

Susan F. Barefoot (219), Departments of Food Science and Microbiology, Clemson University, Clemson, South Carolina 29634

Mark A. Daeschel (63), Department of Food Science and Technology, Oregon State University, Corvallis, Oregon 97331

C. Fremaux (151), Department of Food Science, North Carolina State University, Raleigh, North Carolina 27965

Dale A. Grinstead (219), Department of Food Science, Clemson University, Clemson, South Carolina 29634

Alfred J. Haandrikman (121), Department of Genetics, University of Groningen, 9751 NN Haren, Groningen, The Netherlands

J. Norman Hansen (93), Department of Chemistry and Biochemistry, University of Maryland, College Park, Maryland 20742

Susan K. Harlander (23, 233), Dairy Foods Research and Development, Land O'Lakes, Inc., Arden Hills, Minnesota 55440

Helge Holo (121), Laboratory of Microbial Gene Technology, NLVF, Agricultural University, N-1432 Ås, Norway

Dallas G. Hoover (23, 181), Department of Food Science, University of Delaware, Newark, Delaware 19716

Alan L. Kaiser (1), Joint Graduate Program in Microbiology and Molecular Genetics, New Jersey Agricultural Experiment Station, Rutgers, The State University of New Jersey, New Brunswick, New Jersey 08903-0231

Todd R. Klaenhammer (151), Department of Food Science, North Carolina State University, Raleigh, North Carolina 27965

Jan Kok (121), Department of Genetics, University of Groningen, 9751 NN Haren, Groningen, The Netherlands

John B. Luchansky (41), Food Microbiology and Toxicology, Food Research Institute, University of Wisconsin, Madison, Wisconsin 53706

K. Milton (151), Department of Food Science, North Carolina State University, Raleigh, North Carolina 27965

Thomas J. Montville (1), Department of Food Science and Joint Graduate Program in Microbiology and Molecular Genetics, Rutgers, The State University of New Jersey, New Brunswick, New Jersey 08903-0231

Peter M. Muriana (41), Department of Food Science, Purdue University, West Lafayette, Indiana 47907

Ingolf F. Nes (121), Laboratory of Microbial Gene Technology, NLVF, Agricultural University, N-1432 Ås, Norway

Bibek Ray (181), Department of Animal Science, University of Wyoming, Laramie, Wyoming 82071

Michael E. Stiles (211), Department of Food Science, University of Alberta, Edmonton AB, Alberta, Canada T6G 2P5

Marco J. van Belkum (121), Department of Genetics, University of Groningen, 9751 NN Haren, Groningen, The Netherlands

Willem M. de Vos (249), Department of Biophysical Chemistry, Netherlands Institute for Dairy Research (NIZO), and Department of Microbiology, Wageningen Agricultural University (AUW), 6710 BA Ede, The Netherlands

Foreword

Bacteria exist in complex environments where the continuous cycles of microbial life and death are not visible or obvious. These environments are nutritionally rich and will support the growth of many microbes, but the lactic acid bacteria often dominate because their fermentative metabolism and tolerance of acidic conditions create an advantage. The ability of any bacterium to compete and survive is dependent on a multitude of intrinsic and extrinsic factors, not the least of which is the production of antimicrobial substances that antagonize other microorganisms competing for the same ecological niche.

Antimicrobial proteins and peptides produced by bacteria, termed bacteriocins, are widely acknowledged to be important contributors to those organisms that survive, dominate, or die in microbial ecosystems such as our food supply or digestive tract. Bacteriocins are produced by virtually every known bacterial species and they usually antagonize only those species that occupy the same ecological niche or those that are closely related to the producer organism. Notable exceptions include the lanthionine-containing peptides such as nisin and subtilin, which exhibit bactericidal activity against gram-positive bacteria in general. Because of their long-term and beneficial association with humans and their food supply, there has been increasing interest in examining the lactic acid bacteria for natural protein antimicrobials that could extend the shelf life and improve the margin of safety in our foods. The multitude of bacteriocins available in these bacteria offer great promise that new and improved antimicrobials and preservation systems can be defined.

This book describes a number of well-characterized bacteriocins and, where possible, discusses practical applications for those that have been defined thus far from the lactic acid bacteria. It is foreseeable that, as our knowledge progresses, there will be opportunities to better utilize existing bacteriocins as preservatives and to design improved antimicrobials using genetic and protein engineering. Perhaps we will be able to exert more control over which organisms are slated to die. Then again, we should expect that bacterial adaptation will raise various defenses, such as immunity or resistance, as a means of surviving the microbial killing fields. Then again, maybe combinations of various treatments or antimicrobials could be used to maximize bactericidal activities while minimizing adaptation responses. Then again, then again—but what comes next, if one looks beyond the microbial killing fields?

Over the past decade many fundamental discoveries have been made using bacteriocins as model systems for biochemical and genetic studies. Study of the *ColE1* replicon has elucidated mechanisms of plasmid DNA replication and provided a foundation for the development of sophisticated cloning and expression vectors. Bacteriocin release proteins have been used to direct the excretion of heterologous proteins from recombinant bacteria. Pore-forming colicins and gramicidins generate ion channels that disrupt the electrochemical gradient across bacterial membranes. The discovery of common mechanisms by which large and small bacteriocins act on the cytoplasmic membrane has greatly expanded our understanding of membrane bioenergetics. These notable examples illustrate the enormous potential for contributions in microbial genetics, physiology, and metabolism through investigation of bacteriocin systems. In lactic acid bacteria, the discovery and characterization of bacteriocins and their practical evaluation as bactericidal agents are the current order of the day. But as we look beyond, investigation of bacteriocin systems in lactic acid bacteria will open exciting new research areas and fuel important discoveries about the physiology and genetics of the bacteria which have served humankind for centuries. In light of such developments, one might suspect that the benefits of lactic acid bacteria may far exceed those currently envisioned in the microbial killing fields.

Todd R. Klaenhamer

Preface

Bacteriocins have been known and studied for approximately 65 years. These proteinaceous compounds are commonly produced by a wide variety of bacteria and have counterparts among eukaryotic organisms. In the United States, interest has swelled in recent years, due to the realization that bacteriocins from lactic acid bacteria could be used as additives or "natural preservatives" in food. In 1988, nisin was finally approved for use as a food preservative in the United States after 40 years of commerical use in other countries. (Nisin is presently limited to pasteurized, processed cheese spreads in the United States.) The approval of nisin in food by the FDA has set the stage for possible commercial use of other antibiotic-like compounds in American foods and beverages. Research on nisin and other bacteriocins from lactic acid bacteria has shown that such pervasive and persistant food-borne pathogens as *Listeria monocytogenes* and *Salmonella* species can be specifically inhibited or killed by these compounds. These shelf-life enhancers are now being actively investigated for their use in foodstuffs and personal products. This interest has been the driving force behind this book.

At the 1990 Annual Meeting of the Institute of Food Technologists in Dallas, we co-chaired a symposium on bacteriocins from lactic acid bacteria. It was from that symposium that the seed of this book germinated. Most of the symposium speakers agreed to contribute chapters to this book, and other authors were contacted later to add to areas needed in a book of this magnitude. We feel that this book is the most comprehensive treatise on lactic acid bacteriocins to date.

The subject matter is presented in detail and depth as state of the art in each chapter. Those with interest in the potential for industrial use of bacteriocins as preservative ingredients should benefit from the information contained in this book. Anyone with interest in lactic acid bacteria or the biosynthesis, regulation, and mechanisms of inhibition of these proteinaceous compounds will appreciate the material presented. Such individuals include food scientists, microbiologists, food processors and product development personnel, molecular biologists, bacterial geneticists and physiologists, food toxicologists, and food and personal product regulators.

The book is divided into an introductory overview of naturally occurring antibacterial compounds; methods of detecting bacteriocins and biochemical procedures for extraction and purification; genetics and cellular regulation of bacteriocins (especially nisin); descriptions of bacteriocins based on the genera of lactic acid bacteria *Lactococcus*, *Lactobacillus*, *Pediococcus, and Leuconostoc,* and related bacteria such as *Carnobacterium* and *Propionibacterium*; the regulatory and political aspects for commercial use of these substances; and a concluding chapter on the prognosis for the future of this dynamic area.

Dallas G. Hoover
Larry R. Steenson

CHAPTER 1

Antimicrobial Proteins: Classification, Nomenclature, Diversity, and Relationship to Bacteriocins

THOMAS J. MONTVILLE
ALAN L. KAISER

I. Introduction

A. Current Interest in Bacteriocins

Minimally processed refrigerated foods of extended durability (Notermans et al., 1990) satisfy consumer demand for foods that are "fresh," "natural," and "preservative free" (Lechowich, 1988). Unfortunately, the advent of these foods has coincided with the discovery of psychrotrophic pathogens such as *Listeria monocytogenes* that grow at refrigeration temperatures (Palumbo, 1986; 1987). The growth of psychrotrophic pathogens and the potential for botulinal growth under temperature abuse conditions cast a cloud of doubt over the safety of minimally processed refrigerated foods (Notermans et al., 1990).

Because bacteriocins are proteins and "natural," there is tremendous interest in their use as a novel means to ensure the safety of minimally processed refrigerated foods. Since bacteriocins are biological they may be considered a "biological technology" with potential for use in food systems. This "biotechnology" can be implemented at a relatively low technological level in a variety of food systems described elsewhere in this volume. The

TABLE I
Early Papers Reporting Bacteriocin Inhibition
of *Listeria monocytogenes*

Ahn and Stiles, 1990	*Carnobacterium piscicola*
Berry et al., 1990	Pediococci
Bhunia et al., 1988	*Pediococcus acidilactici*
Carminati et al., 1989	*Lactococcus lactis*
Harding and Shaw, 1990	*Leuconostoc gelidum*
Harris et al., 1989	Lactic acid bacteria
Hoover et al., 1989	Pediococci
Lewus et al., 1991	Lactic acid bacteria
Nielsen et al., 1990	*Pediococcus acidilactici*
Pucci et al., 1988	*Pediococcus acidilactici*
Raccach et al., 1989	Lactic acid bacteria
Schillinger and Lucke, 1989	*Lactobacillus sake*
Spelhaug and Harlander, 1989	*Lactococcus lactis* and *Pediococcus pentosaceus*

tremendous resurgence of interest in antimicrobial proteins is undoubtedly due in part to reports of antilisterial activity in many such proteins. Between 1988 and 1991 more than a dozen papers have described various lactic acid bacteria which inhibit or kill *L. monocytogenes* in test tube and media systems (Table I).

This chapter defines exactly what bacteriocins are (and what they are not), demonstrates the similarities of antimicrobial proteins found throughout the biosphere, and thus provides the context for the more detailed discussion of bacteriocins of lactic acid bacteria found elsewhere in this volume. Because there are already many excellent reviews on the bacteriocins of lactic acid bacteria (Klaenhammer, 1988; Daeschel, 1989; 1990; Schillinger, 1990) and extensive information in other chapters, they will not be discussed here. The following review is not meant to be an exhaustive survey of the literature but an introduction that will serve as a point of departure for further reading.

B. Bacteriocins Defined

1. Classical Definition

As a group, bacteriocins share a number of characteristics. Based on extensive studies of the colicins produced by Gram-negative bacteria, Tagg et al. (1976) suggested six criteria by which to characterize the antimicrobial proteins produced by Gram-positive organisms. Although studies on the antimicrobial proteins produced by lactic acid bacteria frequently cite these criteria (Ahn and Stiles, 1990; Barefoot and Klaenhammer, 1983; Bhunia et al., 1990; Daeschel et al., 1990; Lewus et al., 1991; Lyon and Glatz, 1991; McCormick and Savage, 1983; McGroarty and Reid, 1988; Okereke and

Montville, 1991; Spelhaug and Harlander, 1989), closer reading reveals that these should not be used as inflexible criteria for Gram-positive organisms. It has become increasingly clear that few antimicrobial proteins fit all six criteria. Thus, each of the six criteria is considered separately below.

Bacteriocins must be proteins. This is the one absolute characteristic of bacteriocins and is usually demonstrated by protease negation of the putative bacteriocin's antimicrobial activity. However, since peptides can also be inactivated by proteases, there is some debate about how large a peptide must be before it becomes a protein. Is, for example, the decapeptide gramacidin a protein? We would suggest that if the polymer is made ribosomally via the transcription of a unique gene that is subsequently translated into protein, then it is a protein. If, however, the polymer is made enzymatically by the condensation of amino acids (as is the case for gramacidin), it should be considered a peptide. In the case of larger molecules, inactivation of bacteriocin activity by one or more proteases is sufficient proof that a microbial inhibitor is proteinaceous and therefore, a bacteriocin. Protein synthesis inhibitors should be used to determine whether small amino acid polymers are proteins or peptides.

The criteria suggest that bacteriocins are bactericidal, not just bacteriostatic. While initial characterizations suggested that many bacteriocins are bactericidal, as early as 1977 Lipinska noted that nisin is a weak antimicrobial and is sporostatic or bacteriostatic. Many bacteriocins initially characterized as bactericidal in model systems have later been shown to be static in food applications (Berry et al., 1990; Nielsen et al., 1990; Pucci et al., 1988; Schillinger et al., 1991; Winkowski and Montville, 1992). Since many widely used microbial inhibitors inhibit the growth of pathogens without killing them, and the cidal or static mode of action is often a function of the system under study rather than an inherent characteristic of the molecule, this criterion is not essential to the definition of bacteriocins.

The requirement that bacteriocins have specific binding sites is presumably responsible for the specificity of a bacteriocin against specific pathogens and can be used to distinguish them from other broad spectrum antimicrobials such as organic acids and most antibiotics in general. Recent reports (Sears and Blackburn, 1992; Stevens et al., 1992) on formulations of nisin that inhibit Gram-negative organisms and the effectiveness of bacteriocins against spheroplasts and protoplasts (Andersson, 1986) again suggest that the binding sites need not be very specific.

The plasmid linkage of colicins (Pugsley and Oudega, 1987) suggested that bacteriocins are plasmid mediated. This has been found for *Pediococcus pentosaceus* (Daeschel and Klaenhammer, 1985), *Lactobacillus acidophilus* (Muriana and Klaenhammer, 1987), *Lactococcus lactis* (Steele and McKay, 1986; Kaletta and Entian, 1989), *Carnobacterium piscicola* (Ahn and Stiles, 1990), and *Pediococcus acidilactici* (Ray et al., 1989). However, Joerger and Klaenhammer (1986) have found that the helviticin J gene is located on the *Lactobacillus helveticus* chromosome. Indeed, relatively few bacteriocins produced by lactobacilli are plasmid mediated (Klaenhammer, 1991). The location of the gene that encodes a specific antimicrobial protein is often difficult to determine. For example, evidence has recently been presented that suggests the gene encoding nisin, long believed to be on a plasmid, resides on the chromosome of *Lactococcus lactis* ATCC 11454 (Steen et al., 1991). Thus, gene location is not a useful criterion for its classification as a bacteriocin.

The criterion that bacteriocins must be produced by lethal biosynthesis suggests that the cell producing the bacteriocin should be killed in the process of releasing the bacteriocin. This requirement may have been the result of an orientation toward the secondary metabolism generated by decades of research in antibiotics. However, many bacteriocins are produced as proteins during the growth phase without the lysis of the producing organism (Hurst and Dring, 1968; Lewus, 1991).

The final criterion suggests that bacteriocins are active against a narrow spectrum of closely related bacteria. While this is in many cases true, the observation that many antimicrobial proteins produced by Gram-positive bacteria are effective against many genera of Gram-positive bacteria stretches this criterion to the breaking point.

2. Bacteriocins Redefined

The exceptions to the six criteria are so numerous that Konisky (1982) concluded, and we also hold, that there are only two true requisites for bacteriocins: their proteinaceous nature and their lack of lethality to cells which produce them (i.e., they are not suicide proteins). Given the widespread usage of "bacteriocins" to describe antimicrobial proteins that fit these criteria, we believe that Tagg's proposal (Tagg, 1991) to call them BLIS (Bacteriocin-like Inhibitory Substance) proteins will not be widely accepted.

There are a number of different mechanisms by which bacteria protect themselves from the adverse effect of their own bacteriocins. One is posttranslational modification. Here, after the protein is synthesized as a prepeptide, some amino acid residues are posttranslationally processed to generate the active protein molecule. This is the case for nisin, with the generation of the thioether amino acids lanthionine and 3-methyl-lanthionine (Kaletta and Entian, 1989). Other bacteriocins process a precursor in many cases, splitting the prebacteriocin into an active bacteriocin and an immunity protein (Konisky, 1982). Most colicins work this way. The genes for the immunity protein are closely linked to the genes for bacteriocin production. Compartmentalization can also protect eucaryotic cells from the antimicrobial proteins they produce (see Section V, this chapter).

3. Bacteriocins Are Not Antibiotics

From their very genesis, bacteriocins have been referred to as "antibiotics." Reeves (1965) contrasted bacteriocins to other antibiotics by the fact that bacteriocins are proteinaceous. Since many articles refer to bacteriocins as antibiotics (Kaletta and Entian, 1989; Kordel et al., 1988; Buchman et al., 1988) and it is illegal to use antibiotics as food preservatives, the question should be asked, "Are bacteriocins antibiotics?"

The discussion is, of course, a matter of semantics. It should be framed in the context of nisin's discovery in 1924, before penicillin and colicins, early in the age of antibiotics. Because of the tremendous increase in knowledge about antibiotics during this period, it was reasonable to consider that the newly discovered protein antimicrobials were antibiotics. Technically, antibiotics are made by a restricted group of organisms through the enzymatic packaging of primary metabolites into structurally related secondary metabolites that have no apparent function in the growth of producing cells and are easily secreted from the cell. There are many peptide antibiotics made by fungi and actinomyces that meet this definition. For example, gramacidin S, bacitracin A, and polymixin are all peptide antibiotics. All are made by enzymatic condensation reactions of amino acids to package free amino acids into larger compounds. When the amino acid composition of nisin was discovered to contain several unusual amino acids such as dehydroalanine, dehydrobuterine, and single sulfur lanthione bridges, it was thought that nisin could not be a protein because ribosomes do not process these unusual amino acids. The discovery of the nisin analogues, subtilin and epidermin, further suggested families of structurally related compounds. Only recently has Hansen's elegant work (Hansen et al., 1990) proved that nisin is, in fact, made ribosomally. The dehydration of the amino acids and the lanthione ring formation occur posttranslationally. The presence of discrete genes for bacteriocin synthesis confirms that many other bacteriocins are true proteins.

An alternate approach to the semantic argument is to abandon the semantic issue and examine the reason for prohibiting the use of antibiotics as food preservatives. This prohibition is rooted in the well-justified concern that widespread use of antibiotics in the food supply might compromise the clinical efficacy of antibiotics. The dictionary includes "used to inhibit or treat infectious diseases" in its definition of "antibiotic." Bacteriocins slated for use in foods are not used to inhibit nor treat the clinical progression of an infectious disease. Indeed, this was a major consideration in 1964 when the World Health Organization approved the use of nisin in foods.

C. Nomenclature

Bacteriocin nomenclature is straightforward. Just as "ase" is used in enzyme nomenclature, the suffix "cin" is used to denote bacteriocinogenic activity. The "cin" suffix is appended to either the genus name or (more correctly) to the species name. The colicins were originally isolated from *Escherichia coli*, monocins are antimicrobial proteins made by *Listeria monocytogenes*, subtilin is produced by *Bacillus subtilis*, staphylocin by *Staphylococcus aureus*, and so on. Sequential letters assigned in the order of discovery are used after the bacteriocin name to differentiate unique bacteriocins produced by different strains of the same species. For example, lactacin F was the sixth bacteriocin reported for a lactobacilli species.

D. Occurrence of Bacteriocins

Antimicrobial proteins are produced by many pathogenic and nonpathogenic genera (Table II). The proportion of the bacteriocin-producing strains examined within a given genera ranges from 1 to 100%. Reports of 100% incidence should be examined carefully to insure that adequate controls were run to exclude the possibility of inhibition due to acid, hydrogen peroxide, or bacteriophage. Ortel (1989) reported that listeriocins are produced by 75% of the *Listeria* strains. Geis et al. (1983) reported that only 5% of the lactic streptococci (now classified as lactococci) make bacteriocins. LiPuma et al. (1990) reported that 4% of *Haemophilus influenzae* produce bacteriocins.

The results of screening procedures for the detection of bacteriocinogenic isolates are highly dependent on the screening methods, media, and organisms used to indicate bacteriocin activity. Mayr-Harting et al. (1972) provide an excellent review of the methods used to study bacteriocins. Research in our laboratory provides examples of many of these factors. When agar plates were used to examine 22 strains of lactic acid bacteria previously reported to produce bacteriocins, 19 were confirmed as bacteriocinogenic. When the same bacteriocinogenic cultures were screened in broth media, only 2 strains produced detectable antimicrobial activity (Lewus and Montville, 1991). The choice of indicator (bacteriocin-sensitive) organism used in the screening is extremely important. There is frequently a tradeoff between sensitivity and specificity. Ideally, the ultimate target organisms of the preservation system should be used as the indicators in the screening studies. For example, it was relatively easy for us to isolate a large number of strains with antilisterial activity from meat using listeria as the indictor organism during the primary screening. However, when we examined 19 bacteriocinogenic strains active against *Listeria monocytogenes* (Lewus

Table II
Bacterial Genera That Produce Bacteriocins

Acetobacter	*Haemophilus*	*Salmonella*
Actinobacillus	*Haloferax*	*Propionibacterium*
Bacillus	*Lactobacillus*	*Serratia*
Brevibacterium	*Lactococcus*	*Shigella*
Clostridium	*Listeria*	*Staphylococcus*
Corynebacterium	*Leuconostoc*	*Streptococcus*
Enterococcus	*Pseudomonas*	*Yersinia*
Erwinia	*Pediococcus*	

et al., 1991) for antibotulinal activity, only 4 of them were positive (Okereke and Montville, 1991). Culture collections are a poor source of bacteriocinogenic strains. We have found that only 1 out of 13 randomly selected American Type Culture Collection strains produced bacteriocins (Lewus and Montville, 1991).

II. Colicins

Even though the colicins, antimicrobial proteins of *E. coli*, were discovered in the mid 1920s, research in this area is still strong. The information and ideas generated by this field have been influential in the progress seen in such diverse research interests as bioenergetics, membrane translocation, protein structure and function, protein modeling, and the genetics and molecular biology of membrane proteins. Currently, biophysicists, biochemists, microbiologists, and molecular biologists use the well-defined colicin systems to answer important questions in their respective fields.

A. General Features

The colicins are a fairly large group of antimicrobial proteins produced by *E. coli* and to a lesser extent, other members of the Enterobacteriaceae family (i.e., cloacin from *Enterobacter cloacae*) of Gram-negative bacteria. These fairly large proteins (40–70 kDa) are produced by cells that, in most cases, possess a plasmid that codes for the colicin molecule as well as several other proteins involved in the mode of action of the colicin. The range of susceptible cells is generally fairly narrow but there are exceptions to this. The proteins are produced in large amounts and are secreted into the growth medium, usually during the exponential growth phase of the cell.

Studies on the three-dimensional structure of colicins indicate that the protein may be arranged as three independent structural domains, each of which carries out a separate action that leads to the death of susceptible cells (Yamada et al., 1982; Brunden et al., 1984). Limited proteolysis and deletion analysis have been used to determine the presence of these domains (de Graaf et al., 1978; Ohno-Iwashita and Imahori, 1980; Baty et al., 1988; Shiver et al., 1988). Each of the domains is associated with one of the three steps that lead to cell death. The binding of the colicin to a specific receptor on the surface of the susceptible cell is carried out by the central portion of the molecule. The next step in the mode of action of the colicin involves the translocation of the protein or a portion of it across or into the cell membrane of the susceptible cell. Evidence suggests that the N-terminal region of the molecule interacts with the translocation components of the target cell (Postle and Skare, 1988). The killing action of the colicin is then carried out by the C-terminal portion of the protein.

B. Mode of Action

The lethal action of the colicins is divided into three stages: (1) binding of the colicin to a specific cell surface receptor; (2) insertion into, or transport across the susceptible cell's membrane; and (3) killing of the cell.

Several mechanisms of bacteriocin-induced cell death are seen in the colicins. One mechanism by which this is accomplished is depletion of the proton motive force across the cell membrane. Evidence for this mechanism is also seen in other bacteriocins as well as in other antibacterial proteins (Gould and Cramer, 1977; Schein et al., 1978; Tokuda and Konisky, 1978; Carrasco et al., 1981; Martinac et al., 1990; Lewus, 1991). The second most common mechanism of cell death seen in the colicins is RNase and DNase activity within the susceptible cell. This mechanism has not been seen in other bacteriocins or any of the other antimicrobial proteins discussed in this chapter. A third, but rare, mechanism is lysis of the susceptible cells by the action of the bacteriocin at the cell membrane (Schaller et al., 1981).

The colicins, like all bacteriocins, bind to specific cell surface receptors on susceptible cells (Oudega et al., 1979; Imajoh et al., 1982). These receptors have been highly characterized and usually function in other non-colicin-related translocation pathways. Different cell surface receptors can bind several different types of molecules in many cases. For example, the receptor for colicin A is also involved in the uptake of vitamin B_{12} and has other functions. The receptor for the uptake of cloacin DF13 is also involved in the uptake of the iron chelator aerobactin (van Tiel-Menkveld et al., 1982). Many bacteriocin receptors have also been shown to be receptors for the binding of bacteriophage to cells. In all cases, these receptors are membrane proteins. Some evidence has been shown that other membrane proteins play an important role in the attachment of the colicins and other bacteriocins to susceptible cells. Oudega et al. (1977) present evidence that, in the case of cloacin DF13, the immunity protein is necessary for the binding of the bacteriocin to susceptible cells. Upon binding, the immunity protein is released into the growth medium. The receptors lack protease activity and therefore cell surface proteases in close proximity to the receptor protein must play a role in the fragmentation of the bacteriocin molecule.

Kinetic studies on the interaction of colicin molecules with cell surface receptors indicate that their mode of action follows single-hit inactivation kinetics (Wendt, 1970). A single colicin molecule is able to kill a single cell. Whether this is true for all bacteriocins is not known. Unpublished calculation from our laboratory suggest that several thousand molecules of nisin and other bacteriocins are required to kill an *L. monocytogenes* cell. It is interesting to note that the diphtheria toxin secreted by *Corynebacterium diphtheriae* also displays single-hit inactivation kinetics and has many other similarities with the colicins (Neville and Hudson, 1986).

The mechanism by which the colicin molecule inserts into and/or passes through the cell membrane of the target cell still remains a mystery. A tremendous amount of interest has been invested toward models developed for this mechanism. These models are useful in understanding protein translocation into and through cell membranes. Despite the fact that a mechanism is not known, significant steps have been made in determining the gene products that are involved in getting the colicin molecule into and/or through the cell membrane.

Many colicin molecules contain a region of positively charged amino acids that are thought to interact electrostatically with the negatively charged polar headgroups of the phospholipids of the cell membrane. This interaction occurs only when negatively charged lipids are present (Escuyer et al., 1986; Shiver et al., 1987). A model has been proposed in which large regions of alpha helixes present in many bacteriocins are allowed to spontaneously insert into the membrane as a result of this interaction (Pattus et al., 1990). These alpha-helical regions are present in the C-terminal region of the colicin molecule responsible for the lethal action of the protein. This region also frequently contains long hydrophobic sequences (van den Elzen et al., 1983; Cramer et al., 1990; Song et al., 1991). The alpha-helical conformation apparently solves the inherent conflict that colicins can exist as water-soluble proteins and as membrane-active proteins simultaneously. Whether this model would hold true for the pore-forming as well as the degradative colicins is not known. The shorter antimicrobial proteins, such as the thionins, could not act in this manner due to their size; however, some have been shown to adopt an alpha-helical conformation (Teeter et al., 1981).

Pore-forming colicins include the colicins E1, A, B, N, Ia, Ib, and K. It is well established that these colicins dissipate the membrane potential ($\Delta\psi$) of the sensitive cell and deenergize the cellular membrane (Gould and Cramer, 1977; Schein et al., 1978; Tokuda and Konisky, 1978; Weiss and Luria, 1978). This results in an increase in ion permeability resulting in the leakage of cellular K^+ into the medium. Since the proton motive force is entirely composed of the $\Delta\psi$ component at physiological pH, the colicins eliminate the sole driving force of the active transport systems of the cell. The pore-forming colicins create ion channels in artificial membranes. These channels have been demonstrated *in vitro* but no direct proof exists that pore-forming colicins form ion channels *in vivo*, although there is strong evidence to support this hypothesis (Pattus et al., 1990).

The second major group of the colicins consists of the colicins E2, E3, and cloacin DF13. These colicins, or a fragment of them, enter the cytoplasm of susceptible cells where they exhibit RNase and/or DNase activity (Bowman et al., 1971; Oudega et al., 1975; Schaller and Nomura, 1976). These colicins can therefore be considered enzymes since they possess enzymatic activity. Colicin E3 and cloacin DF13 cause inhibition of protein synthesis by making a single cleavage on the 16S RNA of the 30S ribosomal subunit near the 3' end of the molecule (de Graaf et al., 1973). These bacteriocins are known to interact with the RNA molecule directly.

Therefore, the possibility that they act in an indirect manner has been ruled out.

C. Immunity

The producing cell is conferred immunity to colicin action by the presence of an immunity protein that is cotranscribed from the plasmid with the colicin molecule in a 1:1 ratio. The immunity proteins have a molecular weight of about 10–15 kDa and serve to inhibit the action of the colicin molecule itself (Oudega et al, 1975; Krone et al., 1986). The mechanism by which this happens is not known. Immunity to a given colicin is absolutely specific. The immunity protein for one colicin type will not confer immunity to another colicin type. For example, the immunity protein of colicin E1 will not give immunity to colicins E2 or A. There is some evidence to suggest that the immunity protein may be involved in the uptake and translocation process of the colicin to the sensitive cell (Oudega et al., 1977).

III. Killer Toxins, Yeast Antimicrobial Proteins

Another group of antimicrobial proteins are the killer toxins found in yeast. These toxins are produced by many different genera of yeast and their classification by cross-linking reactions reveals that at least 11 different toxin activities exist (Young and Yagiu, 1978). Each type of activity probably represents a different toxin. Of these different toxins, the K1 and K2 toxins produced by *Saccharomyces cerevisiae* are the best understood.

The similarities between the killer toxins and the bacteriocins are clearly evident. Both are proteinaceous in nature and production is associated with immunity to the producing cell. Toxicity of both is usually restricted to cells of closely related species. Production of both proteins is dependent on carefully controlled growth conditions. Both are more stable in agar medium than in broth. The similarities between these proteins also extend to their proposed mode of action. The postulated mechanism of action for the killer toxin involves the formation of ion channels in the cell membrane with the subsequent release of ions such as K^+, Na^+, and possibly H^+ (Skipper and Bussey, 1977; Bussey, 1981; Martinac et al., 1990). This release would in turn disrupt the energized state of the cell membrane. The mechanism of some colicins involves depleting the membrane proton motive force. This deenergizes the cells. This mode of action has also been proposed for the bacteriocins of the lactic acid bacteria. In addition, in all cases, binding of the toxin or bacteriocin to the cell via a specific receptor is a prerequisite for cell death (Bussey et al., 1979).

The highly characterized K1 toxin is a 20-kDa heterodimeric protein consisting of α and β subunits linked by disulfide bonds (Bostian et al.,

I. Antimicrobial Proteins

1984; Zhu et al., 1987). The α subunit contains 103 amino acids while the smaller β subunit consists of 83 amino acids. The α subunit has been shown by Bostian et al. (1984) to possess two highly hydrophobic regions with a hydrophilic region between them. This topography would suggest that the α subunit may be responsible for the ion channel formation. The β subunit is proposed to bind to the cell wall receptor, which contains a (1-6)-β-D-glucan moiety.

Other factors besides specific receptors on the cell wall may be necessary for the action of the toxin. This can be seen in experiments in which spheroplasts of *Candida albicans* are killed by the K1 toxin but whole cells are not (Zhu and Bussey, 1989). Binding to the cell wall in this case is not enough to result in cell death. Many spheroplasts are killed by killer toxins indicating that the toxins may act nonspecifically at the level of the cell membrane. There are, however, cases in which toxins do not result in death of spheroplasts. This could indicate some processing of the toxin at the level of the cell wall, which would result in a polypeptide that could insert into the cell membrane.

The K1 toxin is only produced by cells that contain the M_1 viruslike particle (VLP). This particle contains a double-stranded RNA molecule that codes for the toxin as well as the immunity function. The VLP is not capable of extracellular transmission and is stably maintained within the yeast cell. It is transferred from generation to generation by mating or budding much as a plasmid would be. The VLPs therefore possess characteristics of both infectious double-stranded RNA viruses and plasmids of eucaryotic cells.

The K1 toxin is extremely potent. A concentration of 2.3 ng/ml is all that is necessary to kill 2.0×10^7 cells/ml (Bussey et al., 1979)! Cell death due to the toxin does not occur immediately after the addition of the toxin to the susceptible cells. During a lag period of one to three hours no lethal effects are manifested (Bussey and Sherman, 1973; de la Pena et al., 1981). This is followed by a shutdown of synthesis of macromolecules, which coincides with the loss of K^+ ion and ATP along with inhibition of amino acid uptake and proton pumping. The killing of the cells is an energy-dependent process (Skipper and Bussey, 1977). This can be demonstrated using 2,4-dinitrophenol (DNP), a proton-conducting uncoupler that acts on the energy generating processes of the cell. DNP prevents toxin-mediated K^+ ion efflux but does not prevent toxin binding to the cell wall receptor (Skipper and Bussey, 1977). As in many of the bacteriocins, cell lysis is not caused by the action of killer toxins on susceptible cells.

IV. Thionins, Plant Antimicrobial Proteins

As a group, the thionins share many structural and biological features. These low molecular weight polypeptides contain 45–47 amino acids and are usually high in cysteine, lysine, and arginine. The high content of lysine and arginine makes these proteins highly charged and basic. They also contain three or four disulfide bonds. The thionins can be divided into two loose groups of toxins based on size, disulfide bonds, and amino acid homology. Research to date reveals that these two groups do not have differences in toxicity or mode of action. In fact, despite their differences these two groups show at least about 40% homology between them.

The first group of thionins consists of the viscotoxins, phoratoxin, and crambin (Figure 1B). The viscotoxins and phoratoxin have been isolated from species of mistletoe found in Europe and America, respectively (Samuelsson, 1973). It is interesting that, historically, mistletoe extracts have been used in folk medicine for the treatment of neoplastic disorders. Crambin, on the other hand, is purified from a plant known as abyssinian cabbage or rapeseed (Teeter et al., 1981). These toxins consist of polypeptides that are 46 amino acids in length and contain three disulfide bonds. The disulfide bonds in all four of these proteins occur between residues 3 and 40, 4 and 32, and 16 and 26 (Samuelsson et al., 1968; Samuelsson and Pettersson, 1971; Mellstrand and Samuelsson, 1974; Teeter et al., 1981). The tertiary structure of crambin has been determined from secondary

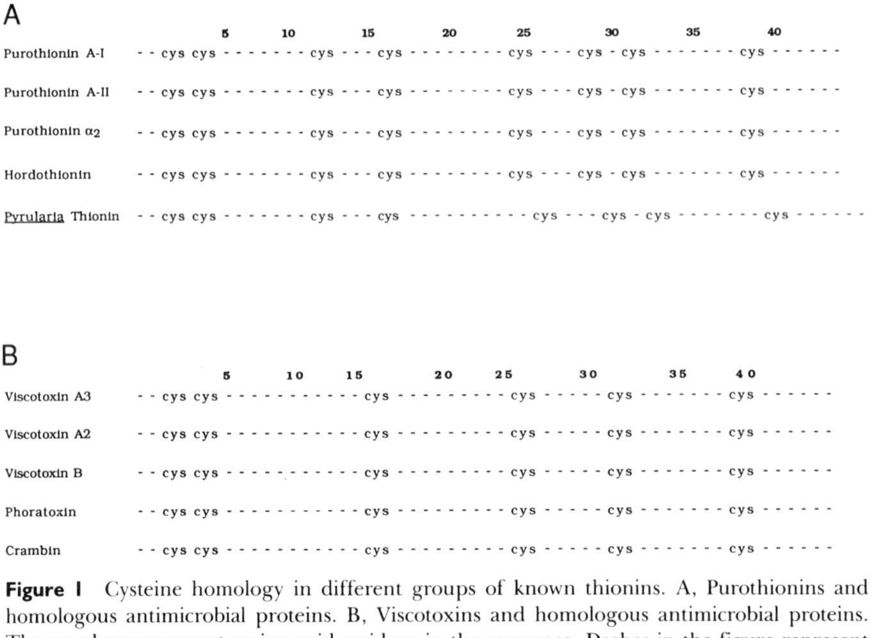

Figure 1 Cysteine homology in different groups of known thionins. A, Purothionins and homologous antimicrobial proteins. B, Viscotoxins and homologous antimicrobial proteins. The numbers represent amino acid residues in the sequence. Dashes in the figure represent noncysteine amino acid residues. Protein name, source, and references: Purothionin A-I, A-II, and α_2, *Triticum aestivum* (Manitoba No. 3 wheat), (Balls et al., 1942; Mak and Jones, 1976); Hordothionin, *Hordeum vulgare* (barley), (Ozaki et al., 1980); Pyrularia Thionin, *Pyrularia pubera* (buffalo nut), (Vernon et al., 1985); Viscotoxin A3, A2, and B, *Viscum album* (European mistletoe), (Samuelsson et al., 1968; Samuelsson and Pettersson, 1971; Samuelsson, 1973); Phoratoxin, *Phoradendron tomentosum* subsp. *macrophyllum* (American mistletoe), (Samuelsson, 1973); Crambin, *Crambe abyssinica* (Abyssinian cabbage), (Teeter et al., 1981). (Adapted from Ozaki et al., 1980; Vernon et al., 1985).

1. Antimicrobial Proteins

structure predictions and x-ray crystallography (Teeter et al., 1981). The high degree of sequence homology seen in this group of thionins may suggest similar three-dimensional structures and perhaps similar functions as well. Crambin may therefore serve as a model for tertiary structure determination in other thionins.

The second group of thionins are the purothionins and other thionins isolated from grains. The first purothionin was isolated from wheat in 1942 (Balls et al., 1942). This protein was a mixture of several highly homologous, but not identical, proteins that have now been separately isolated and designated A-I, A-II, and α_2. These thionins consist of 45 amino acids and have 4 disulfide bonds. In A-I and A-II these bonds are formed between residues 3 and 39, 4 and 31, 12 and 29, and 16 and 25 (Hase et al., 1978). An interesting exception is the thionin isolated from buffalo nut (*Pyrularia pubera*), which has 47 amino acids (Vernon et al., 1985). Except for an addition of two amino acids this protein is about 50% homologous to the purothionins and also contains four disulfide bonds (Figure 1A). The buffalo nut is a parasitic plant in the same order (Santalales) as the mistletoe but in a different family. It therefore appears to be phylogenetically related to the mistletoe but its thionin is more closely homologous to the thionins of the grains.

The function of the thionin proteins is not known at this time; however, a broad range of effects upon various cell types have been reported for the thionin proteins. It has been hypothesized that these plant proteins function in plant immunity. Data obtained by Fernandez de Caleya et al. (1972) show that many phytopathogenic bacteria are susceptible to thionins (Table III). This would seem to support this hypothesis. Thionins are cidal to a broad range of bacteria, yeast, and mammalian cells. Table IV shows the effects of thionins on a wide range of cell types. The lytic or cidal effects of several of the thionins are dependent on the protein binding to the cell membrane of susceptible cells. Kinetic data indicate that the *Pyrularia* thionin probably binds to a protein receptor on the cell membrane (Osorio e Castro et al., 1989). Vernon et al. (1985) report the leakage of K^+ and phosphate ions as well as proteins and nucleotides from thionin-treated cells. Some of the reported effects of thionins upon susceptible organisms have been shown to be secondary effects of the proteins' primary action at the cell membrane. This is the case with the phospholipase activity reported by Osorio e Castro et al. (1989). Other reported effects of thionins on cells include complete

TABLE III
Plant Pathogens Susceptible to Purothionins

Corynebacterium fascians	*Erwinia amylovora*
Corynebacterium flaccumfaciens	*Pseudomonas solanacearum*
Corynebacterium michiganense	*Xanthomonas campestris*
Corynebacterium poinsettiae	*Xanthomonas phaseoli*
Corynebacterium sepedonicum	

TABLE IV
Toxicity of *Pyrularia* Thionin to Various Cell Types

Cell type	ID$_{50}$	Reference
Human Cell Line HeLa	17 µg/ml	Evett et al., 1986
Human Cell Line MRC-5	0.46 µg/ml	Evett et al., 1986
Mouse Cell Line B16 Melanoma	3.0 µg/ml	Evett et al., 1986
Mouse Cell Line P388	0.62 µg/ml	Evett et al., 1986
Mouse Cell Line L1210	3.9 µg/ml	Evett et al., 1986
Monkey Cell Line Vero	9.6 µg/ml	Evett et al., 1986
Live Mice C57Bl/J6 (~20 g)[a]	1.5 µg/kg	Evett et al., 1986
Live Mice (~20 g)[b]	0.5 µg/kg	Samuelsson, 1973
Human Type B RBCs	≥10 µg/ml	Lankisch and Vogt, 1971
Micrococcus luteus	≥125 µg/ml	Fernandez de Caleya et al., 1972

[a] 1.0 mg injected intraperitoneally caused death in 7 min. 0.1 mg injected intraperitoneally caused death in 2 hr.
[b] Reported values are for viscotoxin.

inhibition of sugar incorporation into cells (Okada and Yoshizumi, 1973). Purothionins and viscotoxins inhibit protein synthesis and allow the leakage of Rb$^+$ ions and nucleotides from the cell (Carrasco et al., 1981). Many of these same effects are also seen in the colicins' mode of action. Wada et al. (1982) reported the loss of activity in purothionin by iodination of the tyrosine at position 13. All of the thionins mentioned in this review have a tyrosine at position 13 except crambin. Crambin has been shown by Teeter et al. (1981) to be less toxic than the purothionins. They accounted for this by explaining that the hydrophobicity of crambin may prevent it from getting to its site of activity.

V. Defensins, Animal Antimicrobial Proteins

The tremendous interest in "natural" food ingredients brings this treatise to the defensins, the ultimate in "natural peptide antibiotics" (Lehrer et al., 1990). Elsbach (1990) has called defensins "antibiotics from within." They are proteins made as a special class of human neutrophil proteins in phagocytes. The daily production of defensins in humans is 5–10 mg/kg of body weight, about the same dosage as clinically administered amino glycoside antibiotics.

Figure 2 demonstrates the process by which an infectious bacterium is phagocytized and killed. First, the leukocyte approaches the bacterium and engulfs it through phagocytosis. Simultaneously, thousands of granules containing lipid-encapsulated defensins migrate to and are fused with the membrane of the phagocytic vesicle. Once the bacterium has been successfully phagocytized, the neutrophil has a burst of respiratory activity. The "respiratory burst" produces superoxides, hydrogen peroxides, super-

1. Antimicrobial Proteins

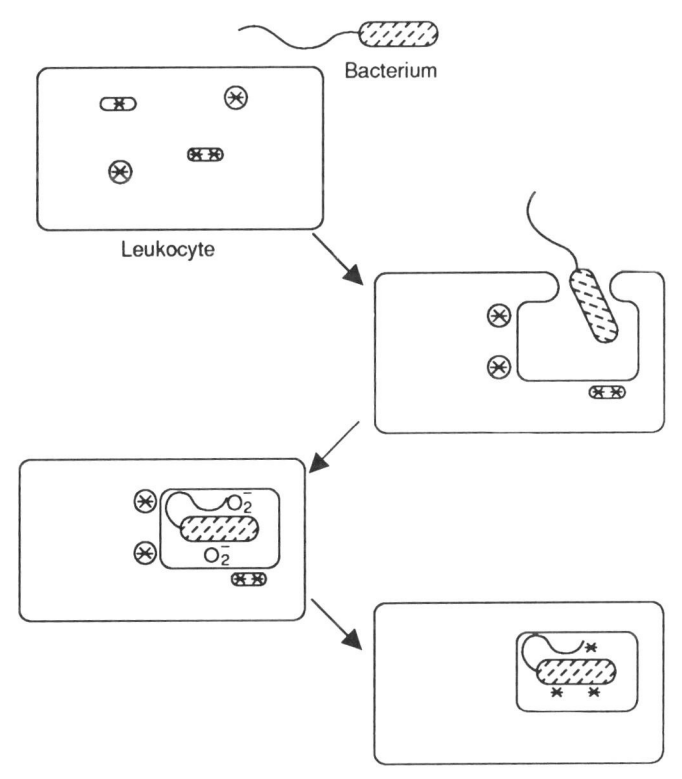

Figure 2 Phagocytosis and destruction of an infectious bacterium by a leukocyte. *, Defensin molecule.

anions, and other bactericidal compounds with known mechanisms of action. The defensins act as part of a "respiratory burst independent" mechanism. The defensin-containing granules fuse to the vesicle membrane and release the defensins into the vesicle lumen. The concentration of defensins inside these vesicles is 1–5 mg/liter. The defensins then target the membrane of the phagocytized pathogen and kill it (Thomas et al., 1988).

The nine different human defensins that have been discovered so far have similar amino acid sequences. Ganz et al. (1990) and Lambert et al. (1989) have reported the amino acid sequences of antimicrobial proteins produced by human, rat, rabbit, mice, guinea pig, and insect cells (Figure 3). Their similarity to plant antimicrobial proteins is also obvious. All of these proteins have molecular weights greater than 4000 Da, are cysteine rich, and share highly conserved amino acid sequences. The cysteines form disulfide bonds that give rise to a cyclic and compact tertiary structure. The disulfide pairs of the human defensin HNP-2 are known to extend between cysteines 1 and 2, 3 and 18, and 8 and 28. The high percentages of glycine are structurally important and result in extensive β-pleated sheeting. Joerger and Klaenhammer (1990) have also suggested that helveticin A has extensive β-pleated sheets at its amino terminus. Hill et al. (1991) resolved

Human	- cys	- cys	- - - - cys	- - - - - - - - - cys	- - - - - - - cys	cys
Rat	- - cys	- cys	- - - - cys	- - - - - - - - - cys	- - - - - - - cys	cys -
Rabbit	- - cys	- cys	- - - - cys	- - - - - - - - - cys	- - - - - - - cys	cys - -
Mouse	- cys	- cys	- - - - cys	- - - - - - - - - cys	- - - - - - - cys	cys -
Guinea Pig	- - cys	- cys	- - - - cys	- - - - - - - - - cys	- - - - - - - cys	cys
Insect	- - cys	- cys	- - - - - cys	- - - - - - - - - cys	- - - cys - - - - - - - - - - - cys	- -

Figure 3 Conservation of cysteine position in defensins and different sources. Dashes represent noncysteine amino acid residues. (Adapted from Lehrer et al., 1990; 1991; Lambert et al., 1989.)

that β-pleated sheets are prominent in the topographical structure of one of the human defensins. The dimers hydrogen bond to form a channel through which small molecules can pass.

VI. Conclusions

This chapter has attempted to emphasize the similarities among colicins, thionins, defensins, and other bacteriocins. This is done with the caveat that antimicrobial proteins, like enzymes, share common general properties, but are each unique. Antimicrobial proteins usually have small molecular weights and are usually processed from larger proteins. At the molecular level, they have both hydrophobic and hydrophilic portions and are thus characterized as amphiphilic. They tend to aggregate and are cysteine rich with sulfhydryl rings. Insertion into or penetration through the membrane of susceptible cells is important to the mode of action of these proteins. In some but not all of the colicins catalytic activity also is important for the mode of action.

There is a normal sequence or unlayering of biological knowledge normally followed in molecular biology, which was particularly a characteristic of the antibiotic era. This sequence starts with a "search and discover" phase. This phase is marked by an initial period of discovery, characterization, and classification. This is followed by a phase in which the chemical structures are examined and related to the functionality of the compound. The final stage is the use of this information to form new "designer" antimicrobials. The field of antimicrobial proteins is particularly exciting because the availability of tools from molecular biology and biochemistry has allowed this sequence to be compressed. All three phases are simultaneously occurring now.

Acknowledgments

This is manuscript F-10564-1-91 of the New Jersey State Agricultural Experiment Station and is based in part on a paper, "An overview of antimicrobial proteins—bacteriocins, colicins, and defensins," given at the symposium "Bacteriocins—Antimicrobial Proteins for Potential Use in Foods" during the Institute of Food Technologists' 1991 annual meeting. Research in the authors' laboratory is supported by state appropriations, U.S. Hatch Act funds, a grant from the Cattlemen's Beef Promotion and Research Board, which is administered in cooperation with the Beef Industry Council, and a grant from the United States Department of Agriculture CSRS NRI food safety program.

References

Ahn, C., and Stiles, M. E. (1990). Antibacterial activity of lactic acid bacteria isolated from vacuum-packaged meats. *J. Appl. Bacteriol.* **69,** 302–310.

Andersson, R. (1986). Inhibition of *Staphylococcus aureus* and spheroplasts of gram-negative bacteria by an antagonistic compound produced by a strain of *Lactobacillus plantarum*. *Int. J. Food Microbiol.* **3,** 149–160.

Balls, A. K., Hale, W. S., and Harris, T. H. (1942). A crystelline protein obtained from a lipoprotein of wheat flour. *Cereal Chem.* **19,** 279–288.

Barefoot, S. F., and Klaenhammer, T. R. (1983). Detection and activity of lacticin B, a bacteriocin produced by *Lactobacillus acidophilus*. *Appl. Environ. Microbiol.* **45,** 1808–1815.

Baty, D., Frenette, M., Lloubes, R., Geli, V., Howard, S. P., Pattus, F., and Lazdunski, C. (1988). Functional domains of colicin A. *Mol. Microbiol.* **2,** 807–811.

Berry, E. D., Liewen, M. B., Mandigo, R. W., and Hutkins, R. W. (1990). Inhibition of *Listeria monocytogenes* by bacteriocin-producing *Pediococcus* during the manufacture of fermented semi-dry sausage. *J. Food Prot.* **53,** 194–197.

Bhunia, A. K., Johnson, M. C., and Ray, B. (1988). Purification, characterization, and antimicrobial spectrum of a bacteriocin produced by *Pediococcus acidilactici*. *J. Appl. Bacteriol.* **65,** 261–268.

Bhunia, A. K., Johnson, M. C., Ray, B., and Belden, E. L. (1990). Antigenic property of pediocin AcH produced by *Pediococcus acidilactici* H. *J. Appl. Bacteriol.* **69,** 211–215.

Bostian, K. A., Elliott, Q., Bussey, H., Burn, V., Smith, A., and Tipper, D. J. (1984). Sequence of the preprotoxin dsRNA gene of type I killer yeast: multiple processing events produce a two-component toxin. *Cell* **36,** 741–751.

Bowman, C. M., Sidikaro, J., and Nomura, M. (1971). Specific inactivation of ribosomes by colicin E3 *in vitro* and mechanism of immunity in colicinogenic cells. *Nature (London) New Biol.* **234,** 133–138.

Brunden, K. R., Cramer, W. A., and Cohen, F. S. (1984). Purification of a small receptor-binding peptide from the central region of the colicin E1 molecule. *J. Biol. Chem.* **259,** 190–196.

Buchman, G. W., Banerjee, S., and Hansen, J. N. (1988). Structure, expression, and evolution of a gene encoding the precursor of nisin, a small protein antibiotic. *J. Biol. Chem.* **263,** 16260–16266.

Bussey, H. (1981). Physiology of killer factor in yeast. *Adv. Microb. Physiol.* **22,** 93–122.

Bussey, H., and Sherman, D. (1973). Yeast killer factor: ATP leakage and coordinate inhibition of macromolecular synthesis in sensitive cells. *Biochim. Biophys. Acta* **298,** 868–875.

Bussey, H., Saville, D., Hutchins, K., and Palfree, R. G. E. (1979). Binding of yeast killer toxin to a cell wall receptor on sensitive *Saccharomyces cerevisiae*. *J. Bacteriol.* **140,** 888–892.

Carminati, D., Giraffa, G., and Bossi, M. G. (1989). Bacteriocin-like inhibitors of *Streptococcus lactis* against *Listeria monocytogenes*. *J. Food Prot.* **52,** 614–617.

Carrasco, L., Vazquez, D., Hernandez-Lucas, C., Carbonero, P., and Garcia-Olmedo, F. (1981). Thionins: plant peptides that modify membrane permeability in cultured mammalian cells. *Eur. J. Biochem.* **116,** 185–189.

Cramer, W. A., Cohen, F. S., Merrill, A. R., and Song, H. Y. (1990). Structure and dynamics of the colicin E1 channel. *Mol. Microbiol.* **4,** 519–526.

Daeschel, M. A. (1989). Antimicrobial substances from lactic acid bacteria for use as food preservatives. *Food Technol.* **43,** 164–166.

Daeschel, M. A. (1990). Applications of bacteriocins in food systems. In "Biotechnology and Food Safety" (D. D. Bills and S.-D. Kung, eds.), pp. 91–94. Butterworth Heinemann, Boston.

Daeschel, M. A., and Klaenhammer, T. R. (1985). Association of a 13.6-megadalton plasmid in *Pediococcus pentosaceus* with bacteriocin activity. *Appl. Environ. Microbiol.* **50,** 1538–1541.

Daeschel, M. A., McKenney, M. C., and McDonald, L. C. (1990). Bacteriocidal activity of *Lactobacillus plantarum* C-11. *Food Microbiol.* **7,** 91–98.

de Graaf, F. K., Niekus, H. G. D., and Klootwijk, J. (1973). Inactivation of bacterial ribosomes *in vivo* and *in vitro* by cloacin DF13. *FEBS Lett.* **35,** 161–165.

de Graaf, F. K., Stukart, M. J., Boogerd, F. C., and Metselaar, K. (1978). Limited proteolysis of cloacin DF13 and characterization of the cleavage products. *Biochemistry* **17,** 1137–1142.

de la Pena, P., Barros, F., Gascon, S., Lazo, P. S., and Ramos, S. (1981). Effect of yeast killer toxin on sensitive cells of *Saccharomyces cerevisiae*. *J. Biol. Chem.* **256,** 10420–10425.

Elsbach, P. (1990). Antibiotics from within: antibacterials from human and animal sources. *Trends Biotechnol.* **8,** 26–30.

Escuyer, V., Boquet, P., Perrin, D., Montecucco, C., and Mock, M. (1986). A pH induced increase in hydrophobicity as a possible step in the penetration of colicin E3 through bacterial membranes. *J. Biol. Chem.* **261,** 10891–10898.

Evett, G. E., Donaldson, D. M., and Vernon, L. P. (1986). Biological properties of *Pyrularia* thionin prepared from nuts of *Pyrularia pubera*. *Toxicon* **24,** 622–625.

Fernandez de Caleya, R., Gonzalez-Pascual, B., Garcia-Olmedo, F., and Carbonero, P. (1972). Susceptibility of phytopathogenic bacteria to wheat purothionins *in vitro*. *Appl. Microbiol.* **23,** 998–1000.

Ganz, T., Selsted, M. E., and Lehrer, R. I. (1990). Defensins. *Eur. J. Haematol.* **44,** 1–8.

Geis, A., Singh, J., and Teuber, M. (1983). Potential of lactic streptococci to produce bacteriocin. *Appl. Environ. Microbiol.* **45,** 205–211.

Gould, J. M., and Cramer, W. A. (1977). Studies on the depolarization of the *Escherichia coli* cell membrane by colicin E1. *J. Biol. Chem.* **252,** 5491–5497.

Hansen, J. N., Banerjee, S., and Buchman, L. W. (1990). Potential of small ribosomally synthesized bacteriocins in the design of new food preservatives. *J. Food Saf.* **10,** 119–130.

Harding, C. D., and Shaw, B. G. (1990). Antimicrobial activity of *Leuconostoc gelidum* against closely related species and *Listeria monocytogenes*. *J. Appl. Bacteriol.* **69,** 648–654.

Harris, L. J., Stiles, M. A., and Klaenhammer, T. R. (1989). Antimicrobial activity of lactic acid bacteria against *Listeria monocytogenes*. *J. Food Prot.* **52,** 384–387.

Hase, T., Matsubara, H., and Yoshizumi, H. (1978). Disulfide bonds of purothionin, a lethal toxin for yeasts. *J. Biochem. (Tokyo)* **83,** 1671–1678.

Hill, C. P., Yee, J., Selsted, M. E., and Eisenberg, D. (1991). Crystal structure of defensin HNP-3, an amphiphilic dimer: mechanism of membrane permeabilization. *Science* **251,** 1481–1485.

Hoover, D. G., Dishart, K. J., and Hermes, M. A. (1989). Antagonistic effect of *Pediococcus* spp. against *Listeria monocytogenes*. *Food Biotechnol.* **3,** 183–196.

Hurst, A., and Dring, S. J. (1968). The relationship between the length of the lag phase of growth to the synthesis of nisin and other basic proteins by *Streptococcus lactis* grown under different cultural conditions. *J. Gen. Microbiol.* **50,** 383–390.

Imajoh, S., Ohno-Iwashita, Y., and Imahori, K. (1982). The receptor for colicin E3. *J. Biol. Chem.* **257,** 6481–6487.

Joerger, M. C., and Klaenhammer, T. R. (1986). Characterization and purification of helveticin J and evidence for a chromosomally determined bacteriocin produced by *Lactobacillus helveticus* 481. *J. Bacteriol.* **167,** 439–446.

Joerger, M. C., and Klaenhammer, T. R. (1990). Cloning, expression, and nucleotide sequence

of the *Lactobacillus helveticus* 481 gene encoding the bacteriocin helveticin J. *J. Bacteriol.* **172,** 6339–6347.
Kaletta, C., and Entian, K.-D. (1989). Nisin, a peptide antibiotic: cloning and sequencing of the nisA gene and posttranslational processing of its peptide product. *J. Bacteriol.* **171,** 1597–1601.
Klaenhammer, T. R. (1988). Bacteriocins of lactic acid bacteria. *Biochimie* **70,** 337–349.
Klaenhammer, T. R. (1991). Regulation and genetics of bacteriocin production by lactic acid bacteria. Abstracts of the Annual Meeting of Institute of Food Technologists. Paper 231, p. 168.
Konisky, J. (1982). Colicins and other bacteriocins with established modes of action. *Annu. Rev. Microbiol.* **36,** 125–144.
Kordel, M., Benz, R., and Sahl, M.-G. (1988). Mode of action of the staphylococcin-like peptide pep 5: voltage-dependent depolarization of bacterial and artificial membranes. *J. Bacteriol.* **170,** 84–88.
Krone, W. J. A., de Vries, P., Koningstein, G., de Jonge, A. J. R., de Graaf, F. K., and Oudega, B. (1986). Uptake of cloacin DF13 by susceptible cells: removal of immunity protein and fragmentation of cloacin molecules. *J. Bacteriol.* **166,** 260–268.
Lambert, J., Keppi, E., Dimarq, J.-L., Wicker, C., Reichhart, J.-M., Dunbar, B., Lepage, P., vanDorsselaer, A., Hoffmann, J., Fothergill, J., and Hoffman, D. (1989). Insect immunity: isolation from immune blood of the dipteran *Phormia terranovae* of two insect antibacterial peptides with sequence homology to rabbit lung macrophage bacterial peptides. *Proc. Natl. Acad. Sci. U.S.A.* **86,** 262–266.
Lankisch, P. G., and Vogt, W. (1971). Potentiation of haemolysis by combined action of phospholipase A and a basic peptide containing S—S bonds (Viscotoxin B). *Experientia* **27,** 122–123.
Lechowich, R. V. (1988). Microbiological challenges of refrigerated foods. *Food Technol.* **42,** 84–85, 89.
Lehrer, R. I., Ganz, T., and Selsted, M. E. (1990). Defensins: natural peptide antibiotics from neutrophils. *ASM News* **56,** 315–318.
Lehrer, R. I., Ganz, T., and Selsted, M. E. (1991). Defensins: endogenous antibiotic peptides of animal cells. *Cell* **64,** 229–230.
Lewus, C. B. (1991). Characterization of bacteriocins produced by lactic acid bacteria isolated from meat. Ph.D. dissertation, Rutgers, The State University, New Brunswick, New Jersey.
Lewus, C. B., and Montville, T. J. (1991). Detection of bacteriocins produced by lactic acid bacteria. *J. Microbiol. Methods* **13,** 145–150.
Lewus, C. B., Kaiser, A., and Montville, T. J. (1991). Inhibition of food-borne bacterial pathogens by bacteriocins from lactic acid bacteria isolated from meat. *Appl. Environ. Microbiol.* **57,** 1683–1688.
Lipinska, E. (1977). Nisin and its application. *In* "Antibiotics and Antibiosis in Agriculture" (M. Woodbine, ed.), pp. 103–130. Butterworth, London.
LiPuma, J. J., Richman, H., and Stull, T. L. (1990). Haemocin, the bacteriocin produced by *Haemophilus influenzae:* species distribution and role in colonization. *Infect. Immun.* **58,** 1600–1605.
Lyon, W., and Glatz, B. A. (1991). Partial purification and characterization of a bacteriocin produced by *Propionibacterium thoenii. Appl. Environ. Microbiol.* **57,** 701–706.
Mak, A. S., and Jones, B. L. (1976). The amino acid sequence of wheat β-purothionin. *Can. J. Biochem.* **54,** 835–842.
Martinac, B., Zhu, H., Kubalski, A., Zhou, X., Culbertson, M., Bussey, H., and Kung, C. (1990). Yeast K1 killer toxin forms ion channels in sensitive yeast spheroplasts and in artificial liposomes. *Proc. Natl. Acad. Sci. U.S.A.* **87,** 6228–6232.
Mayr-Harting, A., Hedges, A. J., and Berkeley, R. C. W. (1972). Methods for studying bacteriocins. *In* "Methods in Microbiology Vol 7A" (J. R. Norris and D. W. Ribbons, eds.), pp. 315–422. Academic Press, Inc., New York, New York.
McCormick, E. L., and Savage, D. C. (1983). Characterization of *Lactobacillus* sp. strain 100-37

from the murine gastrointestinal tract: ecology, plasmid content, and activity toward *Clostridium ramosum* H1. *Appl. Environ. Microbiol.* **45**, 1103–1112.

McGroarty, J. A., and Reid, G. (1988). Detection of a lactobacillus substance that inhibits *Escherichia coli*. *Can. J. Microbiol.* **34**, 974–978.

Mellstrand, S. T., and Samuelsson, G. (1974). Phoratoxin, a toxic protein from the mistletoe *Phoradendron tomentosum* subsp. *macrophyllum* (Loranthaceae). *Acta Pharm. Suec.* **11**, 367–374.

Muriana, P. M., and Klaenhammer, T. R. (1987). Conjugal transfer of plasmid-encoded determinants for bacteriocin production and immunity in *Lactobacillus acidophilus* 88. *Appl. Environ. Microbiol.* **53**, 553–560.

Neville, D. M., and Hudson, T. H. (1986). Transmembrane transport of diphtheria toxin, related toxins, and colicins. *Annu. Rev. Biochem.* **55**, 195–224.

Nielsen, J. W., Dickson, J. S., and Crouse, J. D. (1990). Use of a bacteriocin produced by *Pediococcus acidilactici* to inhibit *Listeria monocytogenes* associated with fresh meat. *Appl. Environ. Microbiol.* **56**, 2142–2145.

Notermans, S., Dufrenne, J., and Lund, B. M. (1990). Botulism risk of refrigerated processes foods of extended durability. *J. Food Prot.* **53**, 1020–1024.

Ohno-Iwashita, Y., and Imahori, K. (1980). Assignment of the functional loci in colicin E2 and E3 molecules by the characterization of their proteolytic fragments. *Biochemistry* **19**, 652–659.

Okada, T., and Yoshizumi, H. (1973). The mode of action of toxic protein in wheat and barley on brewing yeast. *Agric. Biol. Chem.* **37**, 2289–2294.

Okereke, A., and Montville, T. J. (1991). Bacteriocin inhibition of *Clostridium botulinum* spores by lactic acid bacteria. *J. Food Prot.* **54**, 349–353.

Ortel, S. (1989). Listeriocins (monocins). *Int. J. Food Microbiol.* **8**, 249–250.

Osorio E Castro, V. R., Van Kuiken, B. A., and Vernon, L. P. (1989). Action of a thionin isolated from nuts of *Pyrularia pubera* on human erythrocytes. *Toxicon* **27**, 501–510.

Oudega, B., Klaasen-Boor, P., and de Graaf, F. K. (1975). Mode of action of the cloacin DF13-immunity protein. *Biochim. Biophys. Acta* **392**, 184–195.

Oudega, B., Klaasen-Boor, P., Sneeuwloper, G., and de Graaf, F. K. (1977). Interaction of the complex between cloacin and its immunity protein and of cloacin with the outer and cytoplasmic membranes of sensitive cells. *Eur. J. Biochem.* **78**, 445–453.

Oudega, B., van der Molen, J., and de Graaf, F. K. (1979). *In vitro* binding of cloacin DF13 to its purified outer membrane receptor protein and effect of peptidoglycan on bacteriocin-receptor interaction. *J. Bacteriol.* **140**, 964–970.

Ozaki, Y., Wada, K., Hase, T., Matsubara, H., Nakanishi, T., and Yoshizumi, H. (1980). Amino acid sequence of a purthionin homolog from barley flour. *J. Biochem. (Tokyo)* **87**, 549–555.

Palumbo, S. A. (1986). Is refrigeration enough to restrain foodborne pathogens? *J. Food Prot.* **49**, 1003–1009.

Palumbo, S. A. (1987). Can refrigeration keep our foods safe? *Dairy and Food Sanitarian* **7**, 56–60.

Pattus, F., Massotte, D., Wilmsen, H. U., Lakey, J., Tsernoglou, D., Tucker, A., and Parker, M. W. (1990). Colicins: prokaryotic killer-pores. *Experientia* **46**, 180–192.

Postle, K., and Skare, J. T. (1988). *Escherichia coli* TonB protein is exported from the cytoplasm without proteolytic cleavage of its amino terminus. *J. Biol. Chem.* **263**, 11000–11007.

Pucci, M. J., Vedamuthu, E. R., Kunka, B. S., and Vandenbergh, P. A. (1988). Inhibition of *Listeria monocytogenes* by using bacteriocin PA-1 produced by *Pediococcus acidilactici* PAC 1.0. *Appl. Environ. Microbiol.* **54**, 2349–2353.

Pugsley, A. P., and Oudega, B. (1987). Methods for studying colicins and their plasmids. *In* "Plasmids" (M. S. Hardy, ed.), pp. 105–111. IRL Press, Oxford.

Raccach, M., McGrath, R., and Daftarian, H. (1989). Antibiosis of some lactic acid bacteria including *Lactobacillus acidophilus* toward *Listeria monocytogenes*. *Int. J. Food Microbiol.* **9**, 25–32.

Ray, S. K., Johnson, M. C., and Ray, B. (1989). Bacteriocin plasmids of *Pediococcus acidilactici*. *J. Ind. Microbiol.* **4**, 163–171.

1. Antimicrobial Proteins

Reeves, P. (1965). The bacteriocins. *Bacteriol. Rev.* **29,** 24–43.

Samuelsson, G. (1973). Mistletoe toxins. *Syst. Zool.* **22,** 566–569.

Samuelsson, G., and Pettersson, B. (1971). The disulfide bonds of viscotoxin A3 from the European mistletoe *Viscum album* (L. Loranthaceae). *Acta Chem. Scand.* **25,** 2048–2054.

Samuelsson, G., Seger, L., and Olson, T. (1968). The amino acid sequence of oxidized viscotoxin A3 from the European mistletoe *Viscum album* (L., Loranthaceae). *Acta Chem. Scand.* **22,** 2624–2642.

Schaller, K., and Nomura, M. (1976). Colicin E2 is a DNA endonuclease. *Proc. Natl. Acad. Sci. U.S.A.* **73,** 3989–3993.

Schaller, K., Dreher, R., and Braun, V. (1981). Structural and functional properties of colicin M. *J. Bacteriol.* **146,** 54–63.

Schein, S. J., Kagan, B. L., and Finkelstein, A. (1978). Colicin K acts by forming voltage-dependent channels in phospholipid bilayer membranes. *Nature (London)* **276,** 159–163.

Schillinger, U. (1990). Bacteriocins of lactic acid bacteria. *In* "Biotechnology and Food Safety" (D. D. Bills and S.-D. Kung, eds.), pp. 55–74. Butterworth-Heinemann, Boston.

Schillinger, U., and Lucke, F.-K. (1989). Antibacterial activity of *Lactobacillus sake* isolated from meat. *Appl. Environ. Microbiol.* **55,** 1901–1906.

Schillinger, U., McKay, L. L., and Lucke, F. K. (1991). Behaviour of *Listeria monocytogenes* in meat and its control by a bacteriocin-producing strain of *Lactobacillus sake*. *J. Appl. Bacteriol.* **70,** 473–478.

Sears, P. M. and Blackburn, P. (1992). Potential role of antimicrobial proteins in prevention and therapy of mastitis infections. Proceeding of the Empire State Mastitis Council. Biennial Meeting. January 6, Syracuse, NY.

Shiver, J. W., Cramer, W. A., Cohen, F. S., Bishop, L. J., and de Jong, P. J. (1987). On the explanation of the acidic pH requirement for *in vitro* activity of colicin E1. *J. Biol. Chem.* **262,** 14273–14281.

Shiver, J. W., Cohen, F. S., Merrill, A. R., and Cramer, W. A. (1988). Site-directed mutagenesis of the charged residues near the carboxy terminus of the colicin E1 ion channel. *Biochemistry* **27,** 8421–8428.

Skipper, N., and Bussey, H. (1977). Mode of action of yeast toxins: energy requirement for *Saccharomyces cerevisiae* killer toxin. *J. Bacteriol.* **129,** 668–677.

Song, H. Y., Cohen, F. S., and Cramer, W. A. (1991). Membrane topography of ColE1 gene products: the hydrophobic anchor of the colicin E1 channel is a helical hairpin. *J. Bacteriol.* **173,** 2927–2934.

Spelhaug, S. R., and Harlander, S. K. (1989). Inhibition of foodborne bacterial pathogens by bacteriocins from *Lactococcus lactis* and *Pediococcus pentosaceus*. *J. Food Prot.* **52,** 856–862.

Steele, J. L., and McKay, L. L., (1986). Partial characterization of the genetic basis for sucrose metabolism and nisin production in *Streptococcus lactis*. *Appl. Environ. Microbiol.* **51,** 57–64.

Steen, M. T., Chung, Y. J., and Hansen, J. N. (1991). Characterization of the nisin gene as part of a polycistronic operon in the chromosome of *Lactococcus lactis* ATCC 11454. *Appl. Environ. Microbiol.* **57,** 1181–1188.

Stevens, K. A., Sheldon, B. W., Klapes, N. A., and Klaenhammer, T. R. (1992). Effect of treatment conditions on nisin inactivation of Gram-negative bacteria. *J. Food Protect.* **55,** 763–767.

Tagg, J. R. (1991). Bacterial BLIS. *ASM News* **57,** 611.

Tagg, J. R., Dajani, A. S., and Wannamaker, L. W. (1976). Bacteriocins of gram-positive bacteria. *Bacteriol. Rev.* **40,** 722–756.

Teeter, M. M., Mazer, J. A., and L'Italien, J. J. (1981). Primary structure of the hydrophobic plant protein crambin. *Biochemistry* **20,** 5437–5443.

Thomas, E. L., Lehrer, R. I., and Rest, R. F. (1988). Human neutrophil antimicrobial activity. *Rev. Infect. Dis.* **10,** 450–456.

Tokuda, H., and Konisky, J. (1978). Mode of action of colicin Ia: effect of colicin on the *Escherichia coli* proton electrochemical gradient. *Proc. Natl. Acad. Sci. U.S.A.* **75,** 2579–2583.

van den Elzen, P. J. M., Walters, H. H. B., Veltkamp, E., and Nijkamp, H. J. J. (1983).

Molecular structure and function of the bacteriocin gene and bacteriocin protein of plasmid Clo DF13. *Nucleic Acids Research* **11**, 2465–2477.

van Tiel-Menkveld, G. J., Mentjox-Vervuurt, J. M., Oudega, B., and de Graaf, F. K. (1982). Siderophore production by *Enterobacter cloacae* and a common receptor protein for the uptake of aerobactin and cloacin DF13. *J. Bacteriol.* **150**, 490–497.

Vernon, L. P., Evett, G. E., Zeikus, R. D., and Gray, W. R. (1985). A toxic thionin from *Pyrularia pubera*: purification, properties and amino acid sequence. *Arch. Biochem. Biophys.* **238**, 18–29.

Wada, K., Ozaki, Y., Matsubara, H., and Yoshizumi, H. (1982). Studies on purothionin by chemical modifications. *J. Biochem. (Tokyo)* **91**, 257–263.

Weiss, M. J., and Luria, S. E. (1978). Reduction of membrane potential, an immediate effect of colicin K. *Proc. Natl. Acad. Sci. U.S.A.* **75**, 2483–2487.

Wendt, L. (1970). Mechanism of colicin action: early events. *J. Bacteriol.* **104**, 1236–1241.

Winkowski, K., and Montville, T. J. (1992). Use of a meat isolate, *Lactobacillus bavaricus* MN, to inhibit *Listeria monocytogenes* growth in a model meat gravy system. *J. Food Safety* **13**, 19–31.

Yamada, M., Ebina, Y., Miyata, T., Nakazawa, T., and Nakazawa, A. (1982). Nucleotide sequence of the structural gene for colicin E1 and predicted structure of the protein. *Proc. Natl. Acad. Sci. U.S.A.* **79**, 2827–2831.

Young, T. W., and Yagiu, M. (1978). A comparison of the killer character in different yeasts and its classification. *Antonie van Leeuwenhoek* **44**, 59–77.

Zhu, H., and Bussey, H. (1989). The K1 toxin of *Saccharomyces cerevisiae* kills spheroplasts of many yeast species. *Appl. Environ. Microbiol.* **55**, 2105–2107.

Zhu, H., Bussey, H., Thomas, D. Y., Gagnon, J., and Bell, A. W. (1987). Determination of the carboxyl termini of the α and β subunits of yeast K1 killer toxin. *J. Biol. Chem.* **262**, 10728–10732.

CHAPTER **2**

Screening Methods for Detecting Bacteriocin Activity

DALLAS G. HOOVER
SUSAN K. HARLANDER

I. Historical Perspective

The demonstration of inhibitive effects between separate cultures of bacteria is well established. The earliest work includes a recorded observation by Antonie van Leeuwenhoek in 1676 that the product from one microorganism retarded the growth of another. Louis Pasteur, with J. F. Joubert in 1877, documented an antagonistic effect of bacteria from urine against *Bacillus anthracis*. As so often has been the case, such observations of antibiosis may not be elucidated to the point of determining the actual mechanism of action. Inhibition can be due to low molecular weight antibiotics, lytic agents, enzymes, defective bacteriophage, and other metabolic byproducts such as ammonia, organic acids, free fatty acids, carbon dioxide, and hydrogen peroxide. In addition, nutrient depletion and lowering of the redox potential of the medium may antagonize growth of competing strains to some extent. Given the diversity of naturally occurring antagonistic compounds, it is often a challenge to quantify bacteriocin activity while eliminating or measuring other inhibitive effects, some of which may be synergistic with the activity of bacteriocins. Such factors in lactic acid bacteria have been earlier summarized by Klaenhammer (1988).

Total agreement does not exist for the definition of bacteriocin. For this chapter the broader and simpler description will be used, and that is, a bacteriocin is a peptide or protein produced by a bacterium that inhibits the growth of another bacterium.

In spite of numerous inhibitive factors, bacteriocins are ubiquitous in nature. As noted in the milestone review of bacteriocins from Gram-positive bacteria (Tagg et al., 1976), screening of a relatively large number of strains (100 or more) of any one bacterial species will usually offer some evidence of inhibition due to a bacteriocin. Tagg et al. (1976) exemplified this wide occurrence by summarizing 57 papers that surveyed different species of Gram-positive bacteria for strains that produced a bacteriocinlike effect. In eight studies that surveyed the production of bacteriocin activity from cultures of *Staphylococcus aureus,* the occurrence of bacteriocin activity ranged from 1 to 38% of the strains examined with an average of about 13%. The number of strains of staphylococci evaluated in these studies was from 65 to 2035. Five studies that screened for the presence of bacteriocin activity in group D streptococci found a range of 51 to 99% for bacteriocin-positive strains with an overall frequency of occurrence of about 74%. The number examined ranged from 77 to 108 strains.

II. Agar Diffusion Techniques

A. Introduction

The first antimicrobial susceptibility test that utilized diffusion of the antibiotic substance through agar media was done by Fleming in 1924 with penicillin against *Staphylococcus aureus.* Cutting wells into the agar to serve as reservoirs for liquid preparations of antimicrobial agents has been a popular modification (Reddish, 1929). Another approach has been to use sensitivity disks (sterile paper containing diffusible antibiotic) whereby the relationship between diameters of the zones of inhibition in lawns of indicator-seeded agar determines the minimum inhibitory concentration (MIC). Such MIC values are normally derived from serial dilutions of the antibiotic (usually twofold).

There have been many adaptations of the agar zone diffusion technique since it was first used in the testing of antimicrobial compounds (Abraham et al., 1941). Common to these techniques is the measurement of the zones of inhibition in indicator-seeded agar plates (Figure 1). The antimicrobial agent diffuses through the agar to inhibit growth of the indicator organism. With the agar diffusion method, Vesterdal (Cooper, 1964) found the gradient or zone of inhibition to be established when the charge or concentration at the source is well defined and a gradient is established by gradual exhaustion of the diffusing antibiotic (x or $r; x = r - r_d$). There is a linear relationship between the response and the $\log_{10}$ dose. The zone size is a result of diffusion of the antimicrobial compound and the growth rate of the indicator organism (Linton, 1983). An antibiotic compound will diffuse through an agar at a constant rate depending upon its molecular weight, its ionic charge, and the composition of the gel. Temperature and solvent viscosity of the gel will affect this constant. Most microbiological assays are

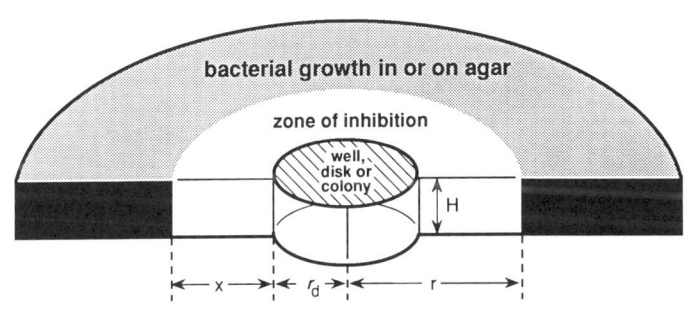

2. Screening Methods for Detecting Bacteriocin Activity

Figure 1 A zone of inhibition formed by radial diffusion in solidified medium.

conducted isothermally, but should two different incubation temperatures be used to either prediffuse the antibiotic or to allow optimum growth temperatures to be used for the producing and indicator strains of a deferred antagonism assay, then the diffusion rate will change and the slope of an assay curve will be altered (Figure 2).

The amount of antibiotic compound will directly affect the zone of inhibition; the greater the concentration, the greater the zone diameter

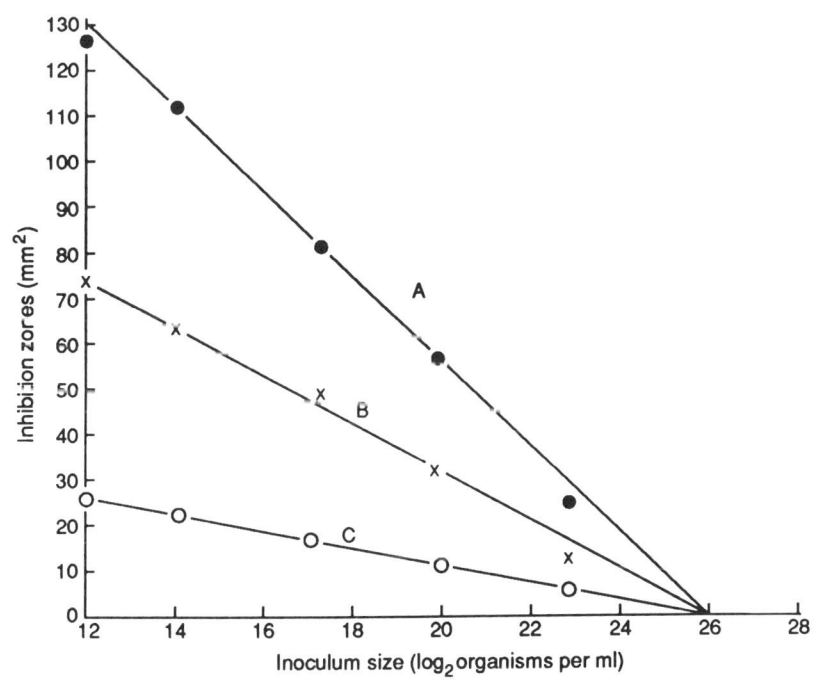

Figure 2 The influence of inoculum size on zones of inhibition. The test organism is *Klebsiella pneumoniae* with three concentrations of streptomycin (1000, 100, and 10 µg/ml: A, B, and C, respectively). The test was carried out at 35°C. The denser the inoculum, the smaller the zone sizes until, with an inoculum of $\log_2 = 26$ and greater, no zones were produced. (From Linton, 1958.)

under standard conditions (Figure 2). The higher the amount of antibiotic substance present, the farther it diffuses out into the agar in a given time period. When there is a significant difference between the critical concentration of the antibiotic that inhibits the indicator organism under conditions of diffusion and the antibiotic concentration at the source (which is usually considered constant) the zone edge tends to be crisp. When they are similar, the zone edges may appear nondescript or diffuse (Linton, 1983).

Overnight incubation is the usual minimum length of time until measurement of zones of inhibition, but the actual location of the zone edge is fixed within a few hours of incubation, after which time the indicator organism grows outside the zone and becomes visible. This length of time for growth is determined by the type of microorganism tested for sensitivity, the nutritional status of the growth medium, the incubation temperature, and the inoculum size (Linton, 1983). Generally, the greater the inoculum, the shorter the required growth time; but too dense an inoculum results in no zones of inhibition because the initial high cell number will mask any subsequent zone formation (Figure 2). Large zones of inhibition are usually formed when the bacteria are slow growing (e.g., due to adverse temperature or minimal media), and small zones are usually formed when bacteria grow rapidly (Piddock, 1990).

In those systems in which the diffusion of the antibiotic and the growth of the indicator organism are simultaneous, all sensitive organisms within the forming zone of inhibition are initially inhibited by the critical inhibitory concentration of the antibiotic diffusing from the source. As incubation continues, cells beyond the zone grow to greater density. At the zone edge, an inhibitory concentration of antibiotic interacts with a cell density that is large enough to absorb antibiotic to an extent large enough to significantly lower the concentration of antibiotic to one that is no longer inhibitory. As the indicator grows in the log phase, the diffusing antibiotic is maintained at a subinhibitory level by the increased density of the indicator organism, and therefore the organism grows uninhibited and becomes visible.

Most antibiotics appear to be diffusible through agar gels, although in some cases this may occur slowly. In those assays in which prediffusion of the antibiotic compound is desirable because the agent diffuses slowly, (e.g., because of large molecular weight or polymeric forms), the initiation of indicator organism growth is deferred. Under these conditions, the zone of inhibition will be a result of the incubation times necessary for prediffusion and growth of the indicator to a level of critical density. Therefore, the longer time will give larger zones, and this is due to the longer time period allowing for a greater distance of antibiotic diffusion (Linton, 1983). When comparing methodologies involving incubation times of therapeutic antibiotics versus bacteriocins, one should keep in mind that bacteriocins are generally of much larger molecular size, so their rate of diffusion will be slower than that of the classical antibiotics. For example, Rogers and Montville (1991) found that the preincubation of plates for 24 hours at 3°C allowed for the diffusion of nisin while the growth of the indicator strain was delayed before subsequent incubation at 30°C. Preincubation increased assay sensitivity by increasing zone size and enhanced assay reproducibility by causing a lesser amount of variability between readings.

B. Plating Methods

1. Deferred Methods

The primary means for detecting bacteriocin activity of the lactic acid bacteria is the agar plate diffusion assay. Agar plate assays are popular for screening of antagonistic activity as well as for monitoring expression, optimization, and inactivation of bacteriocins for their characterization. While these plating methods are straightforward, they are not without certain problems and limitations, especially since lactic acid bacteria can produce other antagonistic compounds and create conditions that mimic bacteriocin activity.

In the screening of bacteriocin activity of Gram-positive bacteria, Tagg et al. (1976) classified the methods of detection as deferred or simultaneous. Fredericq (1948) first used deferred antagonism in study of the colicins. Using solid media in this approach, the producing culture is first grown under optimal conditions. This strain can then be killed by chloroform or heat and subsequently overlaid with indicator-seeded molten agar. The conditions for the second incubation can be optimum for the indicator strain. Killing the producing culture has been used less frequently since Brock et al. (1963) found that chloroform and heat can inactivate bacteriocins. The smearing associated with not killing the producing culture can be minimized by careful application of the soft agar overlay and efforts to control condensation; however, use of chloroform is not uncommon. In their study of diplococcin from *Streptococcus cremoris* 346, Davey and Richardson (1981) stabbed M17 agar plates with *S. cremoris* and incubated for 48 hours at 22°C; surface growth was then killed by exposure to chloroform vapors. The plates were overlaid with 2.5 ml of soft agar inoculated with 0.1 ml of a 24-hour growth of indicator culture and incubated at 22°C with examination for zones of inhibition at 24 and 48 hours. Chloroform was used by Hastings and Stiles (1991) in their adaptation of the "spot-on-lawn" assay that included adjustment of the bacteriocin-producing culture of bacteria (*Leuconostoc gelidum*) to pH 6.5 with 10M NaOH before centrifugation at 6000 g for 5 minutes. Any remaining cells in the supernatant were then inactivated by mixing 1 part chloroform to 4 parts supernatant. After 5 minutes, the aqueous portion was aseptically removed and 20 ml spotted on the surface of an All-purpose Tween (APT) agar plate overlaid with 6 ml soft APT agar (0.75%) inoculated with 1% of an overnight-grown indicator culture. Controls with sterile APT broth were done to assure that no chloroform residues were present.

An alternative approach that omits chloroform treatment is a "slam-transfer" deferred plating method introduced by Kekessy and Piguet (1970). After growth of the producing strain the agar is aseptically dislodged from the bottom half of the petri dish, the dish is closed, and by striking the dish forcefully on the bench upside-down the agar layer will flip onto the top lid. This can then be overlaid with indicator-seeded soft agar and incubated. In this method, the producing cells and indicator cells are separated by a layer of agar. Antagonistic effects by phage and acid are minimized but the bacteriocin must be able to freely diffuse through the solid medium in concentrations adequate for zones of inhibition to form.

The Lutri-Plate™, manufactured by Lutri-Plate, Inc., of Starkville, Mississippi, is a modified petri dish developed for testing bacteria for the production of bacteriocins. It permits the two-sided agar approach of the Kekessy and Piguet (1970) method without the requirement of dislodging and transferring the agar layer that contains the bacteriocin-producing colonies. Instead, the initial agar layer is poured into the dish on top of a fixed, wide-mesh grid that is located above a "false bottom" resting on four plastic posts. Following agar solidification, inoculation, and incubation, the dish is entered from the opposite side and the false bottom is removed from the dish. This newly exposed agar surface of the now inverted plate is overlaid with indicator-seeded soft agar that is incubated and then examined for zones of inhibition in the indicator lawn. The initial agar layer is held in place by the wide-mesh grid.

Agar methods can be adapted to visualize antagonism against an indicator strain from a mixed population sample. The sample can be diluted in molten agar spread over a suitable nutrient agar plate and then overlaid with indicator-seeded soft agar. Bacteriocin-producing colonies can be differentiated from nonproducing clones in this manner. This is applicable for conjugation experiments using bacteriocin production as a plasmid marker whereby bacteriocin-positive transconjugants can be discerned from bacteriocin-negative cells.

Bacteriophage can present an antagonistic effect similar to that of bacteriocins. In fact, it was not uncommon for early studies on lysogeny to be redirected into investigations on colicins when the search for one element uncovered action of another. Dropping dilutions of cell-free supernatant onto an indicator-seeded agar plate with subsequent incubations will distinguish phage from bacteriocins, as phage will continue to cause formation of distinct plaques at high dilutions while the bacteriocin solution will demonstrate gradual decrease in the zones of inhibition with dilution. In addition, bacteriophage can be differentiated from low molecular weight bacteriocins by ultracentrifugation. In extracts, bacteriophage can be eliminated by a 2-hour exposure to ultraviolet light; bacteriocins do not appear to lose any activity from this treatment (Mayr-Harting et al., 1972).

2. Simultaneous Methods

Simultaneous or direct assays are the simplest to do and can be done in a shorter amount of time than deferred assays. Gratia (1946) first published the "spot-on-lawn" procedure. In this approach the producer and indicator cultures are grown concurrently on the same solid media under the same conditions of incubation. The indicator is spread onto the surface of the agar medium and the producing culture is spotted on top of this. While a deferred test works well when antagonistic protein is synthesized and released late in the growth cycle of the producing cell, simultaneous methods require release of the bacteriocin early in the growth cycle or the indicator organism will enter log phase and be established before a diffusible inhibitor can be effective. Adaptations to the spot-on-lawn assay employ dual inoculations in close proximity to one another or overlapping drops on the same agar plate (Barrow, 1963). Wells cut into freshly seeded agars can contain either a cell suspension of the producing strain or a growth extract (Sabine, 1963). A variation by Gagliano and Hinsdill (1970) for facultative anaerobes is stab-inoculation with a producing strain into freshly indicator-seeded pour plates or surface-streaked agar. The slightly diminished oxygen tension enhanced growth of microaerophiles and other facultative types.

C. Media Composition and Conditions of Incubation

1. Media

As noted by Tagg et al. (1976), one must not necessarily assume the conditions for optimal growth of a bacterium coincide with optimal synthesis and release of a bacteriocin. However, it can be anticipated that complex media will support larger bacteriocin yields. The growth medium may also affect the sensitivity of the indicator strain to the bacteriocin. For example, the presence of sucrose in media will cause formation of an extracellular polysaccharide layer that forms a barrier protecting cells of *Streptococcus mutans* and *Streptococcus salivarius* that otherwise are affected by a bacteriocin produced by *S. mutans* (Rogers, 1974). The indicator strains for colicins V, I, and B are much more sensitive when growth media contains low concentrations of iron. The colicin surface receptors in these strains are regulated by exogenous iron concentrations. When grown in iron-deficient media, the number of receptors is significantly diminished (Konisky, 1978). Due to these variations of producer and indicator strains to growth media, it is usually best to screen colonies on several growth agars to select the most suitable. The selection of a medium is often empirical.

Agar predominates as the gelling agent in solid microbiological media. It is a complex of the polysaccharides agarose and agaropectin, and a variety of metallic cations and trace elements. Antibacterial compounds containing cationic molecular structures can be electrostatically bound to the acid or sulfate groups in the agar, which slows the diffusion rate and, as a result, shrinks the size of the zones of inhibition. Barry and Fay (1973) estimated the optimal depth of agar to be about 4 mm for disk susceptibility tests.

In some instances, bacteriocin activity can best be demonstrated by a solid or semisolid matrix. This has been shown primarily with oral streptococci (Hamada and Ooshima, 1975) and staphylococci (Gagliano and Hinsdill, 1970). Using thickening or gelling agents such as starch, agar, dextran, or glycerol in liquid media significantly increases the amount of bacteriocin from oral streptococci (Kelstrup and Gibbons, 1969). Most strains of *Salmonella typhi* require solid media for bacteriocin production (Mayr-Harting et al., 1972). For *Staphylococcus epidermidis*, a semisolid medium increases the yield of bacteriocin 20-fold over the liquid medium (Jetten et al., 1972). Other enhancers for bacteriocin production have included yeast extract for *Streptococcus mutans* (Rogers, 1972) and staphylococci (Mayr-Harting et al., 1972), casein hydrolysate for clostridia (Clarke et al., 1975), and manganese salts for *Bacillus megaterium* (Alfoldi, 1958). Variables in the composition of media for bacteriocin synthesis have included adjustment in the ratios of amino acids for clostridia (Meitert, 1969), adjustment of different concentrations of mannitol for staphylococci (Hale and Hinsdill, 1973; Barrow, 1963), and the concentration of glucose for streptococci (Tagg et al., 1976; 1975).

2. Incubation

Other factors affecting growth and production of bacteriocins are the length of incubation and incubation temperature. The optimum temperature for growth is normally the incubation temperature necessary for observation of the effects of bacteriocin activity. For lactic acid bacteria this is usually a temperature of 32°C. The length of time usually required before examination of plating techniques is at least 24 hours to establish the formation of indicator lawns. In liquid assays and growth medium extractions, enough time is usually allowed for maximum production and release of the bacteriocin before measurement. This is often 48 hours; however, the synthesis and activity of proteases or other inactivators may act to lessen bacteriocin activity with longer incubation times. As noted by Tagg et al. (1976), bacteriocins may be cell surface components that are produced in excess under certain conditions and are released from cells with time. Lactocin LP27 seems to exist in both a cell-associated and an extracellular form (Upreti and Hinsdill, 1973). The maximal time in the growth cycle for production of bacteriocins is dependent of the type of bacteria and can occur from log phase through early stationary, keeping in mind that the bacteriocin may be retained on the surface or held intracellularly. In *Corynebacterium diptheriae*, whether the bacteriocin is released continuously (and constitutively), or produced in bursts is strain dependent (Meitert, 1969). Of course, if a bacteriocin is inducible, it will not be produced in appreciable amounts without the presence of the inducing agent.

3. Inducibility

As noted earlier, many colicins resemble bacteriophage in sharing patterns of inducibility; however, unlike colicins and many other bacteriocins from Gram-negative bacteria, bacteriocins from lactic acid bacteria are not usually inducible by such agents as mitomycin C and ultraviolet (UV) irradiation. With *Listeria monocytogenes,* UV irradiation will increase bacteriocin production a few minutes after exposure until two hours later when the bacteriocin adsorbs to bacteria and antibiotic effects diminish (Mayr-Harting et al., 1972). For some *Bacillus megaterium* strains, mitomycin C will initiate bacteriocin production but UV irradiation will not (Seed, 1979).

To date, the only inducible bacteriocin suggested in lactic acid bacteria is lactacin B, produced by *Lactobacillus acidophilus* N2, which is effective against closely related lactobacilli such as *L. leichmannii* (Hughes and Barefoot, 1990; Chen et al., 1992). With *L. leichmannii* ATCC 4797 present, lactacin B is detectable from *L. acidophilus* N2 grown in Lactobacilli deMan, Rogosa and Sharpe (MRS) broth within 4 to 6 hours, as compared to 8 to 10 hours without *L. leichmannii* present. Lactacin B activity was determined by Hughes and Barefoot (1990) using a critical dilution method whereby serial twofold dilutions of the N2 growth extract were spotted at 1, 10, or 30 μl onto indicator lawns containing approximately 10^6 CFU/ml. These lawns were prepared by adding 100 μl of a tenfold diluted overnight MRS broth culture of *L. leichmannii* to 5 ml MRS soft (0.75%) agar applied over prepared MRS agar plates. Such plates were held at 37°C for approximately 18 hours under carbon dioxide. Bacteriocin titers were expressed as the reciprocal of the highest dilution exhibiting a zone of inhibition and reported as $\log_{10}$ activity units per ml. Lactacin B activity was enhanced by adding washed or unwashed *L. leichmannii* cells, but not spent culture medium, suggesting that a cell-associated moiety stimulated early and increased bacteriocin production in *L. acidophilus* N2.

D. Adaptations of Agar Diffusion Methods Used with Lactic Acid Bacteria

For rapid screening and qualitative assay, Tsai and Sandine (1987) used the flipped agar method of Kekessy and Piguet (1970), while for quantitative assay the method of Tramer and Fowler (1964) was adapted. The latter procedure is an agar plate well diffusion assay that Tsai and Sandine (1987) used to quantify nisin production in their study of conjugally transferring nisin plasmid genes from *Lactococcus lactis* to *Leuconostoc dextranicum*. Overnight cultures of test strains were used at a 1% inoculum in 25 ml of M17-glucose broth and incubated at 30°C into the stationary phase. The culture was then centrifuged and the supernatant filter sterilized. The supernatant samples and nisin standards were added to wells cut into indicator-seeded agar (20 ml/plate) in 90-μl quantities. This method worked quite well in measuring expression of the nisin genes in the donor, recipient, and transconjugant cultures of their study.

For the assay of pediocin AcH activity, Bhunia et al. (1988) used a disk assay method. *Pediococcus acidilactici* H was grown and maintained in casein glucose broth (CGB) containing (in g/l): trypsin-digested casein, 20; glucose, 10; yeast extract, 5; ammonium citrate, 2; disodium phosphate, 2; Tween 80, 1; magnesium sulfate, 0.1; and manganese sulfate, 0.05; in a solution of pH 6.5. For pediocin AcH production, a tenfold concentrate of CGB was dialyzed against distilled water to remove peptides and other macromolecules. The dialysate was autoclaved and used as culture medium. The lawn was made by overlaying CG agar (CGB containing 1.5% agar) prepoured plates with 5 ml of soft CG agar (CGB with 0.8% agar) containing approximately 10^6 CFU of *Lactobacillus plantarum* WSO-39. After 30 minutes at room temperature, 6.25-mm-diameter sterile disks were placed on the agar surface. The disks received 40 µl of serially twofold diluted culture extract. These plates were incubated at 30°C for 18–24 hours and examined for 2 mm or larger clear zones of inhibition around the disks. The antimicrobial titer was reciprocal of the highest dilution producing a definite zone of inhibition.

In an investigation of comparative agar plating methods, Spelhaug and Harlander (1989) examined the production of bacteriocins from different strains of *Lactococcus lactis* subsp. *lactis* and *Pediococcus pentosaceus* effective against a variety of foodborne pathogens. Pediococci were grown in MRS broth, lactococci were grown in M17 broth supplemented with 0.5% glucose or sucrose, and the bacterial pathogens were grown in brain heart infusion (BHI) broth, thioglycolate broth, or Tryptic soy broth plus 0.6% yeast extract. The three plating methods evaluated were the flip streak and flip spot (Kekessy and Piguet, 1970) and the agar spot (Fleming et al., 1975).

For the flip streak method, Spelhaug and Harlander (1989) used an 18-hour growth of the test strain for inoculation as a single streak onto the surface of an M17-glucose agar plate. After a 24-hour incubation at 32°C, a sterile spatula was used to dislodge the agar from the bottom of the petri plate and, with the dish closed, flip it onto the cover of the dish. Spelhaug and Harlander (1989) did not use a seeded soft agar overlay, but instead streaked with a loopful of the indicator organism onto the inverted agar perpendicular to the producer streak (Figure 3). With *Listeria* as indicators, the plates were incubated for 18–20 hours at 30°C. One producing organism and three indicator strains were used for each plate. Inhibition was evident by a clear zone parallel to the producer streak.

In the flip spot method (Spelhaug and Harlander, 1989), 5 µl of a test organism diluted tenfold in pH 7.0 buffer was spot-inoculated onto the surface of an agar plate. As many as 4 cultures, approximately 3 cm apart, were spotted per plate. After 24 hours at 30°C, the agar was detached and transferred onto the lid. A 0.1-ml inoculum of an 18-hour culture of the indicator was then added to 7 ml of soft agar (0.7%) and poured over the surface of the inverted agar. Antagonism was evident by a zone of inhibition extending radially from the test spot.

For the agar spot method (Spelhaug and Harlander, 1989), test organisms were prepared and spot-inoculated as described in the flip spot method; however, after a 24-hour incubation, the agar surface was overlaid with

2. Screening Methods for Detecting Bacteriocin Activity

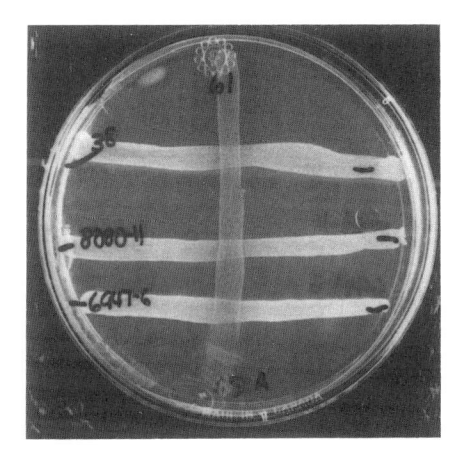

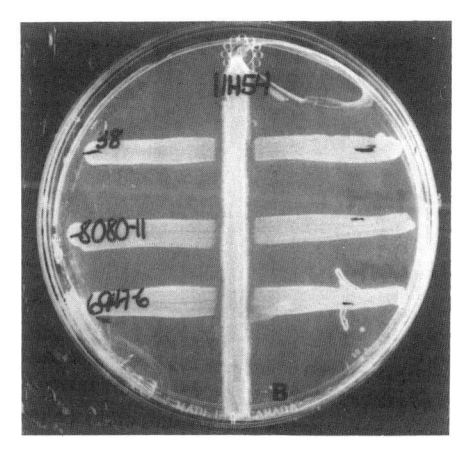

Figure 3 Flip streak assay of *Pediococcus pentosaceus* FBB61 (A) and *Lactococcus lactis* subsp. *lactis* 11454 (B) against *Listeria monocytogenes* 38, *Listeria welshimeri* 8080-11, and *Listeria innocua* 6947-6.

5 ml of BHI soft agar (0.7%) seeded with an 18-hour culture of *Listeria*. A clear zone of inhibition was found after 18–20 hours at 30°C. Using this procedure and the flip spot technique, M17-glucose, BHI, and MRS agars were used as both top and bottom agars. To prevent hydrogen peroxide-caused inhibition, catalase was added to overlay agars at a final concentration of 5 mg/ml.

Although McGroarty and Reid (1988) found MRS broth superior to other media in measuring bacteriocin activity in *Lactobacillus acidophilus* 76 and *L. casei* subsp. *rhamnosus* GR-1, Spelhaug and Harlander (1989) found MRS susceptible to nonspecific inhibition caused by acid production using either the agar spot or flip spot assays. Lewus and Montville (1991) also found the level of glucose to be too high in MRS, resulting in acid inhibition by lactic acid bacteria.

BHI produced smaller zones of inhibition than M17-glucose media (Spelhaug and Harlander, 1989). Using the flip spot method, the bacteriocins produced by *Pediococcus pentosaceus* FBB61 and FBB63-DG2 were not able to diffuse in BHI or M17-glucose agars, thus preventing the formation of any zones of inhibition. Some of the variability caused by plate media composition is shown in Table I. The flip spot method also displayed frequent smearing caused by the pressing of the surface growth of the test organism against the lid; this made precise measurement of the zones of inhibition difficult. The flip streak method was limited by allowing only one test organism per plate so that comparative inhibitory potentials of different strains could not be evaluated on one plate. Because of these disadvantages the authors chose to do the bulk of their testing with the agar spot method because of its greater reproducibility and reliability.

For detection of antagonistic activity from lactobacilli isolated from meat an agar spot test and a well diffusion assay were used by Schillinger and

TABLE I
Effect of Plate Media Composition on Zone Diameter in the Agar Spot Assay

		Zone size (mm)	
Bottom agar	Overlay Agar	*Lactococcus lactis* subsp. *lactis* 11454	*Lactococcus lactis* subsp. *lactis* LM0230
BHI	BHI	0.5	NZD[a]
BHI	M17-Glu	1.5	NZD
BHI	MRS	1.5	1.0
M17-Glu	BHI	3.0	NZD
M17-Glu	M17-Glu	3.0	NZD
M17-Glu	MRS	3.0	NZD
MRS	BHI	6.0	7.0
MRS	M17-Glu	6.0	7.0
MRS	MRS	6.0	7.0

Note: *Listeria monocytogenes* 38 was used as the indicator organism. Zone size was the distance from the edge of the producer spot to the edge of the clearing zone (Spelhaug and Harlander, 1989).
[a] NZD, No zone detected.

Lücke (1989). The agar spot test was an adaptation of that described by Fleming et al. (1975), also used by Spelhaug and Harlander (1989) and described earlier. In the well diffusion assay MRS broth containing 0.2% glucose and 1.2% agar was overlaid with 7 ml of soft MRS (containing 0.7% agar) seeded with 0.3 ml of overnight indicator culture. Wells of 3 mm diameter were cut into these agar plates and 0.03 ml of the filtered, dialyzed, and catalase-treated supernatant of the culture growth was placed into each well. The plates were incubated anaerobically 24 hours at 25°C. For a semiquantitative assay, twofold serial dilutions of the treated supernatant were used.

Lewus and Montville (1991) did a survey using three screening techniques and two media to compare bacteriocin production and sensitivity for 22 strains of lactic acid bacteria. They found the spot-on-lawn assay more reproducible, rapid, and easy to score than the flip plate deferred antagonism method (Kekessy and Piguet, 1970) and the well-diffusion assay (Harris et al., 1989). The flip plate method was found cumbersome and of limited value, while the well-diffusion method gave a large number of false-negative results. As noted earlier, MRS agar was not found to be a satisfactory screening medium because the high levels of acid produced from glucose generated zones of inhibition around all strains. Use of Tryptic soy agar without glucose plus 0.5% yeast extract prevented this occurrence.

Spot-on-lawn agar plates (Lewus and Montville, 1991) were implemented by Lewus et al. (1992) to distinguish cidal versus static mode of action of leuconocin S versus *Lactobacillus sake* 15521. Protease rescue with α-chymotrypsin was used to determine bacteriostatic action rather than bactericidal effect. To do this, the bacteriocin producer, *Leuconostoc* OX was spot-inoculated onto MRS broth or Trypticase soy broth without glucose plus 0.5% yeast extract, each with 2% agar added and incubated anaerobically at 30°C overnight. After incubation the plates were overlaid with BHI containing 1% agar and the indicator, *L. sake*. An agar plug was removed from the zone of inhibition and dissolved in phosphate buffer, pH 6.5, and used to inoculate MRS broth for a 48-hour, 30°C incubation to determine if any viable cells remained. α-Chymotrypsin was used on a second spot-on-lawn plate by spotting a 2-μl (10 mg/ml) amount adjacent to the *Leuconostoc* OX colony in the zone of inhibition. The plates were incubated overnight at 4°C to allow for protease diffusion and then incubated overnight anaerobically at 30°C. If cidal, no cells should be detected, but if static, protease rescue results in viable cells. Appropriate controls were run concurrently.

As evidenced by the descriptions of these procedural adaptations from selected laboratories working with bacteriocins, there are noted differences in approach, but the framework of using plating techniques is essentially the same. The study of bacteriocins from food-related lactic acid bacteria, for the most part and with the exception of noted genetic investigations, is still largely descriptive in nature. Surveys continue to search for more effective bacteriocins with greater potential for use in biocontrol systems. As this area of research continues to mature, the mechanisms of bacteriocin activities and indicator resistances will be closely examined as well.

III. Liquid Media

Liquid methods are not as commonly used for the screening of bacteriocin activity. While plating methods are simpler and quicker to do, the use of liquid methods usually presents greater option for closer examination of bacteriocin activity. One approach is to prepare an extract of the growth medium of a bacteriocin-producing culture. Normally the cells are removed by centrifugation and filtration to present a crude cell-free preparation of the bacteriocin. Dialysis is common to remove organic acids and other dialyzable compounds from the preparation, or the acidity can be neutralized with sodium hydroxide. This extract can be added to a well cut into an agar plate and examined against an indicator strain; or the extract can be inoculated with the indicator strain, with subsequent growth of the indicator monitored by a spectrophotometer or by plate counts. The growth kinetics of the indicator can be followed in this way. Also, the crude bacteriocin extract can be pH adjusted, diluted, amended with supplements or additives, or manipulated in other ways to measure different variables. Two factors to be addressed in such an approach are the appropriateness of the medium for growth of the indicator strain and the depletion of nutrients from the extract by the bacteriocin-producing culture. To correct the latter case, the extract can be dialyzed against sterile growth medium to replenish dialyzable nutrients or additional sterile medium can be added to the extract, although this would dilute the concentration of the bacteriocin, and often the bacteriocin is produced in such low amounts that further dilution is undesirable. Concentration of the bacteriocin in the growth extract is sometimes necessary. For example, the use of a dry filter paper disk method, whereby the paper is wetted with growth extract and then placed on the surface of indicator-seeded agar, may not be effective given the low amounts of bacteriocin produced by many lactic acid bacteria.

IV. Titration of Bacteriocins: Critical Dilution Method

The need frequently occurs for a means of quantitatively estimating the activity of a bacteriocin. Arbitrary units (AU) have been used that are commonly based on the critical dilution of antagonistic activity caused by a bacterial culture. The critical dilution method was developed for the assay of colicins and pyocins. In turn, these assays were derived from earlier methods developed for the titration of bacteriophage. The usual fundamentals are (Mayr-Harting et al., 1972):

1. preparation of a twofold series of dilutions of the bacteriocin sample;
2. use of a standardized plating method for evaluating the antagonistic effect of uniform aliquots from each dilution; and
3. after incubation, selection of an arbitrary endpoint, which is often the last dilution exhibiting total inhibition.

Normally the AU is the reciprocal of the dilution endpoint, that is, 100 AU/ml represents the 10^{-2} dilution that last displayed total inhibition.

The critical dilution method is well suited to preliminary study of a bacteriocin when the protein has not been adequately characterized or purified. The method can become comparative if a known reference sample is used, as would be the case with nisin, where its activity can be standardized.

Joerger and Klaenhammer (1986) adapted the critical dilution assay of Mayr-Harting et al. (1972) to titer helveticin J activity of *Lactobacillus helveticus* 481 against *Lactobacillus bulgaricus* 1489. To measure bactericidal action of helveticin J, growth extract was dialyzed against 0.1M sodium acetate buffer (pH 5.3), filter sterilized, and diluted to give 0, 3.2, and 160 AU/ml. Ten ml of each preparation was added to a sterile cuvette that also received buffer-washed cells of *L. bulgaricus* to yield 10^7 CFU/ml. Optical density at 590 nm and CFU/ml were determined immediately after the indicator was added and after 3 hours at 37°C.

Disadvantages of this method are the subjective judgment of the observer in determining the endpoint, and the vulnerability the method has to inconsistent procedural anomalies existing among laboratories. Variation in techniques when using a critical dilution method will have some effect on the comparability of these data. Day-to-day variation may also be a concern, as will loss in potency of a bacteriocin extract with storage.

V. Survivor Counts

Assessment of the efficacy of a bacteriocin or bacteriocin-producing culture in a model food system is normally conducted by challenge with an indicator organism in the food and subsequent monitoring of the viability of the indicator using plate count methods. Proper controls must be applied to determine the extent to which antibiosis is due to bacteriocins and not to organic acids and other inhibitive compounds and conditions. For example, maintenance of anaerobic conditions for growth of any bacteriocin-producing lactic acid bacteria will preclude formation of hydrogen peroxide, as oxygen will be unavailable for the production of this toxic compound.

As with any enumerative method for microorganisms in foods, the sampling, media, and conditions of incubation should be optimal for the indicator strain to insure colony formation by all viable cells including injured ones. Also, the cells should not form chains or clumps to ensure precision of colony counting.

Application of bacteriocins from lactic acid bacteria as food preservatives has been actively pursued in cultured products where lactic acid bacteria are used in the fermentations. With the notoriety that *Listeria monocytogenes* has received by its presence in animal-derived foods, the use of bacteriocin-positive starter cultures in fermented meat and dairy products has received much attention. Since pediococci are the primary starter cultures for commercial fermented sausage production and can produce bacteriocins active against *L. monocytogenes*, bacteriocin-producing pediococci have been tested for effectiveness in meat emulsions simulating sausage manufacture (Berry et al., 1990; Nielsen et al., 1990; Yousef et al., 1991). These works are described elsewhere in this book. Results indicate that the production of a bacteriocin by a lactic culture represents a significant contribution for inhibiting a sensitive pathogen when present in a fermented meat system, in addition to the antagonistic effects of organic acid production.

References

Abraham, E. P., Chain, E., Fletcher, C. M., Florey, H. W., Gardner, A. D., Heatley, N. G., and Jennings, M. A. (1941). Further observations on penicillin. *Lancet* ii, 177–188.
Alfoldi, L. (1958). La production induite de megacine en milieu synthetique. *Ann. Inst. Pasteur Paris* 94, 474–484.
Barrow, G. I. (1963). Microbial antagonism by *Staphylococcus aureus*. *J. Gen. Microbiol.* 31, 471–481.

Barry, A. L., and Fay, G. D. (1973). The amount of agar in antimicrobic disc susceptibility test plates. *Am. J. Clin. Pathol.* **59**, 196–198.

Berry, E. D., Liewen, M. B., Mandigo, R. W., and Hutkins, R. W. (1990). Inhibition of *Listeria monocytogenes* by bacteriocin-producing *Pediococcus* during the manufacture of fermented semidry sausage. *J. Food Protect.* **53**, 194–197.

Bhunia, A. K., Johnson, M. C., and Ray, B. (1988). Purification, characterization and antimicrobial spectrum of a bacteriocin produced by *Pediococcus acidilactici*. *J. Appl. Bacteriol.* **65**, 261–268.

Brock, T. D., Peacher, B., and Pierson, D. (1963). Survey of the bacteriocins of enterococci. *J. Bacteriol.* **86**, 702–707.

Chen, Y. R., Barefoot, S. F., Hughes, M. D., and Hughes, T. A. (1992). The agent from gram-positive cells that enhances lactacin B production by *Lactobacillus acidophilus* N2 in associative cultures. Institute of Food Technologists Annual Meeting, New Orleans, Louisiana. Abstract 668.

Clarke, D. J., Robson, R. M., and Morris, J. G. (1975). Purification of two *Clostridium* bacteriocins by procedures appropriate to hydrophobic proteins. *Antimicrob. Agents Chemother.* **7**, 256–264.

Cooper, K. E. (1964). The theory of antibiotic inhibition zone. *In* "Analytical Microbiology" (F. Kavanaugh, ed.), pp. 1–86. Academic Press, Inc., New York.

Davey, G. P., and Richardson, B. C. (1981). Purification and some properties of diplococcin from *Streptococcus cremoris* 346. *Appl. Environ. Microbiol.* **41**, 84–89.

Fleming, H. P., Etchells, J. L., and Costilow, R. N. (1975). Microbial inhibition of an isolate of *Pediococcus* from cucumber brines. *Appl. Microbiol.* **30**, 1040–1042.

Fredericq, P. (1948). Actions antibiotiques reciproques chez les Enterobacteriaceae. *Rev. Belge. Pathol. Med. Exp.* **19(Suppl. 4)**, 1–107.

Gagliano, V. J., and Hinsdill, R. D. (1970). Characterization of a *Staphylococcus aureus* bacteriocin. *J. Bacteriol.* **104**, 117–125.

Gratia, A. (1946). Techniques selectives pour la recherche systematique des germes antibiotiques. *C. R. Seances Soc. Biol. Paris* **140**, 1053–1055.

Hale, E. M., and Hinsdill, R. D. (1973). Characterization of a bacteriocin from *Staphylococcus aureus* 462. *Antimicrob. Agents Chemother.* **4**, 634–640.

Hamada, S., and Ooshima, T. (1975). Inhibitory spectrum of a bacteriocin-like substance (mutacin) produced by some strains of *Streptococcus mutans*. *J. Dent. Res.* **54**, 140–145.

Harris, L. J., Daeschel, M. A., Stiles, M. E., and Klaenhammer, T. R. (1989). Antimicrobial activity of lactic acid bacteria against *Listeria monocytogenes*. *J. Food Protect.* **52**, 384–387.

Hastings, J. W., and Stiles, M. E. (1991). Antibiosis of *Leuconostoc gelidum* isolated from meat. *J. Appl. Bacteriol.* **70**, 127–134.

Hughes, M. D., and Barefoot, S. E. (1990). Activity of the bacteriocin lactacin B is enhanced during associative growth of *Lactobacillus acidophilus* N2 and *Lactobacillus leichmannii* ATCC 4797. American Society for Microbiology Annual Meeting, Anaheim, California. Abstract O-12.

Jetten, A. M., Vogels, G. D., and De Windt, F. (1972). Production and purification of a *Staphylococcus epidermidis* bacteriocin. *J. Bacteriol.* **112**, 235–242.

Joerger, M. C., and Klaenhammer, T. R. (1986). Characterization and purification of helveticin J and evidence for a chromosomally determined bacteriocin produced by *Lactobacillus helveticus* 481. *J. Bacteriol.* **167**, 439–446.

Kekessy, D. A., and Piguet, J. D. (1970). New method for detecting bacteriocin production. *Appl. Microbiol.* **20**, 282–283.

Kelstrup, J., and Gibbons, R. J. (1969). Bacteriocins from human and rodent streptococci. *Arch. Oral Biol.* **14**, 251–258.

Klaenhammer, T. R. (1988). Bacteriocins of lactic acid bacteria. *Biochimie* **70**, 337–349.

Konisky, J. (1978). The bacteriocins. *In* "The Bacteria" (L. N. Ornston and J. R. Sokatch, eds.), Vol. 6, pp. 71–136. Academic Press, Inc., New York.

Lewus, C. B., and Montville, T. J. (1991). Detection of bacteriocins produced by lactic acid bacteria. *J. Microbiol. Meth.* **13**, 145–150.

Lewus, C. B., Sun, S., and Montville, T. J. (1992). Production of an amylase-sensitive bacteriocin by an atypical *Leuconostoc paramesenteroides* strain. *Appl. Environ. Microbiol.* **58,** 143–149.

Linton, A. H. (1958). Influence of inoculum size on antibiotic assays by the agar diffusion technique with special reference to *Klebsiella pneumoniae* and streptomycin. *J. Bacteriol.* **76,** 94–103.

Linton, A. H. (1983). Theory of antibiotic inhibition zone formation, disc sensitivity methods and MIC determinations. *In* "Antibiotics: Assessment of Antimicrobial Activity and Resistance" (A. D. Denver and L. B. Quesnel, eds.), pp. 19–30. Academic Press, Inc., London.

Mayr-Harting, A., Hedges, A. J., and Berkeley, R. C. W. (1972). Methods for studying bacteriocins. *In* "Methods in Microbiology" (T. Bergen and J. R. Norris, eds.), Vol. 7A, pp. 315–422. Academic Press, Inc., London.

McGroarty, J. A., and Reid, G. (1988). Detection of a lactobacillus substance that inhibits *Escherichia coli. Can. J. Microbiol.* **34,** 974–978.

Meitert, E. (1969). Sur les bacteriocines de *Corynebacterium diphtheriae.* II. Etude des bacteriocines produites par *C. diphtheriae, C. ulcerans, C. atypique, C. hoffmani* and *C. xerose. Arch. Rouni. Pathol. Exp. Microbiol.* **28,** 1086–1097.

Nielsen, J. W., Dickson, J. S., and Crouse, J. D. (1990). Use of bacteriocin produced by *Pediococcus acidilactici* to inhibit *Listeria monocytogenes* associated with fresh meat. *Appl. Environ. Microbiol.* **56,** 2142–2145.

Pasteur, L., and Joubert, J. F. (1877). Charbon et septicemie. *C. R. Soc. Biol. Paris* **85,** 101–115.

Piddock, L. J. V. (1990). Techniques used for the determination of antimicrobial resistance and sensitivity in bacteria. *J. Appl. Bacteriol.* **68,** 307–318.

Reddish, G. F. (1929). Methods for testing antiseptics. *J. Lab. Clin. Med.* **14,** 649–658.

Rogers, A. H. (1972). Effect of the medium on bacteriocin production among strains of *Streptococcus mutans. Appl. Microbiol.* **24,** 294–295.

Rogers, A. H. (1974). Bacteriocin production and susceptibility among strains of *Streptococcus mutans* grown in the presence of sucrose. *Antimicrob. Agents Chemother.* **6,** 547–550.

Rogers, A. M., and Montville, T. J. (1991). Improved agar diffusion assay for nisin quantification. *Food Biotechnol.* **5,** 161–168.

Sabine, D. B. (1963). An antibiotic-like effect of *Lactobacillus acidophilus. Nature (London)* **199,** 811.

Schillinger, U., and Lücke, F.-K. (1989). Antibacterial activity of *Lactobacillus sake* isolated from meat. *Appl. Environ. Microbiol.* **55,** 1901–1906.

Seed, H. D. (1970). Ph.D. Thesis, University of Bristol, UK.

Spelhaug, S. R., and Harlander, S. K. (1989). Inhibition of foodborne bacterial pathogens by bacteriocins from *Lactococcus lactis* and *Pediococcus pentosaceus. J. Food Protect.* **52,** 856–862.

Tagg, J. R., Dajani, A. S., and Wannamaker, L. W. (1975). Bacteriocin of a group B streptococcus: partial purification and characterization. *Antimicrob. Agents Chemother.* **7,** 764–772.

Tagg, J. R., Dajani, A. S., and Wannamaker, L. W. (1976). Bacteriocins of gram-positive bacteria. *Bacteriol. Rev.* **40,** 722–756.

Tramer, J., and Fowler, G. G. (1964). Estimation of nisin in foods. *J. Sci. Food Agric.* **15,** 522–528.

Tsai, H.-J., and Sandine, W. E. (1987). Conjugal transfer of nisin plasmid genes from *Streptococcus lactis* 7962 to *Leuconostoc dextranicum* 181. *Appl. Environ. Microbiol.* **53,** 352–357.

Upreti, G. C., and Hinsdill, R. D. (1973). Isolation and characterization of a bacteriocin from a homofermentative *Lactobacillus. Antimicrob. Agents Chemother.* **4,** 487–494.

Yousef, A. E., Luchansky, J. B., Degnon, A. J., and Doyle, M. P. (1991). Behavior of *Listeria monocytogenes* in wiener exudates in the presence of *Pediococcus acidilactici* H or pediocin AcH during storage at 4 or 25°C. *Appl. Environ. Microbiol.* **57,** 1461–1467.

CHAPTER 3

Biochemical Methods for Purification of Bacteriocins

PETER M. MURIANA
JOHN B. LUCHANSKY

I. Introduction

Lactic acid bacteria (LAB) have been used for centuries in the preparation and preservation of foods of meat, milk, and vegetable origin and are generally recognized as safe (GRAS). These bacteria produce a heterologous array of functional products including organic acids, flavor compounds, proteases, and various antimicrobials, most notably bacteriocins. Bacteriocins are bactericidal proteins that inhibit other, usually closely related, bacteria (Tagg et al., 1976). More recently, Tagg (1991) has argued that the term "bacteriocin" be restricted to inhibitors that more closely resemble the bacteriocins produced by *Escherichia coli* and related Gram-negative bacteria (colicins), and that the term bacteriocinlike inhibitory substances (BLIS) be used for other proteinaceous inhibitors. We retain the term bacteriocin herein to describe bactericidal proteins produced by LAB because this descriptor has been in use for decades.

General properties shared by most bacteriocins of LAB origin include sensitivity to various proteases, bactericidal mode of action, insensitivity to heat, and narrow inhibitory spectrum (Klaenhammer, 1988). However, some bacteriocins (e.g., nisin, pediocins, and sakacins) produced by LAB exhibit more broad-spectrum inhibitory activity toward Gram-positive bacteria. Thus, there is considerable interest in exploiting bacteriocins and LAB (i.e., biopreservation systems) as supplemental barriers to preclude the growth/survival of pathogenic and spoilage organisms in foods. Pursuant to the use of bacteriocins for (bio-) control of undesirable bacteria in food and/or other applications, research efforts have been directed to obtain information on their physical properties. Invariably, this has involved the purification and characterization of the bioactive moieties and the genetic sequences that encode for them. In this chapter, we review recent advances in the purification, detection, production, and physical characterization of bacteriocins of LAB origin and briefly discuss their applications.

II. Detection and Assay of Bacteriocin Activity

Methods to detect and assay bacteriocin activity from LAB have been "borrowed" from prior studies with Gram-negative bacteriocins, primarily the colicins (Mayr-Harting et al., 1972). Direct or deferred antagonism (i.e., either simultaneous growth of bacteriocin-producing and indicator cells or inhibition of indicator lawns by preformed bacteriocin, respectively) are methods frequently used to detect bacteriocin activity. These methods may involve (1) spotting culture supernatants on indicator lawns; (2) crossstreaking bacteria; (3) overlaying colonies of the producer strain with an indicator lawn; (4) agar well diffusion of culture supernatant; and/or (5) the flip plate method (Table I).

Although various methods have been devised to approximate bacteriocin activity, the most common method used is the "critical dilution method" (Mayr-Harting et al., 1972). Briefly, culture supernatant (bacteriocin) and dilutions thereof are spotted on an indicator lawn and activity is "quantitated" subjectively in arbitrary units (AU) of activity (i.e., the reciprocal of the last dilution demonstrating discernible activity). Since the titration of bacteriocin activity is subject to error, depending on the reproducibility of the indicator cell concentration and the ability of the investigator to determine the last dilution showing inhibition, values determined for AU merely approximate rather than precisely quantitate bacteriocin activity. As described below, detection of antimicrobial activity is critical for subsequent studies to optimize growth conditions for the production of bacteriocins.

III. Production of Bacteriocins

A. Production on Agar

Most studies demonstrating bacteriocinogenesis by a producing organism are initiated by detection of inhibitory activity on an agar medium (usually by an overlay of indicator cells on spotted culture or on plated colonies of the producer strain). However, LAB characteristically produce organic acids

TABLE I
Methods of Testing Bacteria for Production of Bacteriocins

Application	Method	Procedure	Reference
Culture Supernatants			
	1. Spot-on-lawn test (spot test)	Spot neutralized and filter-sterilized supernatant onto an indicator lawn	Tagg et al., 1976
	2. Agar well diffusion	Add neutralized and filter-sterilized supernatant into wells bored into an indicator lawn	Tagg and McGiven, 1971
	3. Activity assays	Spot serial dilutions of neutralized and filter-sterilized supernatant onto indicator lawns	Mayr-Harting et al., 1972
Bacterial Colonies			
	1. Flip plate method	Grow colonies on agar, "flip" agar onto inside of petri plate cover, and overlay with indicator	Kekessy and Piguet, 1970
	2. Sandwich overlay	Plate bacterial dilution, cover with sterile medium, incubate until colonies are visible, and overlay with indicator lawn	Mayr-Harting et al., 1972
	3. Lutri-Plates	Special two-sided agar plates: grow colonies on one side, and overlay other side with indicator lawn	Lutri-Plate™ [a]

[a] Lutri-Plate, Inc., Starkville, Mississippi.

(e.g., lactic, acetic) and other antimicrobials (e.g., hydrogen peroxide) that can also inhibit indicator cells. The problem is exacerbated when screening for bacteriocin susceptibility by nonlactic organisms, which are often more sensitive to acid than LAB indicators. Modifications to exclude inhibition by acid include buffering of the agar medium, reduction of fermentable carbohydrate in the agar medium, or use of an indicator organism not affected by acid. Potential phage interactions can be eliminated by detection of bacteriocin activity after diffusion through agar, and inhibition due to hydrogen peroxide can be eliminated by incorporation of catalase into the agar medium. In general, the probability of initially observing antimicrobial activity is potentially greater using agar media rather than broth due to continual production and localization/concentration of bacteriocin around producer cells. Also, the use of agar-based media permits facile screening for antagonism of several (different) producer cells against a single indicator.

B. Production in Broth

The production of inhibitory compounds is readily visualized on agar, but further characterization is facilitated by their production in liquid media. Liquid media are also easier to manipulate and monitor than agar-based media, thus allowing for maintenance of optimal growth conditions and maximum bacteriocin production. Although media optimization to increase yields of bacteriocin in culture supernatants at the onset of a purification scheme may not be necessary, it can provide relief via the tenet that "it is always better to start with more bacteriocin than with less." The ability to obtain a concentrated source of crude bacteriocin through optimization of growth parameters greatly simplifies recovery of bacteriocin in subsequent purification steps because dramatic losses of activity may be incurred during the course of protein purification.

Several studies of antimicrobials produced by LAB have used growth media similar to the products the cultures are associated with, namely milk. Hamdan and Mikolajcik (1974), Shahani et al. (1977), and Reddy et al. (1983) reported the production of acidolin, acidophilin, and bulgarican, respectively, in milk by lactobacilli; however, because of variability in proteolytic ability, many LAB do not grow well in milk-based media. For this reason, both commercially prepared dehydrated culture media and defined synthetic media are used to grow LAB for production of bacteriocins.

Most studies undertaken to optimize for the production of bacteriocins by LAB have utilized commercial media (or modified commercial media) to provide a rich supply of growth nutrients (Table II). For example, Upreti and Hinsdill (1973) obtained maximum levels of lactocin 27 from *Lactobacillus helveticus* LP27 cultured in APT broth, and found that substitution of APT with either BHI, trypticase soy, or thioglycollate broth yielded inferior results. Similarly, McGroarty and Reid (1988) examined skim milk, BHI broth, BHI plus 2% yeast extract (BYE), nutrient broth (1% and 2.5%), Mueller Hinton broth, tryptose broth, and MRS broth for the ability of *Lactobacillus casei* GR-1 and *Lactobacillus acidophilus* 76 to produce an inhibitor against *E. coli* Hu734. They found that both strains of lactobacilli produced maximum amounts of inhibitor when grown in MRS broth. Additionally, de Klerk and Smit (1967) dialyzed MRS broth prior to inoculation with bacteriocinogenic LAB to remove high molecular weight components from the growth medium, which could interfere in subsequent purification steps. Dialyzed MRS broth has also facilitated the purification of pediocin AcH (produced by *Pediococcus acidilactici* H). Bhunia et al. (1987; 1988) prepared casein glucose broth as a 10× concentrate, which was dialyzed with a 14-kDa membrane against deionized water; the dialysate was then autoclaved and used as the growth medium for *P. acidilactici* H.

Bacteriocin production is also influenced by growth conditions. Production of lactacin B (Barefoot and Klaenhammer, 1984), helveticin J (Joerger and Klaenhammer, 1986), and lactacin F (Muriana and Klaenhammer,

TABLE II
Conditions for Production and Purification of Bacteriocins of Lactic Acid Bacteria

Organism	Bacteriocin	Medium[a]	Purification scheme[a]	Molecular weight (kDa)	Reference
Carnobacterium					
C. piscicola LV17	NN[b]	APT, pH 6.5	AS, dialysis	ND[c]	Ahn and Stiles, 1990a
C. piscicola UAL26	NN	APT, pH 6.5	UF	ND	Ahn and Stiles, 1990b
C. piscicola LV61	NN	D-MRS (MRS w/o acetate, pH 8.5)	NP[d]	ND	Schillinger and Holzapfel, 1990
Lactobacillus					
L. acidophilus 11088	Lactacin F	MRS held at pH 7.0	AS, GF, HPLC, SDS-PAGE	6.3	Muriana and Klaenhammer, 1987, 1991a
L. acidophilus N2	Lactacin B	MRS held at pH 6.0	IEX, UF, GF	6.0	Barefoot and Klaenhammer, 1984
L. acidophilus LAPT1060	Acidophilucin A	MRS (0.5% gluc/0.5% lact)	NP	ND	Toba et al., 1991c
L. brevis B37	Brevicin 37	TJM & SM-2$_p$ (synth. med.)	NP	ND	Rammelsberg and Radler, 1990
L. brevis B155	NN	TSA (w/o gluc; w/0.5% YE)	NP	ND	Lewus and Montville, 1991
L. casei B80	Caseicin 80	TJM & SM-2 (synth. med.)	UF, CEX, GF-FPLC	~41	Rammelsberg et al., 1990
L. delbrueckii JCM1106	Lacticin A	MRS (0.5% gluc/0.5% lact)	NP	ND	Toba et al., 1991b
L. delbrueckii JCM1248	Lacticin B	MRS (0.5% gluc/0.5% lact)	NP	ND	Toba et al., 1991b
L. fermenti 466	NN	MRS dialysate	Vacuum concentrated, dialysis, GF, CEX	ND	de Klerk and Smit, 1967
L. gasseri strains	Gassericin A	MRS	NP	ND	Toba et al., 1991d
L. helveticus 481	Helveticin J	MRS held at pH 5.5	AS, GF, SDS-PAGE	~37	Joerger and Klaenhammer, 1986

(*continued*)

TABLE II (*Continued*)

Organism	Bacteriocin	Medium[a]	Purification scheme[a]	Molecular weight (kDa)	Reference
L. helveticus LP27	Lactocin 27	APT w/o Tween 80	GF, CHCl$_3$ precipitation, SDS-PAGE	12.4	Upreti and Hinsdill, 1973
L. plantarum C-11	Plantaricin A	MRS	Concentrated by dialysis	ND	Daeschel et al., 1990
L. plantarum NCDO 1193	Plantacin B	MRS (0.5% gluc)	NP	ND	West and Warner, 1988
L. reuteri LA6	Reutericin 6	MRS	UF, dialysis	ND	Toba et al., 1991a
L. sake Lb706	Sakacin A	MRS-0.2 (0.2% gluc)	Dialysis, vacuum concentrated	ND	Schillinger and Lucke, 1989
L. sake L45	Lactocin S	MRS	AS, GF, IEX, HIC, HPLC	ND	Mortvedt et al., 1991
L. sake 148	NN	MRS	NP	ND	Sobrino et al., 1991
Lactobacillus sp. 100-37	NN	MRS	AS, dialysis	ND	McCormick and Savage, 1983
Lactococcus					
L. cremoris 346	Diplococcin	M17 broth; M17 agar	AS, CEX, dialysis, SDS-PAGE	5.3	Davey and Richardson, 1981
L. cremoris LMG2130	Lactococcin A	M17 broth; M17G agar	AS, CEX, HIC, FPLC	5.8	Holo et al., 1991
L. diacetylactis S50	Bacteriocin S50	M17G	NP	ND	Kojic et al., 1991
L. lactis ADRIA-85LO30	Lactococcin	CG broth; M17G agar	Dialysis, CEX, GF, SDS-PAGE	2.3	Dufour et al., 1991
L. lactis DRC 1	Dricin	SL1 broth	NP	ND	Powell et al., 1990
L. lactis CNRZ 481	Lacticin 481	Elliker (w/o gelatin, pH 5.5)	AS, GF, HPLC, SDS-PAGE	1.7	Piard et al., 1992
L. lactis	Nisin	Commercial preparation	GF, ultracentrifugation	3.5	Jarvis et al., 1968

Leuconostoc				
L. gelidum UAL187	Leucocin A-UAL 187	CAA broth (semidefined)	AS, AP, HIC, GF, HPLC, SDS-PAGE	3.9	Hastings et al., 1991
L. gelidum IN139	NN	BM broth (MRS w/o citrate/acetate)	Dialysis	ND	Harding and Shaw, 1990
L. mesenteroides UL5	Mesenterocin 5	MRS	AS, dialysis, UF, SDS-PAGE	4.5	Daba et al., 1991
L. paramesenteroides OX	Leuconocin S	APT	SDS-PAGE	ND	Lewis et al., 1992
Pediococcus				
P. acidilactici H	Pediocin AcH	TGE	AS, GF, IEX, SDS-PAGE	2.7	Biswas et al., 1991; Bhunia et al., 1988
P. acidilactici PAC1.0	Pediocin PA-1	MRS	AS, dialysis, GF, IEX	16.5	Gonzalez and Kunka, 1987
P. acidilactici PO2	NN	APT	NP	ND	Hoover et al., 1989
P. acidilactici JD1-23	NN	MRS + 2% YE (MRS-YE)	NP	ND	Berry et al., 1991
P. pentosaceus FBB61	Pediocin A	MRS	NP	ND	Daeschel and Klaenhammer, 1985

[a]Abbreviations: MRS, APT, M17, and Elliker broths are commercially available media; TJM, tomato juice medium; TSA, tryptic soy agar; YE, yeast extract; M17G, M17-glucose; SM-2, CG, SL1, CAA, BM, TGE, and D-MRS broths are modified media; AS, ammonium sulfate precipitation; AP, acid precipitation; GF, gel filtration; IEX, ion exchange; CEX, cation exchange; HPLC, high performance liquid chromatography; UF, ultrafiltration; FPLC, fast protein liquid chromatography; HIC, hydrophobic interaction chromatography; SDS-PAGE, sodium dodecyl sulfate-polyacrylamide gel electrophoresis.
[b]NN, Not named.
[c]ND, Not determined.
[d]NP, Not purified.

1987) was examined in MRS-based media maintained at pH levels ranging from 5.0 to 7.5 by controlled addition of ammonium hydroxide. Maximum levels of helveticin J (produced by *Lactobacillus helveticus* 481), lactacin B (produced by *L. acidophilus* N2), and lactacin F (produced by *L. acidophilus* 11088) were detected at pH 5.5, 6.0, and 7.0, respectively, demonstrating that conditions affecting bacteriocin synthesis can vary even among closely related strains. As another example, Biswas et al. (1991) adjusted MRS broth (TGE broth: trypticase, yeast extract, glucose, Tween 80, Mg^{2+}, Mn^{2+}) to achieve maximal production of pediocin AcH by *Pediococcus acidilactici* H. However, a pH effect was observed whereby negligible levels of pediocin AcH were produced when TGE was maintained above pH 5.0. These data indicate that optimization of growth media can potentiate the production of bacteriocins and potentially contribute to greater yields of the purified product.

IV. Bacteriocin Purification

A. Biochemical Methods

After detecting antagonism of producer cells against indicator cells on microbiological media and perhaps identifying conditions that lead to enhanced production of the inhibitory substance(s), research efforts are usually directed to recover the bacteriocin in purified form. A variety of techniques (or combinations thereof) have been used to obtain purified, or partially purified, bacteriocins of LAB origin (Table II). The final purification regimen may be determined following consideration of the targeted application of the desired end product. For example, some applications (e.g., determination of amino acid sequence) would require highly purified bacteriocin preparations, whereas in other applications (e.g., use as a biopreservation system) it may be preferable to forego purity to recover high yields of active bacteriocin. Cost, reproducibility, and logistics may also influence the selection of a purification regimen. A factor of paramount importance to the overall success of any purification scheme is the ability to assay for the protein being purified at each step in the process. In this regard, the presence of bacteriocins can be determined by titering for biological activity and/or by other analytical methods (e.g., SDS-PAGE, staining methods). The progress of a purification scheme can also be followed at each step in the process by monitoring arbitrary units (total AU or AU/ml), protein concentration (mg/ml), and/or specific activity (AU/mg). Empirical observations of fluctuations in the aforementioned parameters often are used to tailor methodologies for enhanced purification of bacteriocins.

Because bacteriocins are secreted into the growth medium, most approaches to purification are initiated with a method to concentrate bac-

3. Biochemical Methods for Purification of Bacteriocins

teriocins from culture supernatants (i.e., reduce the working volume). Vacuum concentration has been used as a method to concentrate bacteriocins in culture broth, but components of the growth medium and other cellular products are also retained (de Klerk and Smit, 1967; Schillinger and Lucke, 1989). Other methods, such as precipitation and size exclusion, can simultaneously concentrate bacteriocins and eliminate undesirable (contaminating) proteins. Precipitation of bacteriocins from culture supernatants has been accomplished by salt fractionation (e.g., ammonium sulfate, NaCl), acid precipitation (Hastings et al., 1991), or organic solvents (Upreti and Hinsdill, 1973). These procedures are well suited for an initial step in the purification process because they accommodate large volumes of starting materials. Ammonium sulfate precipitation is often the method of choice because it is gentle and simple to perform. More specifically, as ammonium sulfate molecules become hydrated, the salt ions "tie up" water molecules that would otherwise be associated with maintaining proteins (i.e., bacteriocins) in solution. As the number of stabilizing water molecules is reduced, hydrophobic regions of proteins associate to generate aggregates that precipitate. Ammonium sulfate precipitation is an effective method to fractionate proteins, since the variable distribution of hydrophobic and hydrophilic regions allows specific proteins to precipitate over a narrow range of salt concentrations.

Other methods that have proven valuable to simultaneously concentrate and purify bacteriocins include dialysis and ultrafiltration. Dialysis is generally performed in dialysis "bags" in which a molecular equilibrium causes an exchange of diffusible molecules, whereas ultrafiltration utilizes positive pressure to force the equilibrium in favor of the filtrate. However, both methods are similar in their use of membranes of specific pore size to retain proteins above a particular size and allow smaller proteins (or other molecules) to escape. Concomitant loss of water allows concentration of retained proteins. Detergents may then be used to "free" bacteriocins from associated constituents, such as media components and cellular debris, followed by extensive dialysis to remove salts and/or detergents from protein samples prior to subsequent purification steps.

Although the aforementioned procedures play an important role in bacteriocin purification, they typically do not provide for a high degree of resolution from the plethora of contaminating proteins contributed by growth media and/or cellular metabolism. Therefore, several modes of chromatography including size exclusion (gel filtration), adsorption (ion exchange), and/or hydrophobic interaction (reversed-phase) chromatography have been used to achieve significant purification of bacteriocins. High performance liquid chromatography (HPLC) has also been used, albeit less frequently, than low-pressure column systems to obtain more highly purified bacteriocin preparations. As a final step, the purity of the bacteriocin preparation should be confirmed; gel electrophoresis (SDS-PAGE) is commonly employed for this purpose. This chapter will not focus on the details of the theory and/or practice of protein purification procedures because they are reviewed in sufficient clarity and depth elsewhere (Deutscher, 1990; Scopes, 1987). However, the utility of protein purification methods will be highlighted, as they have been specifically exploited to purify LAB bacteriocins.

B. Purified LAB Bacteriocins

1. Lactocin, Helveticin, Lactacin, and Caseicin

Bacteriocins of *Lactobacillus* spp. are typically produced as large native complexes. Lactocin 27, produced by *L. helveticus* LP27, was purified to homogeneity by sequential rounds of organic extractions, lyophilization, and gel filtration (Upreti and Hinsdill, 1973). Lactocin 27 is produced as a large complex (>200 kDa) containing protein, lipid, carbohydrate, and phosphorous; however, the purified adduct dissociated by SDS is a glycoprotein estimated at 12.4 kDa (Upreti and Hinsdill, 1973). Another bacteriocin produced by *L. helveticus*, helveticin J, was shown by ultrafiltration to exist in culture supernatants as large aggregates in excess of 300 kDa (Joerger and Klaenhammer, 1986). Despite the use of semidefined media for improved recovery of the bacteriocin and the application of ammonium sulfate precipitation and denaturing gel filtration, helveticin J was not purified to homogeneity. However, extraction of helveticin J activity from SDS-PAGE gel slices correlated with a stained protein band of 37 kDa (Joerger and Klaenhammer, 1986). In later work, helveticin J was isolated from SDS-PAGE gels, and antibodies to this protein were produced in rabbits for use as immuno-probes to subsequently clone, sequence, and express the *L. helveticus* 481 gene (*hlv*) that encoded for helveticin J (Joerger and Klaenhammer, 1990).

Two bacteriocins produced by *Lactobacillus acidophilus* have also been purified. Lactacin B (strain N2; Barefoot and Klaenhammer, 1984) and lactacin F (strain 11088; Muriana and Klaenhammer, 1987) are produced as complexes in excess of 100 kDa and 180 kDa, respectively. Lactacin B was purified by ion-exchange chromatography followed by ultrafiltration and gel filtration in the presence of denaturants ($8M$ urea, 0.1% SDS). Although insufficient levels of lactacin B were obtained for visualization on protein-stained SDS-PAGE gels, molecular weight determinations from gel-extracted activity (6 kDa) correlated with estimates obtained by gel filtration (6.5 kDa) (Barefoot and Klaenhammer, 1984). The other *L. acidophilus* bacteriocin, lactacin F, was effectively fractionated from culture supernatants as a surface pellicle by using relatively low ammonium sulfate saturation levels (Muriana and Klaenhammer, 1991a). The presence of lactacin F in the pellicle was attributed to its association with hydrophobic globular micellelike structures in culture supernatants. Lactacin F was resolved from the associated material by reversed-phase HPLC and shown by SDS-PAGE to have a molecular weight of 2.5 kDa. However, amino acid composition analysis indicated that lactacin F was composed of approximately 57 amino acids, indicating a molecular weight of about 6.2 kDa (Muriana and Klaenhammer, 1991a). Subsequent cloning and sequencing of the lactacin F structural gene (*laf*) confirmed that the mature bacteriocin is 6.3 kDa (Muriana and Klaenhammer, 1991b). The size discrepancy on SDS-PAGE gels may be a consequence of the small (<10 kDa) and hydrophobic nature of lactacin F, which (presumably) allowed for disproportionate binding to SDS and/or excessive charge density and spurious migration.

Caseicin 80, a bacteriocin produced by *Lactobacillus casei* B80, has been purified by ultrafiltration, cation exchange, and gel filtration-FPLC (Rammelsberg et al., 1990). Despite excessive losses during ultrafiltration, the molecular mass of caseicin 80 was estimated at 40–42 kDa by gel filtration (Rammelsberg et al., 1990). However, the molecular weight of caseicin 80 was not obtained under denaturing conditions, nor was it confirmed by another method such as SDS-PAGE. As a final example of a bacteriocin purified from lactobacilli, Mortvedt et al. (1991) recently purified (by ammonium sulfate precipitation, anion and cation exchange, gel filtration, HIC, and reversed-phase FPLC) and partially sequenced lactocin S (produced by *L. sake* L45). Lactocin S eluted well behind the smallest protein standard (RNase A, 13.7 kDa) when analyzed by gel filtration, suggesting it is a relatively small peptide. Amino acid composition analysis indicated lactocin S consists of 33 amino acids with a projected molecular weight of about 3 kDa; however, the molecular weight was not confirmed by another method.

2. Nisin, Diplococcin, Lactococcin, and Lacticin

Nisin is the most well-characterized bacteriocin of LAB origin, having first been isolated in 1951 (Hirsch, 1951). Numerous reviews have since been written that detail the biology, chemistry, biosynthesis, and applications of this bacteriocin (Hurst, 1981; 1983; Lipinska, 1976). To our knowledge nisin remains the only bacteriocin that is commercially available (Nisaplin™, Aplin and Barrett Ltd., Trowbridge, England). For a detailed account of the purification of nisin the reader is referred to Cheeseman and Berridge (1957). In recent years, considerable progress has been made in the purification of other bacteriocins produced by LAB (Table II). As an example, Davey and Richardson (1981) purified diplococcin from *Streptococcus cremoris* 346 primarily by precipitation and cation-exchange chromatography. Gel filtration analysis of purified diplococcin gave two peaks with corresponding molecular weights of 6 and 9 kDa. Although attempts to size diplococcin by SDS-PAGE were obscured by diffuse banding, amino acid composition analysis indicated a molecular weight of 5.3 kDa. Several other bacteriocins from lactococci have also been purified. Lactococcin A, produced by *Lactococcus cremoris* LMG2130, was purified by ammonium sulfate precipitation, cation exchange, hydrophobic interaction, and reversed-phase HPLC (Holo et al., 1991). Protein sequence analysis of lactococcin A revealed 54 amino acids with a projected molecular weight of 5.8 kDa. The molecular weight of lactococcin A was confirmed following the cloning and sequencing of the structural gene (*lcnA;* Holo et al., 1991). A bacteriocin produced by *L. lactis* ADRIA 85L030, also designated as lactococcin, was purified by dialysis, cation exchange, and gel filtration (Dufour et al., 1991). The molecular weight of lactococcin estimated by gel filtration (2.3 kDa) was confirmed by urea-SDS-PAGE (Dufour et al., 1991). However, the smallest LAB bacteriocin identified to date is lacticin 481 (Piard et al., 1992). Lacticin 481 (produced by *L. lactis* CNRZ 481) was initially subjected to two rounds of ammonium sulfate precipitation (60% and 80% saturation) whereby total activity increased 455-fold over that detected in the original culture supernatant; presumably, active monomers were released from bacteriocin complexes. Lacticin 481 was ultimately purified by reversed-phase HPLC and shown by SDS-PAGE to exist predominately as a 1.7-kDa peptide; however, low levels of a dimer (3.4 kDa) were also detected (Piard et al., 1992). Protein sequence analysis of purified lacticin 481 revealed the presence of β-methyllanthionine, dehydroalanine, and dehydrobutyrine, indicating that lacticin 481 is a member of the "lantibiotic" group of antimicrobial peptides. Aside from nisin, lacticin 481 is the only other LAB bacteriocin that has been determined to contain lanthionine residues.

3. Pediocins

Considerable work has also been conducted to characterize and purify bacteriocins produced by pediococci. Briefly, pediocin PA-1 (produced by *Pediococcus acidilactici* PAC.1; Gonzalez and Kunka, 1987) and pediocin AcH (produced by *P. acidilactici* H; Bhunia et al., 1988) were crudely purified from (filtered) culture supernatants by precipitation with ammonium sulfate (60 and 70%, respectively). Dialyzed pediocin PA-1 was further purified using ion-exchange chromatography (IEX) on DEAE-Sephadex A-25 and then CM-Sephadex C-25. Fractions recovered following IEX were concentrated in dialysis tubing by removing water with Carbowax 20 resulting in a tenfold concentration of the partially purified bacteriocin. Although values for percent recovery of total protein, antimicrobial activity, or specific activity were not reported, the molecular weight of pediocin PA-1 was estimated at 16.5 kDa by ascending gel filtration. In contrast, crude pediocin AcH preparations were dialyzed to remove residual ammonium sulfate and further purified on a Superose-12 HR 10/30 FPLC column (Bhunia et al., 1988). Following passage through a gel filtration column and elution from an anion-exchange column, recovery of pediocin AcH was monitored by SDS-PAGE and by determining protein concentration and biological activity. Using this approach, a 98-fold increase in pediocin AcH activity was observed in the final preparation compared to the original culture supernatant. Other pediocins, including those produced by *P. acidilactici* PO2 (Hoover et al., 1989), *P. acidilactici* JD1-23 (Berry et al., 1990), and *P. pentosaceus* FBB61 (pediocin A; Daeschel and Klaenhammer, 1985), have not as yet been purified. Lastly, comparison of native plasmids and inhibitory spectra (Hoover et al., 1989), as well as the genomic fingerprints of producer strains generated by pulsed-field gel analysis (J. B. Luchansky, unpublished data), suggest a common parentage for several strains of *P. acidilactici* (e.g., H, PO2, PAC.1, and several commercial starter cultures) and their associated pediocins. However, neither the amino acid sequence of the various pediocins nor the nucleotide sequence of the corresponding genes has been published.

4. Leucocin, Mesenterocin, and Leuconocin

There have been several recent reports on the detection and concentration or purification of bacteriocins produced by *Leuconostoc* spp. Harding and Shaw (1990) described bacteriocinogenesis among leuconostocs, but attempts were not made to specifically purify and characterize the antimicrobial peptide. Bacteriocins have also been identified in *L. gelidum* (leucocin A-UAL187; Hastings et al., 1991), *L. mesenteroides* (mesenterocin 5; Daba et al., 1991), and *L. paramesenteroides* OX (leuconocin S; Lewus et al., 1992). Perhaps the most thoroughly characterized bacteriocin of this group is leucocin A-UAL187; both the protein (37 amino acids, 3.93 kDa) and the gene (localization on a 2.9-kb *Hpa*II fragment of a native *L. gelidum* plasmid) encoding for this protein have been sequenced. Approximately 60% (approximately 4500-fold purification) of leucocin A-UAL187 activity was recovered by ammonium sulfate or acid precipitation followed by hydrophobic interaction chromatography, gel filtration, and reversed-phase HPLC. Regarding other bacteriocins of *Leuconostoc*, Daba et al. (1991) concentrated mesenterocin 5 from culture supernatants of *L. mesenteroides* UL5 using ammonium sulfate precipitation or ultrafiltration, but did not further purify the antimicrobial peptide. Likewise, efforts were not made to specifically purify leuconocin S from culture supernatants of *L. paramesenteroides* OX; however, SDS-PAGE analysis of crude leuconocin S intimated that 2 small (approximately 2 and 10 kDa) glycoproteins may be involved in antimicrobial activity (Lewus et al., 1992). Further work is required to more precisely and comparatively characterize bacteriocins produced by leuconostocs.

V. Applications of Purified Bacteriocins

A. Biopreservation Systems

Bacteriocins have been evaluated for their ability to inhibit food-related pathogens and spoilage organisms. Although the efficacy of these biopreservation systems has been proven in synthetic media, it has not been firmly established that bacteriocins are equally effective in foods (Delves-

Broughton, 1990; Eckner, 1991). Nisin, by far the most well-characterized LAB bacteriocin, is the only bacteriocin approved for use in foods. In the United States, nisin is used in certain pasteurized cheese spreads to inhibit the outgrowth of *Clostridium botulinum* spores (Federal Register, 1988). Worldwide, nisin is used in a variety of products including pasteurized, flavored, and long-life milks, aged and processed cheeses, and canned vegetables and soups (Delves-Broughton, 1990). More recent applications have utilized nisin to inhibit undesirable LAB in wine (Daeschel et al., 1991; Radler, 1990a; 1990b) and beer (Ogden, 1986; Ogden et al., 1988). Widespread application of nisin in foods is perhaps limited because of its poor solubility, cost, variations in sensitivity among target strains, and the strong influence of environmental parameters (e.g., pH, NaCl, temperature) on its activity (Delves-Broughton, 1990). Therefore, research efforts have been directed to purify and characterize other bacteriocins of LAB origin for application in foods.

With the exception of studies evaluating nisin, to date, relatively little has been published describing the use of (partially) purified bacteriocins to inhibit undesirable organisms directly in food. However, recent studies have revealed the potential of partially purified pediocins as bioactive agents. For example, Pucci et al. (1988) demonstrated that a crude preparation of pediocin PA-1 delayed growth or inhibited *Listeria monocytogenes* in dairy products (e.g., cottage cheese, cheese sauce, and half-and-half) during refrigerated storage. In another study, pediocin PA-1 retained listericidal activity on pediocin-saturated beef pieces for 28 days at 5°C; soaking beef in a concentrated pediocin preparation prior to exposure to *L. monocytogenes* was more effective than application of the pediocin after exposure to the pathogen (Nielsen et al., 1990). Yousef et al. (1991) demonstrated that counts of *L. monocytogenes* decreased by 0.74 $\log_{10}$ CFU/ml in ≤2 hours after the addition of pediocin AcH to wiener exudate during storage at 4°C, and by 3.6 $\log_{10}$ CFU/ml in 6.6 days during storage of exudates at 25°C. These examples demonstrate that purification and subsequent characterization of LAB bacteriocins may lead to the identification and application of bacteriocins with properties superior to nisin.

The question of whether bacteriocins retain antimicrobial activity in foods has been the topic of much recent research. Although bacteriocins are generally less effective in foods than in microbiological media (M. A. Daeschel, Chapter 4, this volume; J. B. Luchansky, unpublished data; Schillinger et al., 1991), preliminary studies indicate that bacteriocins (added directly or produced *in situ* by LAB) do provide an additional barrier against growth or survival of undesirable bacteria in foods (Berry et al., 1990, 1991; Daeschel et al., 1991; Degnan et al., 1992; Delves-Broughton, 1990; Nielsen et al., 1990; Pucci et al., 1988; Sabel et al., 1991; Schillinger et al., 1991; Yousef et al., 1991). It is likely that several factors (food components, poor diffusion, indigenous flora, pH, and storage temperature of foods) may influence the activity of biopreservation systems in foods (Daeschel, 1990; Schillinger et al., 1991). As more detailed information accumulates concerning the structure, function, and *modus operandi* of LAB bacteriocins, future research should be directed to employ genetic or protein engineering to "design" bacteriocins with enhanced or diversified capabilities (i.e., greater killing spectrum, less susceptibility to association/inactivation by food components) and to develop novel delivery systems to afford protection and ensure more even distribution of these molecules in foods. Advances in the purification of bacteriocins will augment efforts to fully exploit their potential as biocontrol agents for extending the shelf life of foods and precluding microbial hazard. At present, this remains an active area of research.

B. Characterization of Physical Properties

Although the structure of nisin was determined in 1971 (Gross and Morell, 1971) and a commercial preparation (Nisaplin™) has been available since 1960, detailed information concerning its mode of action on prokaryotic, eukaryotic, and artificial membranes was only recently obtained (Gao et al.; 1991; Kordel and Sahl, 1986; Kordel et al., 1989; Ruhr and Sahl, 1985). The use of highly purified nisin preparations for plasma desorption mass spectrometry (PDMS) analyses determined the precise mass of nisin as 3354.3 Da (Nielsen and Roepstorff, 1988). Similarly, nuclear magnetic resonance (NMR) analyses confirmed the structure (Chan et al., 1989b; Palmer et al., 1989; Slijper et al., 1989), molecular reactiveness (Liu and Hansen, 1990; Slijper et al., 1989), and degradative products (Chan et al., 1989a) of nisin. Studies designed to elaborate the physical properties, cellular interactions, and genetic relatedness of other LAB bacteriocins remain in their infancy, presumably because the majority of these antimicrobial peptides were only recently identified (Table II). One exception is Hastings et al.'s (1991) study, which employed mass spectrometry to determine the precise mass (3930.3 Da) of leucocin A-UAL187. In another example, van Belkum et al. (1991) reported that lactococcin A induced membrane permeability in membrane vesicles and whole cells in a voltage-independent manner. In contrast, the ability of nisin to promote membrane permeability is contingent upon the presence of a sufficient membrane potential (i.e., voltage-dependent process) (Kordel et al., 1989). More detailed studies on the physical properties of LAB bacteriocins are warranted to precisely determine structure-function attributes of these bioactive peptides.

C. Immunological Studies

Purified bacteriocin preparations have also found utility in immunological studies. For example, antibodies isolated from antisera raised against nisin (conjugated to egg albumen) fostered the development of an enzyme-linked immunosorbent assay (ELISA) to quantify nisin in foods (Falahee et al., 1990). Similar studies performed with albumin-conjugated and nonconjugated pediocin AcH failed to elicit an immune response in mice or rabbits

(Bhunia et al., 1990), suggesting that pediocin AcH would not be toxic to humans if used as a food biopreservative. In another example of an immunologic application, purified helveticin J was injected into rabbits to obtain antihelveticin J antibodies, which were subsequently used as immunoprobes to clone the structural gene (*hlv*) for this bacteriocin (Joerger and Klaenhammer, 1990). Collectively, these examples illustrate that antibodies prepared against purified bacteriocins can be used to localize, quantify, sequester, and analyze bacteriocins and the corresponding genes that encode for these proteins.

D. Protein Sequence Determinations and "Reverse Genetics"

The recovery of purified bacteriocins facilitated efforts to elucidate the primary amino acid sequence of several LAB bacteriocins. Highly purified bacteriocin preparations were used to determine the first 13 and 25 N-terminal amino acids of leucocin A-UAL187 (Hastings et al., 1991) and lactacin F (Muriana and Klaenhammer, 1991b), respectively, and all 54 amino acids of lactococcin A (Holo et al., 1991). Amino acid sequence information and protein composition data indicated that lactacin F, lactococcin A, and leucocin A-UAL187 did not contain unusual or modified amino acids (e.g., lanthionine, dehydroalanine) and revealed that these bacteriocins were different from nisin and the other lantibiotics. In contrast, the presence of lanthionine-related residues (confirmed by composition analysis) restricted protein sequence analysis of lacticin 481 to the first 7 N-terminal amino acids (Piard et al., 1992). Although the presence of lanthionine-related residues has not been demonstrated among other LAB bacteriocins, attempts to sequence the helveticin J protein were also impeded due to unspecified N-terminal blockage (Joerger and Klaenhammer, 1990). N-terminal blockage encountered during sequence analysis of lactocin S was circumvented by cyanogen bromide cleavage and sequencing was initiated from an internal methionine residue (Mortvedt et al., 1991). In another application, knowledge of the amino acid sequence of bacteriocins can be extrapolated to obtain the corresponding sequence of nucleotides for the synthesis of oligomeric probes (i.e., "reverse genetics"). Probes derived in this manner can then be used to screen genomic libraries to sequester bacteriocin genes. Indeed, cloning the structural gene for lactacin F (Muriana and Klaenhammer, 1991b), lactococcin A (Holo et al., 1991), and leucocin A-UAL 187 (Hastings et al., 1991) was accomplished by "reverse genetics."

VI. Summary

During the past decade bacteriocins have quite arguably become a primary focus of research efforts involving lactic acid bacteria. Although detailed studies to understand the mode of action or structure–function relationships of these bioactive peptides have not been widely performed, it is clear from preliminary studies that several biological and physical characteristics are shared among LAB bacteriocins. It is also clear from seminal studies that bacteriocins possess many salient features for application as biopreservatives in foods or as tools (e.g., immuno-probes, food grade markers, "reverse genetics") to further increase our understanding of LAB and their bioactive peptides. Future efforts directed toward molecular characterization of the structure, function, and regulation of highly purified LAB bacteriocins will accelerate efforts to engineer innovative antimicrobial peptides with enhanced capabilities and diversified applications.

References

Ahn, C., and Stiles, M. E. (1990a). Plasmid-associated bacteriocin production by a strain of *Carnobacterium piscicola* from meat. *Appl. Environ. Microbiol.* **56,** 2503–2510.

Ahn, C., and Stiles, M. E. (1990b). Antibacterial activity of lactic acid bacteria isolated from vacuum-packaged meats. *J. Appl. Bacteriol.* **69,** 302–310.

Barefoot, S. F., and Klaenhammer, T. R. (1984). Purification and characterization of the *Lactobacillus acidophilus* bacteriocin lactacin B. *Antimicrob. Agents Chemother.* **26,** 328–334.

Berry, E. D., Liewen, M. B., Mandigo, R. W., and Hutkins, R. W. (1990). Inhibition of *Listeria monocytogenes* by bacteriocin-producing *Pediococcus* during the manufacture of fermented semidry sausage. *J. Food Prot.* **53,** 194–197.

Berry, E. D., Hutkins, R. W., and Mandigo, R. W. (1991). The use of bacteriocin-producing *Pediococcus acidilactici* to control postprocessing *Listeria monocytogenes* contamination of frankfurters. *J. Food Prot.* **54,** 681–686.

Bhunia, A. K., Johnson, M. C., and Ray, B. (1987). Direct detection of an antimicrobial peptide of *Pediococcus acidilactici* in sodium dodecyl sulfate-polyacrylamide gel electrophoresis. *J. Ind. Microbiol.* **2,** 319–322.

Bhunia, A. K., Johnson, M. C., and Ray, B. (1988). Purification, characterization and antimicrobial spectrum of a bacteriocin produced by *Pediococcus acidilactici*. *J. Appl. Bacteriol.* **65,** 261–268.

Bhunia, A. K., Johnson, M. C., Ray, B., and Belden, E. L. (1990). Antigenic property of pediocin AcH produced by *Pediococcus acidilactici* H. *J. Appl. Bacteriol.* **69,** 211–215.

Biswas, S. R., Ray, P., Johnson, M. C., and Ray, B. (1991). Influence of growth conditions on the production of a bacteriocin, pediocin AcH, by *Pediococcus acidilactici* H. *Appl. Environ. Microbiol.* **57,** 1265–1267.

Chan, W. C., Bycroft, B. W., Lian, L.-Y., and Roberts, G. C. K. (1989a). Isolation and characterization of two degradation products derived from the peptide antibiotic nisin. *FEBS Lett.* **1,2,** 29–36.

Chan, W. C., Lian, L.-Y., Bycroft, B. W., and Roberts, G. C. K. (1989b). Confirmation of the structure of nisin by complete ^{1}H N.M.R. resonance assignment in aqueous and dimethyl sulphoxide solution. *J. Chem. Soc. Perkin Trans.* **1,** 2359–2367.

Cheeseman, G. C., and Berridge, N. J. (1957). An improved method of preparing nisin. *Biochem. J.* **65,** 603–608.

Daba, H., Pandian, S., Gosselin, J. F., Simard, R. E., Huang, J., and Lacroix, C. (1991). Detection and activity of a bacteriocin produced by *Leuconostoc mesenteroides*. *Appl. Environ. Microbiol.* **57,** 3450–3455.

Daeschel, M. A. (1990). Applications of bacteriocins in food systems. *In* "Biotechnology and Food Safety" (D. D. Bills and S. Kung, eds.), pp. 91–104. Butterworth-Heinemann, Boston.

Daeschel, M. A., and Klaenhammer, T. R. (1985). Association of a 13.6-megadalton plasmid in *Pediococcus pentosaceus* with bacteriocin activity. *Appl. Environ. Microbiol.* **50,** 1538–1541.

Daeschel, M. A., McKenney, M. C., and McDonald, L. C. (1990). Bacteriocidal activity of *Lactobacillus plantarum* C-11. *Food Microbiol.* **7,** 91–98.

Daeschel, M. A., Jung, D.-S., and Watson, B. T. (1991). Controlling wine malolactic fermentation with nisin and nisin-resistant strains of *Leuconostoc oenos. Appl. Environ. Microbiol.* **57,** 601–603.

Davey, G. P., and Richardson, B. C. (1981). Purification and some properties of diplococcin from *Streptococcus cremoris* 346. *Appl. Environ. Microbiol.* **41,** 84–89.

Degnan, A. J., Yousef, A. E., and Luchansky, J. B. (1992). Use of *Pediococcus acidilactici* to control *Listeria monocytogenes* in temperature-abused vacuum-packaged wieners. *J. Food Prot.* **55,** 98–103.

de Klerk, H. C., and Smit, J. A. (1967). Properties of a *Lactobacillus fermenti* bacteriocin. *J. Gen. Microbiol.* **48,** 309–316.

Delves-Broughton, J. (1990). Nisin and its uses as a food preservative. *Food Technol.* **44**(11), 100–117.

Deutscher, M. P. (ed.) (1990). Guide to protein purification. *In* "Methods in Enzymology," Vol. 182. Academic Press, Inc., New York.

Dufour, A., Thuault, D., Boulliou, A., Bourgeois, C. M., and Le Pennec, L.-P. (1991). Plasmid-encoded determinants for bacteriocin production and immunity in a *Lactococcus lactis* strain and purification of the inhibitory peptide. *J. Gen. Microbiol.* **137,** 2423–2429.

Eckner, K. F. (1991). Bacteriocins and food applications. *Scope (Silliker Labs Technical Bulletin)* **6,** 1–5.

Falahee, M. B., Adams, M. R., Dale, J. W., and Morris, B. A. (1990). An enzyme immunoassay for nisin. *Int. J. Food Sci. Technol.* **25,** 590–595.

Federal Register. (1988). Nisin preparation: affirmation of GRAS status as a direct human food ingredient. *Fed. Regist.* **58,** 11247–11250.

Gao, F. H., Abee, T., and Konings, W. N. (1991). Mechanism of action of the peptide antibiotic nisin in liposomes and cytochrome c oxidase-containing proteoliposomes. *Appl. Environ. Microbiol.* **57,** 2164–2170.

Gonzalez, C. F., and Kunka, B. S. (1987). Plasmid-associated bacteriocin production and sucrose fermentation in *Pediococcus acidilactici. Appl. Environ. Microbiol.* **53,** 2534–2538.

Gross, E., and Morell, J. L. (1971). The structure of nisin. *J. Am. Chem. Soc.* **93,** 4634–4635.

Hamdan, I. Y., and Mikolajcik, E. M. (1974). Acidolin: an antibiotic produced by *Lactobacillus acidophilus. J. Antibiot.* **27,** 631–636.

Harding, C. D., and Shaw, B. G. (1990). Antimicrobial activity of *Leuconostoc gelidum* against closely related species and *Listeria monocytogenes. J. Appl. Bacteriol.* **69,** 648–654.

Hastings, J. W., Sailer, M., Johnson, K., Roy, K. L., Vederas, J. C., and Stiles, M. E. (1991). Characterization of leucocin A-UAL187 and cloning of the bacteriocin gene from *Leuconostoc gelidum. J. Bacteriol.* **173,** 7491–7500.

Hirsch, A. (1951). Growth and nisin production of a strain of *Streptococcus lactis. J. Gen. Microbiol.* **5,** 208–221.

Holo, H., Nilssen, O., and Nes, I. F. (1991). Lactococcin A, a new bacteriocin from *Lactococcus lactis* subsp. *cremoris:* isolation and characterization of the protein and its gene. *J. Bacteriol.* **173,** 3879–3887.

Hoover, D. G., Dishart, K. J., and Hermes, M. A. (1989). Antagonistic effect of *Pediococcus* spp. against *Listeria monocytogenes. Food Biotech.* **3,** 183–196.

Hurst, A. (1981). Nisin. *Adv. Appl. Microbiol.* **27,** 85–123.

Hurst, A. (1983). Nisin and other inhibitory substance from lactic acid bacteria. *In* "Antimicrobials in Foods" (A. L. Brenen and P. M. Davidson, eds.), pp. 327–351. Marcel Dekker, Inc., New York.

Jarvis, B., Jeffcoat, J., and Cheeseman, G. C. (1968). Molecular weight distribution of nisin. *Biochim. Biophys. Acta* **168,** 153–155.

Joerger, M. C., and Klaenhammer, T. R. (1986). Characterization and purification of helveticin J and evidence for a chromosomally determined bacteriocin produced by *Lactobacillus helveticus* 481. *J. Bacteriol.* **167,** 439–446.

Joerger, M. C., and Klaenhammer, T. R. (1990). Cloning, expression, and nucleotide sequence

3. Biochemical Methods for Purification of Bacteriocins

of the *Lactobacillus helveticus* 481 gene encoding the bacteriocin helveticin J. *J. Bacteriol.* **172,** 6339–6347.

Kekessy, D. A., and Piguet, J. D. (1970). New method for detecting bacteriocin production. *Appl. Microbiol.* **20,** 282–283.

Klaenhammer, T. R. (1988). Bacteriocins of lactic acid bacteria. *Biochimie* **70,** 337–349.

Kojic, M., Svircevic, J., Banina, A., and Topisirovic, L. (1991). Bacteriocin-producing strain of *Lactococcus lactis* subsp. *diacitilactis* S50. *Appl. Environ. Microbiol.* **57,** 1835–1837.

Kordel, M., and Sahl, H.-G. (1986). Susceptibility of bacterial, eukaryotic and artificial membranes to the disruptive action of the cationic peptides Pep 5 and nisin. *FEMS Microbiol. Lett.* **34,** 139–144.

Kordel, M., Schuller, F., and Sahl, H.-G. (1989). Interaction of the pore-forming peptide antibiotics Pep 5, nisin and subtilin with non-energized liposomes. *FEBS Lett.* **244,** 99–102.

Lewus, C. B., and Montville, T. J. (1991). Detection of bacteriocins produced by lactic acid bacteria. *J. Microbiol. Meth.* **13,** 145–150.

Lewus, C. B., Sun, S., and Montville, T. J. (1992). Production of an amylase-sensitive bacteriocin by an atypical *Leuconostoc paramesenteroides* strain. *Appl. Environ. Microbiol.* **58,** 143–149.

Lipinska, E. (1976). Nisin and its applications. In "Antibiotics and Antibiosis in Agriculture" (M. Woodbine, ed.), pp. 103–131. Butterworths, Boston.

Liu, W., and Hansen, J. N. (1990). Some chemical and physical properties of nisin, a small-protein antibiotic produced by *Lactococcus lactis*. *Appl. Environ. Microbiol.* **56,** 2551–2558.

Mayr-Harting, A., Hedges, A. J., and Berkley, R. C. W. (1972). Methods for studying bacteriocins. In "Methods in Microbiology" (J. R. Norris and D. W. Ribbons, eds.), Vol. 7A, pp. 315–422. Academic Press, Inc., New York.

McCormick, E. L., and Savage, D. C. (1983). Characterization of *Lactobacillus* sp. strain 100-37 from the murine gastrointestinal tract: ecology, plasmid content, and antagonistic activity toward *Clostridium ramosum* H1. *Appl. Environ. Microbiol.* **46,** 1103–1112.

McGroarty, J. A., and Reid, G. (1988). Detection of a lactobacillus substance that inhibits *Escherichia coli*. *Can. J. Microbiol.* **34,** 974–978.

Mortvedt, C. I., Nissen-Meyer, J., Sletten, K., and Nes, I. F. (1991). Purification and amino acid sequence of lactocin S, a bacteriocin produced by *Lactobacillus sake* L45. *Appl. Environ. Microbiol.* **57,** 1829–1834.

Muriana, P. M., and Klaenhammer, T. R. (1987). Conjugal transfer of plasmid-encoded determinants for bacteriocin production and immunity in *Lactobacillus acidophilus* 88. *Appl. Environ. Microbiol.* **53,** 553–560.

Muriana, P. M., and Klaenhammer, T. R. (1991a). Purification and partial characterization of lactacin F, a bacteriocin produced by *Lactobacillus acidophilus* 11088. *Appl. Environ. Microbiol.* **57,** 114–121.

Muriana, P. M., and Klaenhammer, T. R. (1991b). Cloning, phenotypic expression, and DNA sequence of the gene for lactacin F, an antimicrobial peptide produced by *Lactobacillus* spp. *J. Bacteriol.* **173,** 1779–1788.

Nielsen, P. F., and Roepstorff, P. (1988). Sample preparation dependent fragmentation in 252-CF plasma desorption mass spectrometry of the polycyclic antibiotic, nisin. *Biomed. Environ. Mass Spectromet.* **17,** 137–141.

Nielsen, J. W., Dickson, J. S., and Crouse, J. D. (1990). Use of a bacteriocin produced by *Pediococcus acidilactici* to inhibit *Listeria monocytogenes* associated with fresh meat. *Appl. Environ. Microbiol.* **56,** 2142–2145.

Ogden, K. (1986). Nisin: a bacteriocin with a potential use in brewing. *J. Inst. Brew.* **92,** 379–383.

Ogden, K., Waites, M. J., and Hammond, J. R. M. (1988). Nisin and brewing. *J. Inst. Brew.* **94,** 233–238.

Palmer, D. E., Mierke, D. F., Pattaroni, C., and Goodman, M. (1989). Interactive NMR and computer simulation studies of lanthionine-ring structures. *Biopolymers* **28,** 397–408.

Piard, J.-C., Muriana, P. M., Desmazeaud, M. J., and Klaenhammer, T. R. (1992). Purification

and partial characterization of lacticin 481, a lanthionine-containing bacteriocin produced by *Lactococcus lactis* subsp. *lactis* CNRZ 481. *Appl. Environ. Microbiol.* **58,** 279–284.
Powell, I. B., Ward, A. C., Hillier, A. J., and Davidson, B. E. (1990). Simultaneous conjugal transfer in *Lactococcus* to genes involved in bacteriocin production and reduced susceptibility to bacteriophages. *FEMS Microbiol. Lett.* **72,** 209–214.
Pucci, M. J., Vedamuthu, E. R., Kunka, B. S., and Vandenbergh, P. A. (1988). Inhibition of *Listeria monocytogenes* by using bacteriocin PA-1 produced by *Pediococcus acidilactici* PAC1.0. *Appl. Environ. Microbiol.* **54,** 2349–2353.
Radler, F. (1990a). Possible use of nisin in winemaking. I. Action of nisin against lactic acid bacteria and wine yeasts in solid and liquid media. *Am. J. Enol. Vitic.* **41,** 1–6.
Radler, F. (1990b). Possible use of nisin in winemaking. II. Experiments to control lactic acid bacteria in the production of wine. *Am. J. Enol. Vitic.* **41,** 7–11.
Rammelsberg, M., and Radler, F. (1990). Antibacterial polypeptides of *Lactobacillus* species. *J. Appl. Bacteriol.* **69,** 177–184.
Rammelsberg, M., Muller, E., and Radler, F. (1990). Caseicin 80: purification of characterization of a new bacteriocin from *Lactobacillus casei*. *Arch. Microbiol.* **154,** 249–252.
Reddy, G. V., Shahani, K. M., Friend, B. A., and Chandan, R. C. (1983). Natural antibiotic activity of *Lactobacillus acidophilus* and *bulgaricus*. III. Production and partial purification and bulgarican from *Lactobacillus bulgaricus*. *Cult. Dairy Prod. J.* (May), 15–19.
Ruhr, E., and Sahl, H.-G. (1985). Mode of action of the peptide antibiotic nisin and influence on the membrane potential of whole cells and on cytoplasmic and artificial membrane vesicles. *Antimicrob. Agents Chemother.* **27,** 841–845.
Sabel, G., Yousef, A. E., and Marth, E. H. (1991). Behavior of *Listeria monocytogenes* during fermentation of beaker sausage made with or without a starter culture and antioxidant food additives. *Lebensm.-Wiss. u.-Technol.* **24,** 252–255.
Schillinger, U., and Holzapfel, W. H. (1990). Antibacterial activity of carnobacteria. *Food Microbiol.* **7,** 305–310.
Schillinger, U., and Lucke, F.-K. (1989). Antibacterial activity of *Lactobacillus sake* isolated from meat. *Appl. Environ. Microbiol.* **55,** 1901–1906.
Schillinger, U., Kaya, M., and Lucke, F.-K. (1991). Behaviour of *Listeria monocytogenes* in meat and its control by a bacteriocin-producing strain of *Lactobacillus sake*. *J. Appl. Bacteriol.* **70,** 473–478.
Scopes, R. K. (ed.). (1987). "Protein Purification. Principles and Practice." Springer-Verlag, New York.
Shahani, K. M., Vakil, J. R., and Kilara, A. (1977). Natural antibiotic activity of *Lactobacillus acidophilus* and *bulgaricus*. II. Isolation of acidophilin from *L. acidophilus*. *Cult. Dairy Prod. J.* (May), 8–11.
Slijper, M., Hilbers, C. W., Konings, R. N. H., and van de Ven, F. J. M. (1989). NMR studies of lantibiotics. Assignment of the ^{1}H-NMR spectrum of nisin and identification of inter-residual contacts. *FEBS Lett.* **1,2,** 22–28.
Sobrino, O. J., Rodriguez, J. M., Moreira, W. L., Fernandez, M. F., Sanz, B., and Hernandez, P. E. (1991). Antibacterial activity of *Lactobacillus sake* isolated from dry fermented sausages. *Int. J. Food Microbiol.* **13,** 1–10.
Tagg, J. R. (1991). Bacterial BLIS. *ASM News* **57,** 611.
Tagg, J. R., and McGiven, A. R. (1971). Assay system for bacteriocins. *Appl. Microbiol.* **21,** 943.
Tagg, J. R., Dajani, A. S., and Wannamaker, L. W. (1976). Bacteriocins of Gram-positive bacteria. *Bacteriol. Rev.* **40,** 722–756.
Toba, T., Samant, S. K., Yoshioka, E., and Itoh, T. (1991a). Reutericin 6, a new bacteriocin produced by *Lactobacillus reuteri* LA6. *Lett. Appl. Microbiol.* **13,** 281–286.
Toba, T., Yoshioka, E., and Itoh, T. (1991b). Lacticin, a bacteriocin produced by *Lactobacillus delbrueckii* subsp. *lactis*. *Lett. Appl. Microbiol.* **12,** 43–45.
Toba, T., Yoshioka, E., and Itoh, T. (1991c). Acidophilucin A, a new heat-labile bacteriocin produced by *Lactobacillus acidophilus* LAPT 1060. *Lett. Appl. Microbiol.* **12,** 106–108.
Toba, T., Yoshioka, E., and Itoh, T. (1991d). Potential of *Lactobacillus gasseri* isolated from infant faeces to produce bacteriocin. *Lett. Appl. Microbiol.* **12,** 228–231.

Upreti, G. C., and Hinsdill, R. D. (1973). Isolation and characterization of a bacteriocin from a homofermentative *Lactobacillus*. *Antimicrob. Agents Chemother.* **4,** 487–494.

van Belkum, M. J., Kok, J., Venema, G., Holo, H., Nes, I. F., Konings, W. N., and Abee, T. (1991). The bacteriocin lactococcin A specifically increases permeability of lactococcal cytoplasmic membranes in a voltage-independent, protein-mediated manner. *J. Bacteriol.* **173,** 7934–7941.

West, C. A., and Warner, P. J. (1988). Plantacin B, a bacteriocin produced by *Lactobacillus plantarum* NCDO 1193. *FEMS Microbiol. Lett.* **49,** 163–165.

Yousef, A. E., Luchansky, J. B., Degnan, A. J., and Doyle, M. P. (1991). Behavior of *Listeria monocytogenes* in wiener exudates in the presence of *Pediococcus acidilactici* H or pediocin AcH during storage at 4 or 25°C. *Appl. Environ. Microbiol.* **57,** 1461–1467.

CHAPTER **4**

Applications and Interactions of Bacteriocins from Lactic Acid Bacteria in Foods and Beverages

MARK A. DAESCHEL

To those that devote their lives to science nothing can give more happiness than making discoveries, but their cups of joy are full only when the results of their studies find practical application.
—LOUIS PASTEUR

I. Introduction

The scope of current investigations on bacteriocins from lactic acid bacteria (LAB) is quite extensive, ranging from basic studies on genetic regulation to applications in food preservation. One can easily justify scientific study on LAB bacteriocins from an applied viewpoint. First, the LAB are indispensable in the production of certain foods and beverages where their fermentation properties endow a myriad of desirable and unique sensory attributes to foods. Second, the preservative properties of LAB when used as fermentation agents in food was historically and still is an important means of food preservation. Two relatively recent factors accelerating interest in LAB bacteriocins are the increasing incidence and detection of food-borne disease and the emerging consumer resistance to highly processed foods. Most food-borne illness (> 80%) can be directly attributed to microbial infection or intoxication (Todd, 1989). Estimates of the number of food-borne illness cases per year and their associated costs in the United States were given (Oblinger, 1988) as 24–81 million cases and 5–17 billion dollars. The relatively recent recognition of *Campylobacter jejuni, Escherichia coli* O157:H7, and *Listeria monocytogenes* as major food-borne pathogens has increased awareness of the extent and costs of food-borne disease.

The recent trend of food processors to develop foods that are minimally processed (believed by some to be more "natural") may in fact compromise or eliminate previous product safeguards. Food processors are no doubt responding to a growing consumer preference for such foods and capitalizing on a new lucrative market with high growth potential. Minimally processed foods are attractive to the consumer because they are convenient, have a natural, fresh appearance, are viewed as nutritionally correct, and are generally devoid of added preservatives. More and more consumers now read food ingredient labels and will tend to select foods that do not contain preservatives if given a choice. The "contains no preservatives" label syndrome is quite acute with obvious abuse by marketing strategists. Examples such as the labeling of bottled water with statements that it contains no preservatives and is fat and cholesterol free, have no doubt fueled legislation of the Food and Drug Administration's (FDA) new labeling guidelines. On the other hand, one cannot say that justifiable concerns about preservatives do not exist, sulfite sensitivity being an example. Minimally processed foods may carry risks that are not readily apparent to the consumer. The historical introduction and use of chemical preservatives were developed on a trial-and-error basis when it was observed that people did not sicken or die after ingesting perishable foods treated with certain chemicals. Elimination of preservatives from perishable foods and the subsequent necessary dependence on refrigeration as the primary preservation mechanism may be a risk. It has generally been assumed that refrigeration is adequate to prevent the growth of pathogenic and toxigenic microorganisms associated with foods. This assumption is no longer valid; pathogens such as *L. monocytogenes* and *Aeromonas hydrophila* can grow at temperatures used for refrigerated storage of foods. The probability of temperature abuse during commercial handling compounds the problem. Thus, food processors are faced with the dilemma of how to best manufacture attractive, minimally prepared foods without chemical preservatives, and ensure a profitable shelf life and a safe product.

Consumer bias against chemical preservatives and preference for minimally processed foods reflect consumers' perception of what constitutes a natural and healthy food. To meet the demand and expectation of today's consumer, any preservative or additive under consideration in a product formulation must be carefully evaluated in terms of the consumer's perception of its validity. Lactic acid bacteria produce metabolites that are inhibitory toward other microorganisms, and historically, lactic fermentation as a preservation process is based in part upon such metabolites accumulating to inhibitory levels in certain foods and beverages. Fermented foods have a positive image among consumers as healthy and natural substances, with yogurt and sour dough bread, and more recently red wine, being examples. The inhibitory microbial metabolites produced during fermentation include lactic and acetic acids, ethanol, hydrogen peroxide, and bacteriocidal proteins (bacteriocins). Much recent interest, even in the popular press (Cowley, 1988), has centered on the potential use of LAB bacteriocins as possible replacements or adjuncts to preservatives that have fallen out of favor with consumers. Some of the factors fueling the interest are summarized in Table I. The use of LAB bacteriocins is also attractive because the use of LAB in foods has withstood the test of time concerning safe use. However, it would be misleading to assume that all primary and secondary metabolites produced by LAB are nontoxic to humans. There simply is not enough information to support such a claim. Nevertheless, the safe and efficacious use of the LAB bacteriocin nisin for the past 30 years is encouraging and provides the precedent for continued research on and evaluation of LAB bacteriocins as an emerging class of food preservatives. In addition to the use of bacteriocins to control food-borne pathogens, they have potential application in controlling bacterial spoilage of foods and in directing pure culture fermentations in foods and beverages.

It is the objective of this chapter to briefly review the past and present uses of nisin in food/beverage processing and preservation, to discuss factors that may enhance or diminish bacteriocin efficacy in food applications, and to review applied studies with bacteriocins other than nisin. The reader is referred to Chapter 5, this volume, by N. Hansen for discussion on the chemistry and molecular biology of nisin.

TABLE I
Events Contributing to the Increasing Number of Applied Investigations on Bacteriocins and Bacteriocinogenic LAB

Safe and efficacious use of nisin during the past 30 years.
Recent FDA approval of nisin as a "generally regarded as safe" (GRAS) substance in certain applications.
Consumer resistance to traditional "chemical" preservatives.
Justifiable concerns over the safety of existing food preservatives such as sulfites and nitrites.
Realization that bacteriocinogenicity is not a rare occurrence within the lactic acid bacteria.
Feasibility of using bacteriocin production and immunity as selectable genetic markers in starter culture bacteria.
Availability of molecular biology tools to transfer, clone, and sequence the genetic determinants and to engineer genetic variants of bacteriocins.
Willingness of federal funding agencies, food commodity groups, and food processing corporations to fund both basic and applied researches.

Source: Daeschel, 1992.

II. Using Bacteriocinogenic Lactic Acid Bacteria and Bacteriocins to Control Food-Borne Pathogens

LAB bacteriocins are primarily active against Gram-positive species and generally are not inhibitory toward Gram-negative bacteria, yeasts, or molds. However, studies (Blackburn et al., 1989; Stevens et al., 1991) have described treatments that can make Gram-negative species susceptible to the action of nisin. Furthermore, very high concentrations of nisin (10,000 IU on filter disks) have been reported (Henning et al., 1986a) to be active against yeast and, to a lesser extent, fungi. As LAB bacteriocins are characterized, it is apparent that activity spectra can range from narrow (same species or genus, e.g., diplococcin) to broad (most Gram-positive species, e.g., nisin). Summaries and discussion of the activity spectra of LAB bacteriocins including nisin are given by Klaenhammer (1988), Schillinger (1990), Delves-Broughton (1990), and Ray (1992a, b). Therefore, the ensuing discussion is limited to a review of recent studies pertaining to LAB bacteriocins and their activity against *Listeria monocytogenes* and *Clostridium botulinum*.

A. Sensitivity of Listeria monocytogenes to LAB Bacteriocins

Much recent interest has been focused on evaluating the sensitivity of *L. monocytogenes* to LAB bacteriocins and it has become the model target bacterium in many studies of food preservatives. In particular, the ability of *L. monocytogenes* to grow at refrigeration temperatures, its propensity for chilled meats and dairy foods, and its Gram-positive nature are all factors that have spurred the many investigations (Table II) addressing its sensitivity to LAB bacteriocins. *Listeria monocytogenes* is sensitive to many of the broader spectra bacteriocins. The effect of nisin on *L. monocytogenes* is depicted in Figure 1, demonstrating the loss of cellular material following lysis. It is interesting that the LAB bacteriocin mesenterocin 5 appears to be specific for *Listeria* species and not active against other LAB (Daba et al., 1991). Strategies to use LAB or LAB bacteriocins to protect refrigerated foods from *Listeria* contamination may include direct addition of purified bacteriocin preparations, addition of fermentation by-products containing active bacteriocin, or formulation of foods to include bacteriocinogenic LAB either as starter cultures or as safeguards against temperature abuse by their subsequent outgrowth and bacteriocin production. However, these approaches should not be viewed as substitutes for appropriate sanitation and food handling practices but rather as enhancements or safeguards to preservation.

TABLE II
Studies Addressing LAB Bacteriocins and Their Action on *Listeria monocytogenes*

Bacteriocin and/or bacteriocinogenic LAB	*L. monocytogenes* strains	Study summary	Reference
Pediocin PA-1	LMO 1	Inhibition of *L. monocytogenes* demonstrated in dairy products	Pucci et al. (1988)
Lactobacillus sake	Scott A, NCTC 5105, NCTC 7973	Concentrated supernatants active against *L. monocytogenes*	Sobrino et al. (1991)
Pediococcus acidilactici P. *pentosaceus* P. *damnosus*	CA, Scott A, 19111 19113, 19115 F5027, F5069	28% of 296 pairings were inhibition positive against *L. monocytogenes*	Hoover et al. (1989)
Nisin	Scott A 35152, 15313, 7644 V7	MIC values ranged from 740 to 10^5 IU/ml in TSB; *L. monocytogenes* completely inhibited in cottage cheese (2500 IU/g)	Benkerroum and Sandine (1988)
Nisin	Scott A, V7, 20A$_2$, Ohio, California	Lysis of *L. monocytogenes* monitored by absorbance; 100 IU/ml nisin reduced *L. monocytogenes* 10^6 to 10^3 in 24 hours	Monticello and O'Connor (1990)
Nisin	Scott A, 19115, UAL 500	Nisin-resistant mutants detected at 10^{-6} to 10^{-8} frequency	Harris et al. (1991)
Nisin	Scott A, Jalisico	Influence of milk fat and emulsifiers on nisin activity against *L. monocytogenes*	Jung et al. (1992)
14 Bacteriocinogenic LAB (lactobacilli, leuconostoc, pediococci, lactococci)	F5069, Scott A, F-5027, 19115, 675-3, DA-1, UAL 501, 15313	7 LAB antagonistic to all *L. monocytogenes* strains tested	Harris et al. (1989)
L. lactis 11454, *P. pentosaceus* FBB-61, FBB-63-DG2	14 Strains including Scott A and V7	Survey of bacteriocinogenic LAB ability to inhibit food-borne pathogens; *L. monocytogenes* sensitive to all LAB tested	Spelhaug and Harlander (1989)
Pediococcus acidilactici JBL 1095, produces Pediocin AcH	Scott A, V7, 101M	Inhibition of *L. monocytogenes* by production of Pediocin AcH in packaged wieners held at 25°C	Degnan et al. (1992)

(*continued*)

TABLE II (Continued)

Bacteriocin and/or bacteriocinogenic LAB	*L. monocytogenes* strains	Study summary	Reference
Pediococcus acidlactici, Pediocin AcH containing supernatants	Scott A, V7, 101M	Addition of cells or pediocin AcH enhanced inhibition of *L. monocytogenes* in wiener fluids	Yousef et al. (1991)
Lactobacillus sake lb706	17a	Survey of bacteriocinogenic LAB from meat; *L. monocytogenes* inhibition with supernatants	Schillinger and Lucke (1989)
Lactobacillus sake lb 706, produces Sakacin A	4a, 16e, 17a, 17b	Decline in populations of *L. monocytogenes* in Mettwurst compared to nonbacteriocinogenic *L. sake*	Schillinger (1990)
Pediococcus acidilactici PAC 1.0	Scott A, F5069, 19115, NCF-U2K3, NCF-F1KK4	Pediocin enhanced inhibition of *L. monocytogenes* in fermented sausage; novel enumeration of cultures	Foegeding et al. (1992)
Pediococcus sp. JD1-23	Scott A	Bacteriocin-producing culture more effective in inhibiting *L. monocytogenes* compared to non-bacteriocinogenic strain in fermented sausage	Berry et al. (1990)
Pediococcus acidilactici LAC-TACEL 110	Scott A	Inhibition of attachment of *L. monocytogenes* to fresh meat by addition of culture supernatants	Nielsen et al. (1990)
Leuconostoc gelidium	NCTC 9863, 10537, 11994	Characterization of a broad spectrum bacteriocin from *Leuconostoc gelidium* active against *L. monocytogenes*	Harding and Shaw (1990)
Leuconostoc mesenteroides UL5, mesenterocin 5	12 Strains	Characterization of bacteriocin essentially specific for *Listeria* sp.	Daba et al. (1991)

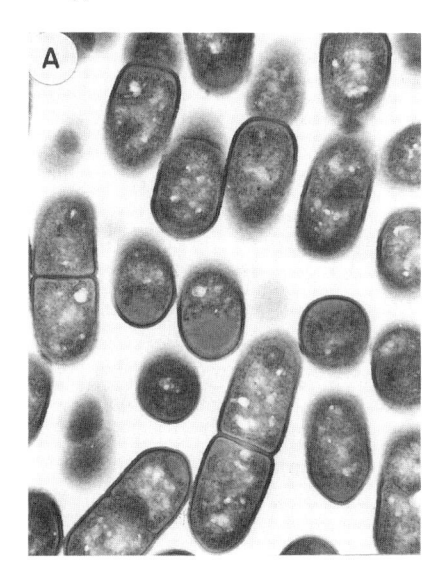

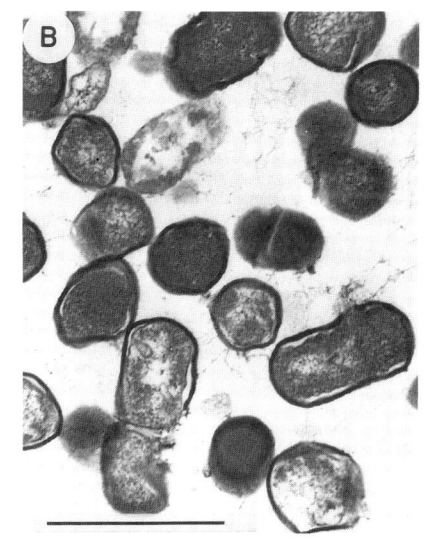

Figure 1 Lytic activity of nisin on *Listeria monocytogenes* Scott A. Cells were washed and suspended in $0.1M$ phosphate buffer (pH 6.0) with and without 300 IU/ml nisin for 4 hours at 37°C. A, Without nisin; B, with nisin. Bar = 1 μm. (From the authors' laboratory, X. Ming and M. A. Daeschel.)

B. Sensitivity of Clostridium botulinum to LAB Bacteriocins

Nisin has anticlostridial activity (Campbell et al., 1959; Denny et al., 1961; Ramseier, 1960; Scott and Taylor, 1981a,b). Hurst (1972) cited *C. botulinum* as one of the more nisin-resistant species among the clostridia. Scott and Taylor (1981a,b) studied the nisin sensitivity of *C. botulinum* according to inoculum level, nisin level, pH, and strain. They found resistance to nisin to be greatest in type A strains, followed by type B and then type E. Type 56A was most resistant to the two type A strains tested. Scott and Taylor (1981b) demonstrated activity of 2000 IU/ml nisin at pH 6.0 with an inoculum level of 10^4 spores *C. botulinum* 56A, but observed no activity at pH 7.0 or 8.0. At pH 6.0, activity was observed at inoculum levels of 10^4 and 10^3. Hurst (1981) and Lipinska (1977) noted the irreversible inactivation of nisin above pH 7.0 at room temperature. At pH 2.5, nisin is stable; at pH 5.0, it loses 40% activity, and it loses 90% above pH 6.8 after autoclaving at 121°C for 15 minutes. Fowler (1981) also reported that spores of the proteolytic strains were more nisin resistant than the spores of the nonproteolytic strains. Rayman et al. (1983) reported that proteolysis was not related to nisin resistance. The sensitivity of 19 strains of *Clostridium botulinum* and 2 strains of *C. sporogenes* (13 of 19 strains were type B, and of those, 7 were proteolytic; 6 strains were type A) to nisin was investigated by Montville et al. (1992). Type A strains exhibited the most resistance, although 62A and 25763A were least resistant among type A strains. Of all *Clostridium botulinum* strains, 56A, 17409A, 213B, and 2129B were the most resistant. This information enhances the conclusion drawn by Scott and Taylor (1981a) that strain 56A is the most resistant strain. Montville et al. (1992) also demonstrated that no one biotype was more resistant to nisin than another, concurring with Rayman et al.'s (1983) observations in pork slurries.

The sensitivity of *Clostridium botulinum* to the bacteriocin (Pediocin A) of *Pediococcus pentosaceus* FBB-61 and L 7230, first observed by Daeschel and Klaenhammer (1985), was further investigated by Okereke and Montville (1991a). Their investigation determined the minimum inhibitory cell concentration (MICC) of the inhibitor producer strain required for observing inhibition in deferred antagonism assays. Eleven strains of *C. botulinum* representing both type A and B strains were all inhibited by the two pediococcal strains. The MICC (spot inoculum) required to produce an inhibition zone against a 10^4 composite cell population of *C. botulinum* strains seeded into agar was 1×10^5 CFU/ml for *P. pentosaceus* FBB-61 and 4×10^6 for *P. pentosaceus* L-7230. Similarly, a MICC of 7×10^5 CFU/ml was required for the nisin producer *Lactobacillus lactis* 11454. The investigators concluded that *P. pentosaceus* FBB-61 was the "most promising strain for further research on bacteriocin-mediated protection against *C. botulinum* hazards." In a later study, Okereke and Montville (1991b) used bacteriocinogenic LAB in deferred and simultaneous antagonism assays to demonstrate bacteriocin-mediated suppression of *C. botulinum* spores at temperatures (4, 10, 15, and 30°C) that minimally processed refrigerated foods may encounter during handling. They observed that *Lactobacillus plantarum* BN, *L. lactis* 11454, and *P. pentosaceus* 43200 and 43201 were bacteriocinogenic and inhibitory toward *C. botulinum* spores at 4, 10, 15, and 30°C. Furthermore, bacteriocinogenic effects were enhanced by the addition of sodium chloride, with 3–4% found as optimal.

III. Using Bacteriocinogenic Lactic Acid Bacteria and Bacteriocins to Direct Food and Beverage Fermentations

Many if not all food and beverage fermentations are conducted under nonsterile conditions. Fermentation processes rely on natural or modified environments to select against spoilage and pathogenic microorganisms and to promote the growth of desirable microorganisms, either those naturally occurring or intentionally added. In practice, most fermentations progress through a sequence of microbial species, where the desirable species usually predominates and imparts the characteristic identity attributes to the particular food or beverage.

The desirable microbial sequence can be easily perturbed, depending

4. Applications and Interactions of Bacteriocins from Lactic Acid Bacteria

on maintenance and monitoring of the physical and chemical parameters determined to be critical for reproducible, high-quality finished products. Examples of disturbances that can obstruct the fermentation process and compromise the product are phage infection of dairy starters, growth of lactobacilli in wines, staphylococcal growth in fermented meats, and pediococcal growth in beer. Stuck fermentations, undesirable sensory attributes, and growth of pathogenic and toxigenic species can all occur in fermentations that have had their intended microbial ecology altered. Fermentation processing aids, such as sulfur dioxide, nitrites and nitrates, potassium sorbate, and various organic acids have been used with some success to control and direct fermentation processes. The use of bacteriocins as fermentation aids has also been studied and presents a unique feature, that is, the desirable microorganisms used in fermentation can also be the source of bacteriocin. This negates the need to include the bacteriocin as a direct food additive.

In one review (Lipinska, 1977), the use of nisin-producing and nisin-resistant dairy lactococci as starters in cheese was discussed in detail. Hurst (1981) also reviewed this topic. Briefly, the first applied evaluation of nisin was the study of Hirsch et al. (1951) in which they made a swiss-style cheese with nisin-producing starters, which prevented butyric blowing faults from occurring. This observation generated considerable interest in the 1950s and 1960s when much research was centered on using nisin-producing starters in Swiss and Dutch cheeses to prevent the undesirable growth of clostridia, the cause of butyric blowing. However, several observations came to light during cheesemaking trials with nisin-producing starters that did not encourage their further development and application. As summarized by Lipinska (1977) and Hurst (1981), it was confirmed that nisin-producing starters were effective in controlling the incidence of clostridial spoilage. However, it was also observed that cheese made with such starters was not of as high a quality as cheeses made with regular starters. Moreover, it appeared that nisin-producing starters were sensitive to phage attack and that "nisinase"-producing organisms may compromise the efficacy of nisin. A second approach was to use nisin-producing starters with nisin-resistant starters with good cheesemaking properties. Work cited by Lipinska (1977) indicated that a desirable nisin-resistant strain should be resistant to 300 IU nisin/ml, its resistance should be stable, and it should not be resistant to nisin by means of nisin-degrading mechanism, that is, nisinase. The nisin-resistant starters were observed to possess some significant differences in terms of starter performance and stability when compared to the regular parent starters. Longer lag phase and lower cell densities were observed as well as lowered heat resistance. Nevertheless, Lipinska (1977) cited several studies in which nisin-producing and nisin-resistant strains when used together significantly reduced the incidence of butyric spoilage and made cheese of acceptable quality. Hurst (1981) made the point that the second approach of using nisin-producing and nisin-resistant starters is an alternative to consider as a substitute to the use of nitrate (a suspected carcinogen) to control butyric spoilage in cheese. However, in the United States, nitrate is not permitted in cheese.

The antagonistic substance diplococcin isolated from *Lactococcus lactis* sp. *cremoris* (Whitehead, 1933; Oxford, 1944) has a narrow spectrum of activity, being restricted to within the *Lactococcus* genus (Davey and Pearce, 1980). In some situations bacteriocin-producing starters may not be desirable, such as with diplococcin-producing *L. lactis* sp. *cremoris* strains, as they tend to dominate or inhibit other cultures in multiple starter systems (Davey and Richardson, 1981). Davey and Richardson obtained nondiplococcin-producing variants of *L. lactis* subsp. *cremoris* 346 by subjecting the strain to elevated growth temperatures. Subsequently, these variants were demonstrated to be suitable for cheese production having retained desirable properties in that regard.

During beer fermentation operations, the wort may often become contaminated with lactobacilli or pediococci that are tolerant of the antibacterial properties of hops (Hough et al. 1982). Growth of these lactic acid bacteria can result in several serious sensory defects, the most important being the appearance of diacetyl. In addition, certain LAB strains can produce polysaccharides that can impart a ropy or slimy character to beer. To prevent LAB contamination, breweries need to pay special attention to sanitation procedures. However, not all contamination stems from unclean equipment. Many times yeast stocks as well as pitching yeast is contaminated with LAB. Several procedures have been developed to rid yeast of LAB, including washing treatments with sulfur dioxide, organic acids, acidified ammonium persulfate, and antibiotics such as polymixin and penicillin. Several problems and concerns became evident with these treatments, including the likelihood of generating contaminants resistant to clinical antibiotics and the loss of viability of yeasts exposed to the chemical washes (Hough et al., 1982). The use of nisin to control the growth of lactic acid bacteria in fermenting beer was investigated by Ogden and co-workers (Ogden and Tubb, 1985; Ogden, 1986; Ogden and Waites, 1986; Ogden, 1987). Tubb and Ogden (1986) applied for patents on technologies using nisin to prevent spoilage of alcoholic beverages. A review on the topic recently appeared (Ogden et al., 1988). To summarize, Ogden and co-workers conducted experiments that defined a level of 100 IU of nisin/ml wort as adequate to kill 10^5 cells/ml of added lactobacilli within 6 hours. Furthermore, nisin did not affect the populations of yeast reached during fermentation or the fermentation performance (time to decrease specific gravity of the wort from 1.040 to 1.020) of 9 ale and lager yeasts. Three grades of nisin were evaluated for their effect on the sensory attributes of finished beer. Grade A, Nisaplin, 1000 IU/mg; Grade B, 2500 IU/mg; and Grade C, 40,000 IU/mg. Using triangular taste testing procedures it was determined that Grade C, the purest form of nisin, contributed no discernible taste to the beer. Grades A and B did elicit significant taste differences compared to control beer (no nisin added). The results are interesting: with Grade A-treated beer, of the 22 taste panelists, 7 preferred the nisin-treated beer, none preferred the untreated beer, and 15 had no preference.

Henning et al. (1986b), investigated the use of nisin to control the incidence of LAB during the fermentation of fruit mashes for the production of brandy. Added nisin was observed to decrease populations of LAB during apple mash fermentations from an initial count of 1.8×10^6/ml to a low of 1.0×10^2/ml 3 days after initiation of fermentation. However, populations rose to 6.4×10^4/ml after 17 days and to 2.7×10^7/ml after 75 days, and it was suggested by the authors that nisin-resistant LAB populations may have arisen. Significant decreases in the levels of ethyllactate and ethylacetate and increases in alcohol in fermented mash containing nisin were reported. Furthermore, none of the yeast strains used to ferment the mashes was adversely affected by the presence of nisin.

The use of nisin in wine processing was recently evaluated by Radler (1990a, b), Daeschel et al. (1991), and Daeschel and Watson (1991). Wine, depending on grape variety and vinification procedures, can spoil as the result of undesirable LAB growing and producing compounds detrimental to wine quality (Wiwobo et al., 1985). For example, metabolites such as diacetyl, acrolein (2-propenal), acetic acid, and lactic acid are undesirable at certain concentrations. In addition, certain lactobacilli are capable of imparting a ropy or oily character to wine because of polysaccharide synthesis. On the other hand, certain wines benefit from the growth of LAB that can decarboxylate malic acid [malolactic fermentation (MLF)], resulting in the reduction of wine acidity, microbial stability, and the addition of complex sensory overtones. Thus, LAB in wine can be desirable or not depending on the particular wine and style. Traditionally, winemakers have relied on indigenous LAB (primarily *Leuconostoc oenos*) to carry out MLF, and more recently the use of commercial LAB starters to conduct MLF has become commonplace. To prevent MLF and the growth of spoilage LAB, winemakers will use sulfur dioxide, low temperatures, frequent racking, and even filtration. The use of nisin or other bacteriocins may provide wine processors with an alternative way to prevent MLF and the growth of spoilage LAB. Nisin in particular may be attractive because it is quite stable at pH values encountered in wine (pH 3–4), is active against LAB, and has no effect against yeast. Radler (1990a) evaluated the nisin sensitivity of 83 strains of LAB (*Pediococcus*, *Lactobacillus*, and *Leuconostoc*) and found most strains were inhibited by less than 10 IU/ml, with *L. oenos* being sensitive to 5 IU/ml. However, some strains of lactobacilli (*L. plantarum* and *L. casei*) were able tolerate up to 1000 IU/ml. Again, as observed previously by Henning et al. (1986b) and Ogden (1986) fermentation yeasts were not affected by nisin.

In a companion study (Radler, 1990b), winemaking trials were conducted to evaluate the sensitivity of LAB to nisin under commercial conditions. Several varietal grape musts were prepared with the inclusion of 1000 IU/ml nisin along with yeast cultures. After primary yeast fermentation, the wines were racked and an additional 100 IU/ml nisin was added. It was observed that even the most nisin-resistant LAB (*L. casei*) was unable to grow (evaluated at 49 days). However, in MRS medium with the same concentrations of nisin *L. casei* was able to grow. These results indicate that either the wine environment potentiated the action of nisin or the summation of low pH, and high acid, ethanol, and nisin provided a cumulative inhibitory effect. By whatever mechanism, nisin was clearly shown to be effective in suppression of LAB during wine fermentation. Sensory analysis of wines prepared with and without nisin indicated that tasters more often were unable to detect organoleptic differences. Like the results in Ogden's study (1986) with beer, nisin-treated samples were preferred by panelists more often than the controls.

Often, MLF in wine is difficult to promote when it is most needed and difficult to prevent when unwanted, such as in wines of high pH. Daeschel et al. (1991) addressed this problem by investigating the use of nisin to specifically prevent or promote MLF. They used nisin to inhibit naturally occurring LAB capable of MLF and developed and used nisin-resistant mutants of *L. oenos* to conduct MLF in wines containing levels of nisin that are inhibitory to naturally occurring LAB. They observed that nisin (100 IU/ml) when added to wine was able to prevent MLF and the growth of intentionally added spoilage (LAB (*Pediococcus damnosus*) or the naturally occurring LAB. Commercial strains of *L. oenos* (sensitive to < 1 IU/ml nisin) possessing desirable MLF properties were made resistant to 100 IU/ml nisin by incremental exposure to nisin. Wines made with nisin-resistant *L. oenos* strains and added nisin had greater than 90% of malate decarboxylated within three months with no other LAB growing. Nisin-resistant strains of *L. oenos* were tested for the ability to inactivate nisin. No significant differences in nisin activity levels were observed when nisin was added to medium supernatants in which nisin-resistant or nisin-sensitive strains of *L. oenos* were grown, nor were any significant differences observed in wines inoculated with the sensitive or resistant strains. This was an important observation; it would be counterproductive to have nisin degraded when nisin-resistant mutants are used to promote a pure-culture MLF. These observations demonstrated that nisin and nisin-resistant strains of desirable MLF bacteria can be used to promote a pure-culture MLF in the presence of other LAB, a situation representative of a commercial winery environment. Daeschel and Watson (1991) observed that the wine protein fining agent, bentonite, was able to remove nisin by binding-precipitation reactions. This may be useful if it is desired to remove residual nisin from wine before bottling in order to preclude the necessity to declare it on the label. In the same study it was determined that 100 IU/ml nisin could effectively substitute for 50 mg/liter SO_2 in preventing induction of MLF. A summary of applications of bacteriocins in alcoholic beverages is given in Table III.

In vegetable fermentations, a variable succession of LAB growth occurs, which in most cases results in a desirable lactic acid fermentation (Daeschel et al., 1987). Approaches to controlling and directing vegetable fermenta-

4. Applications and Interactions of Bacteriocins from Lactic Acid Bacteria

TABLE III
Potential Applications of Bacteriocins in Alcoholic Beverage Fermentations

Product	Function
Distilled spirit mashes	Inhibit LAB, resulting in acid and ester reduction
Beer	Prevent contamination of pitching yeast with LAB
	Prevent growth of LAB in fermenting wort
Wine	Prevent growth of spoilage LAB
	Use bacteriocins with bacteriocin-resistant malolactic starters to establish pure-culture MLF
	Partial substitute for sulfites

tions have involved the selective addition of defined and optimal amounts of salt, acid, and buffer. More recently, the use of starter cultures (Daeschel and Fleming, 1987) and controlling exposure of fermentation vessels to oxygen (Fleming et al., 1988) have furthered efforts to precisely direct fermentation. The use of bacteriocinogenic LAB as starter cultures to achieve pure-culture vegetable fermentations was discussed by Daeschel and Fleming (1987). This approach was explored and developed by Harris et al. (1992a) using a model cabbage juice system (representing sauerkraut fermentation), where a paired starter culture consisting of nisin-resistant *Leuconostoc mesenteroides* and a nisin-producing *Lactococcus lactis* (isolated from sauerkraut; Harris et al., 1992b) were used as fermentation strains. The strategy was to have the desirable heterolactic *L. mesenteroides* ferment cabbage sugars in the presence of levels of nisin produced by the *L. lactis* strain adequate to suppress undesirable homolactic LAB. Data presented by these investigators indicated that nisin levels produced by the *L. lactis* strain were sufficient to inhibit *L. plantarum* when the two strains were co-cultured. Co-culture of the nisin-resistant *L. mesenteroides* with nisin-producing *L. lactis* did not affect the *L. mesenteroides* strain in terms of growth and acid production (pH) in cabbage juice, and coculture of all 3 strains with specifically defined initial numbers resulted in the elimination of *L. plantarum* and growth of *L. mesenteroides*.

Foegeding et al. (1992) used a strain of *Pediococcus acidilactici* that produces Pediocin A1 (Gonzales and Kunka, 1987) as a lactic starter for production of dry fermented sausage. The objective of their study was to evaluate the ability of Pediocin produced *in situ* to inhibit intentionally introduced *Listeria monocytogenes*. An isogenic strain of *P. acidilactici* lacking pediocin-producing ability was used as a control. Selective enumeration of all strains was accomplished with antibiotic resistance markers. It was concluded that the lactic fermentation by itself was sufficient to control *L. monocytogenes* if enough acid was produced. However, pediocin production was shown to provide a safeguard against *L. monocytogenes* if acid production was insufficient.

The use of bacteriocins and bacteriocinogenic LAB to control microbial populations in food and beverage fermentations has been demonstrated with various commodities. Whether this approach becomes commonplace technology will depend on a proven track record under commercial production conditions. Certain problems or hurdles that may emerge are the appearance of bacteriocin-resistant populations of spoilage-type LAB, increased likelihood of phage infection because of dependence on one or two starters as opposed to the heterogenous indigenous LAB, and inactivation of inhibitory properties by interaction with product components.

IV. Factors Affecting the Efficacy of Bacteriocins in Foods and Beverages

The transition from bacteriocin study in the laboratory environment to application in the food processing arena is one that few LAB bacteriocins will successfully accomplish. Demonstration of bacteriocin activity under controlled laboratory conditions may be straightforward, whereas proving their efficacy as antimicrobials in food systems poses a much greater challenge. Foods and beverages are essentially undefined mixtures of compounds and biological structures that are processed with a variety of physical and chemical treatments, with the expectant result of an attractive, stable, safe, good-tasting, nutritious, and marketable product. The latter are limitations imposed by government, industry, consumers and the basic laws of chemistry and physics, which in total make the application of bacteriocins a challenge. Nisin is the one bacteriocin that has surmounted these hurdles through the efforts of basic and applied researches. Clearly, much effort and support is needed to get other promising LAB bacteriocins into applications. The ensuing discussion is primarily about studies addressing the interactions of nisin with food components and should be viewed as initial guidelines for investigations with other LAB bacteriocins. Some of the factors that may compromise the efficacy of bacteriocins in food applications are summarized in Table IV.

A. Interactions with Food Components

Nisin antibotulinal activity has been investigated in food systems, with its activity being dependent on the particular type of food. Somers and Taylor (1987) found nisin in cheese spreads to inhibit *Clostridium botulinum* at levels of 4000–16,000 IU/g. The addition of hurdles such as low moisture (54–57%) and salt (2%) interactions could enhance the inhibition due to nisin at 30°C. Rayman et al. (1983) demonstrated the inability of nisin to inhibit a cocktail of five strains of *C. botulinum* type A at a level of 12,000 IU/g in pork slurries at pH 5.0 and 25°C. Reduction of the pH of pork slurries to a value of 5.1 increased nisin effectiveness in this system. Scott and Taylor (1981a,b)

TABLE IV
Factors That May Compromise the Efficacy of Bacteriocins and Bacteriocinogenic LAB in Product Applications

Bacteriocins
1. Emergence of bacteriocin-resistant pathogens or spoilage bacteria
2. Conditions that may destabilize biological activity of proteins (bacteriocins)
    a. Nonspecific proteolytic enzymes
    b. Oxidation
    c. Heavy metals
    d. Excessive agitation, foaming
    e. Freeze-thaw, shearing
3. Binding to food components
4. Inactivation by other food additives
5. Partitioning into polar or nonpolar food components
5. pH effects
    a. Solubility
    b. Activity dependent on narrow pH range

Bacteriocinogenic LAB
1. Inadequate environment for growth or bacteriocin production
2. Spontaneous loss of bacteriocin-producing ability
3. Phage infection
4. Antagonism by other flora
5. Development of bacteriocin-resistant microflora

found that nisin activity against *C. botulinum* decreased in cooked meat media, in comparison to tryptone/peptone/yeast extract/glucose broth (TPYG). They reported that 2000 IU/ml nisin inhibited *C. botulinum* type A in TPYG, but 5000 IU/ml could not inhibit the same organism in cooked meat medium at pH 6.0 and 35°C. They postulated that protein surfaces interacted with nisin to neutralize its inhibitory action, but did not evaluate the percentage lost.

Bell and De Lacy (1986) found that when 100 IU/g nisin was added to meat, residual nisin concentration ranged from 26% in the presence of 3% beef fat to 75% in the presence of 83% beef fat. Other investigators found that lipid such as butterfat (Jones, 1974) or phospholipid (Henning et al., 1986a) interfered with nisin activity. Jung et al. (1992) observed that the activity of nisin against *L. monocytogenes* in fluid milk was directly dependent upon the fat content with increasing concentrations decreasing the efficacy of nisin. Moreover, it was shown that the addition of Tween 80 could restore nisin activity in high-fat milks. These investigators hypothesized that nisin absorbed to milk-fat globules, reducing its availability to inhibit cells, and that the addition of Tween 80 released absorbed nisin from globule surfaces. Daeschel (1990) demonstrated with sterile homogeneously prepared foods that nisin lost significantly more initial activity in high-fat foods (e.g., turkey puree, 7.8% fat) than in low-fat foods (e.g., green bean puree, 0.2% fat).

Taylor et al. (1985) used the procedure of Tramer and Fowler (1964) for assay of residual nisin and found at day zero approximately 10 to 25% nisin remaining after the addition of nisin to chicken frankfurter emulsions at 27°C, fat 13–20%. Taylor and Somers (1985) again used the procedure of Tramer and Fowler (1964) and detected approximately 60% residual nisin at 100 IU/ml in pickled pork bellies and bacon at 27°C (pH not mentioned).

Somers and Taylor (1987) used the Tramer and Fowler (1964) method for nisin determination in cheese and found residual nisin to be 25 to 300% of target nisin. They attributed nisin variability to the presence of unknown factors in the cheese spread being investigated, which may have affected the growth of the indicator organism, *Micrococcus luteus*. They discussed the possibility of inconsistent extraction of nisin from cheese emulsion caused by nisin binding to the cheese spread ingredients as a source of variability. Bell and De Lacy (1986) investigated different concentrations of salt, nitrite, and fat and subsequent effects on the recovery of nisin from the sample using the procedure of Tramer and Fowler (1964). They found that increasing fat levels increased the nisin recovered; however, they could not explain it.

Residual nisin and its detection have generated controversy, and appear to be dependent on the temperature, outgrowth of organism, and media constituents. Once the nisin is added to a simple or complex matrix, such as a food system, only a percentage of nisin remains to be detected. This detectable level may remain the same or change with storage, depending on the matrix. Rogers (1991) monitored nisin levels during 35°C storage in 10% albumen solution, 10% lecithin solution, 30% soluble starch solution, and a basal medium consisting of 0.5% proteose peptone, 0.5% yeast extract, and 1.0% glucose. Nisin levels remained the same for 37 days in all but the egg albumen. Levels of approximately 400 IU/ml nisin dropped to less than 5 IU/ml in less than 48 hours in the 10% albumen solution. At lower initial nisin levels, the outgrowth of either *Clostridium botulinum* or *Lactobacillus sake* in the albumen-, starch-, or lecithin-containing media with nisin resulted in progressively decreasing nisin levels over time, ultimately reaching zero after approximately three weeks at 35°C.

Daeschel et al. (1991) observed that nisin retained activity in white wines but decreased to less than 90% of its original activity in red wines within 4 months. They hypothesized that the decrease in nisin activity seen in red wines may be the result of nisin interactions with phenolic compounds that are absent in white wines. Further investigation (Bower et al., 1992) using individual phenolic components (gallic acid, catechin, myricetin, quercetin, malvidin 3,5-diglucoside, and crude grape tannin) in a model wine system with added nisin revealed significant nisin activity decreases in the presence of myricetin and, in particular, tannin (Figure 2). Tannin interaction with nisin was further evaluated using a series of red wines aged for 2, 4, and 6 months before nisin addition. A greater loss of nisin activity was observed in

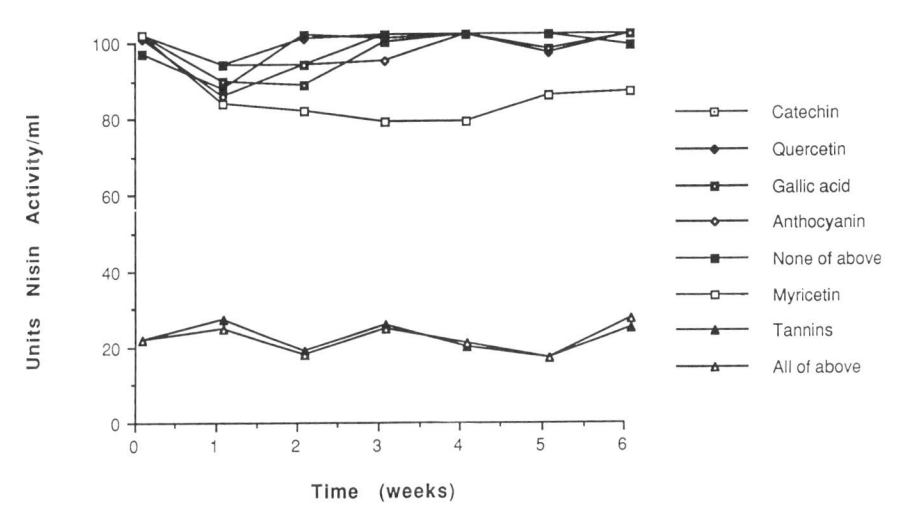

4. Applications and Interactions of Bacteriocins from Lactic Acid Bacteria

Figure 2 Effect of phenolic components at concentrations found in wine on the activity of nisin (100 IU/ml) in a model wine system (Aqueous solution of 13% v/v ethanol, tartaric acid; 2 g/L and pH 3.5) with phenolic components added singly and all together. Storage temp. 23°C. (From Bower et al., 1992.)

wine aged for 6 months than in the younger wines. The authors suggest this result is linked to the ratio of polymeric to monomeric phenols increasing as wine ages; polymeric phenols may have a greater affinity for binding or inactivating nisin. Moreover, they felt it was reasonable to assume that the binding and precipitation reactions (Yokotsuka and Singleton, 1987) for peptide–tannin interactions in wines may be applicable to nisin–tannin interactions. Tannins are sometimes used specifically as fining agents in wines to remove unstable proteins that may precipitate out and cause cloudiness. In fact, any protein fining agent is likely to interact with nisin and inactivate it by precipitation. Daeschel and Watson (1991) observed that the commonly used fining agent, bentonite, was able to inactivate nisin at concentrations commonly used in wine processing. However, this should not be necessarily viewed as a disadvantage, as it provides a way, if desired, to selectively remove nisin and likely most other bacteriocins from beverage products before packaging and distribution.

Again, using nisin as an example, it has been observed that pH conditions and thermal processing of foods may compromise the activity of added bacteriocins. Research at Aplin & Barrett Ltd. (Tech. Info. Sheets, 1/88, 10/91) has determined the percent retention of nisin after heating at different F values in various foods. Examples given were sour cherries (pH 3.28), 69.5% nisin retention at F = 3 and 37.8% at F = 15; and cream-style corn (pH 6.88), 47.9% retention at F = 3 and 7.6% at F = 15. The heat stability of nisin is well documented (Hurst, 1981) and is also evident with many other LAB bacteriocins (summarized by Schillinger, 1990). However, nisin becomes increasingly ineffective in foods that approach pH neutrality. The effect of pH on nisin is twofold, the first being decreasing solubility with increasing pH. Nisin solubility values were given by Liu and Hansen (1990) as 57 mg/ml at pH 2.0 and 0.25 mg/ml at pH 8–12. The second pH effect on nisin, also elaborated by Liu and Hansen (1990) occurs under alkaline conditions where nisin inactivation was believed to be a result of hydroxyl ion addition to the dehydro residues. The latter effect is likely to be of little consequence in foods with pH values below 6.0.

It is clear that LAB bacteriocins by their nature are subject to inactivation by various environmental conditions. Successful bacteriocin applications will result if our understanding of how these inactivating mechanisms work is enhanced. In turn, this will allow strategies to be developed that circumvent bacteriocin inactivation, perhaps by protein engineering or chemical modification.

B. Enhancement of Bacteriocin Activity

Several observations of enhanced bacteriocin activity have been reported in the literature. A common observation has been the effect of sodium chloride in increasing the efficacy of bacteriocins against target microorganisms. Harris et al. (1991) reported a decrease in survivability of *Listeria monocytogenes* in the presence of nisin with NaCl added (2.5 and 3.5%); however, they were unable to conclude whether the effect was due to additional environmental stress or other effects. Okereke and Montville (1991b) evaluated the effects of sodium chloride in deferred antagonism assays using bacteriocinogenic LAB (*Lactobacillus lactis* 11454, *Pediococcus pentosaceus* 43200, 43201, and *L. plantarum* BN) as antagonists against *Clostridium botulinum* spores. Their results indicated that sodium chloride at 3 and 4% levels either enhanced bacteriocin activity or somehow made spores more sensitive. The latter possibility was addressed by evaluating the sole effect of sodium chloride on spore viability. The effect observed was much less than one that could be explained by subtracting the salt effect from the nisin–salt effect, leading to the conclusion that a synergistic effect occurred with sodium chloride and bacteriocin. However, essentially the opposite effect with salt was observed by Bell and De Lacy (1985) with heat-activated *Bacillus lichenformis* spore. The inclusion of various combinations of salt (1–5% w/v) and nisin (1–100 IU/ml) in spore suspensions (10^6/ml) consistently showed a protective effect of salt on inhibition of spore outgrowth by nisin. These studies led the authors to conclude that salt somehow prevents nisin from absorbing to the spores.

Taylor et al. (1985) observed an "adjuvant effect" when combinations of nisin and nitrite were used to inhibit *Clostridium botulinum* spores in TPYG broth. In one example, in the absence of nitrite, 1500 IU/ml nisin was needed to inhibit spore outgrowth; however the addition of 20 ppm nitrite lowered the nisin requirement to 100 IU/ml. Rayman et al. (1981, 1983) also described nisin-nitrate combinations and conditions for inhibiting outgrowth of clostridial spores. Their initial experiments (Rayman et al., 1981) with *C. sporogenes* indicated that nisin and nitrate had an additive inhibitory effect. However, in later experiments with *C. botulinum* and pork slurries it was observed that levels up to 550 ppm (22,000 IU/ml) with 60 ppm nitrite were inadequate to prevent outgrowth of *C. sporogenes*. They concluded that high pH (5.8) was the main cause for the failure of nisin to inhibit spore outgrowth. At pH 5.1, nisin (300 ppm) without nitrate was able to prevent spore outgrowth.

Oscroft et al. (1990) investigated the inhibition of thermally stressed *Bacillus* spores by various combinations of nisin, pH, and food-grade organic acids. Synergistic inhibition effects were reported between nisin and acetic acid, lactic acid, citric acid, or glucono-delta-lactone. With some treatments, synergy was more evident at higher pH values (6.0) than at lower pH values (5.4–5.7).

Several studies have addressed the influence of food emulsifiers on the activity of nisin. Observations reported by Henning et al. (1986a) indicated an antagonistic effect of emulsifiers on the antimicrobial activity of nisin in a tomato juice system. Specifically, Henning et al. (1986a) demonstrated 70% inactivation of nisin in the presence of approximately 2.8 μmol/ml (0.081%) phosphatidylcholine. However, other experiments with Tween 80 (Jung et al., 1992) clearly showed an enhancement effect of the emulsifier on nisin activity in high-fat milks against *L. monocytogenes;* neither an enhancement nor an antagonistic effect was seen with lecithin. The latter data is in agreement with that of Blackburn et al. (1989) who demonstrated similar nisin enhancement effects against *Streptococcus agalactiae* and *L. monocytogenes* in milk with monoacylglycerols (lauric and oleic acids), which function as nonionic emulsifiers.

An alternative approach to increase the efficacy and activity spectra of bacteriocins is to make target cells more sensitive by eliminating some of their natural defenses. Blackburn et al. (1989) and Stevens et al. (1991) described experiments showing that EDTA in combination with nisin could effectively inhibit Gram-negative food-borne pathogens previously insensitive to nisin. As discussed by Stevens et al. (1991), the most likely mechanism is a disruption of the Gram-negative outer membrane by EDTA chelation of membrane-stabilizing magnesium ions, thus exposing the cell to the action of nisin.

Advances in genetic technologies are rapidly providing new molecular tools that may be used to enhance the activity of existing antimicrobials and to create new superior antimicrobials. Bacteriocins such as nisin are ribosomally synthesized and posttranslationally modified. When the modification systems bacterial cells use to process these proteins are understood, the likelihood of engineering and enhancing their biological activity could be realized. This is a current frontier in LAB bacteriocin research and one with much potential reward.

C. Resistance of Target Microorganisms to Bacteriocins

The appearance of populations of pathogenic bacteria resistant to clinical antibiotics has resulted from indiscriminate use and is of serious medical concern. The large current research effort directed toward isolating and synthesizing new antibiotics has in part stemmed from the problem of resistance. Early concerns on the use of nisin included possible cross-resistance to clinical antibiotics developing in bacteria exposed to nisin. The recognition of bacteriocin-sensitive bacteria becoming resistant was first observed during early studies on nisin by Hirsch (1950). He observed that cells of *Streptococcus agalactiae* that survived exposure to nisin were able to grow in concentrations of nisin forty times higher than those previously tolerated. Carlson and Bauer (1957) were able to increase the resistance of *Staphylococcus aureus* from 2.5 to 2000 IU/ml through successive stepwise transfers. Mention was made of a substance being isolated from the nisin-resistant strain that destroyed nisin. The acquired resistance was reported lost following growth in nisin-free media. *S. aureus* strains described as resistant to clinical antibiotics were also resistant to nisin, but not vice versa. Kooy (1952) described strains of *Lactobacillus plantarum* isolated from cheese and raw milk that could degrade the antimicrobial activity (presumably nisin) in *L. lactis* ssp. *lactis* culture filtrates. Galesloot (1956) also described similar observations with other LAB. However, neither observation was pursued further. Alifax and Chevalier (1962) described nisinase activity in a strain of *Streptococcus thermophilus* and partially purified preparations of the enzyme. Optimum conditions for degradation of nisin by their preparations were pH 7.0 (no activity at pH 3 or 4) and a 6-hour incubation at 37°C. Furthermore, the nisinase preparation was described as being inactive against other antibiotics, including the polypeptide antibiotics, polymixin, subtilin, and bacitracin, suggesting it was specific for nisin and not a nonspecific protease. Jarvis (1967) described strains of *Bacillus cereus* and *Bacillus polymyxa,* which produced enzymes that inactivated nisin and subtilin but not other polypeptide antibiotics. Further investigation (Jarvis, 1970; Jarvis and Farr, 1971) provided details on substrate specificity, thermal stability, and mechanism of nisin inactivation. Briefly, nisin inactivation by the nisinase preparation was shown not to be a proteolytic degradation process but was suggestive of a reductase reaction of the dehydroalanine residue adjacent to the C-terminal. The inactivation of subtilin by the nisinase preparation was consistent with their hypothesis that nisin and subtilin have a similar C-terminal sequence. Nisin is also labile to enzymatic degradation by chymotrypsin (Heineman and Williams, 1966; Jarvis and Mahoney, 1969). The appearance of spontaneous nisin-resistant variants or the purposeful exposure of strains to increasing amounts to obtain variants with high resistance has been frequently observed. Lipinska (1977) "trained" cheese starters to become nisin resistant but not produce nisinase. Daeschel et al. (1991) obtained strains of

Leuconostoc oenos 100-fold more resistant to nisin than the wild-type for use in malolactic fermentation of wine that were not resistant by means of a secreted nisinase. Harris et al. (1991) described the appearance of *Listeria monocytogenes* mutants resistant to 2000 IU/ml nisin at frequencies between 10^{-6} and 10^{-8}. On the other hand, Collins-Thompson et al. (1985) were unable to increase the resistance of certain lactobacilli to nisin (resistant to < 10 IU/ml) by repeated passage through media containing increasing concentrations of nisin. However, mechanisms to explain nisin resistance other than "nisinase" have not been forthcoming. Scenarios such as an alteration of nisin-binding receptors and intracellular modification (inactivation) of bound nisin by resistant cells have circulated among investigators in the field. Genes that encode for nisin resistance have been described and are transferrable by conjugation (Gasson, 1984; Klaenhammer and Sanosky, 1984; McKay and Baldwin, 1984), have been cloned (Forseth et al., 1988), and have been used in the construction of vectors (von Wright et al., 1990). Nevertheless, the mechanism by which the proteins encoded by these genes confer resistance is not yet understood.

Specific resistances to other LAB bacteriocins have not to this author's knowledge been described, likely because adequate quantities of purified bacteriocin have been lacking. However, the appearance of spontaneous resistant colonies in inhibition zones from various LAB bacteriocin preparations has been observed by this author and others, suggesting that bacteriocin resistance is not an uncommon phenomenon. This brings up the specter of the bane of clinical antibiotics: the appearance of resistant strains. Although the potential compromise of human health in using LAB bacteriocins as food preservative agents is not of the same magnitude as that of clinical antibiotic-resistant pathogens, the appearance of bacteriocin-resistant food-borne pathogens can pose problems. As pointed out by Harris et al. (1991), when high levels of microbial contamination are present, the likelihood of spontaneous (nisin)-resistant mutants arising should be expected. In addition, other characteristics of food environments may interfere with the efficacy of LAB bacteriocins (discussed in a previous section) and thus increase the likelihood of resistant populations developing. A hypothetical scenario that might play out is as follows. A food product such as bologna, which can be subject to inadvertent postprocessing contamination has a bacteriocin such as nisin added to it to prevent the growth of *Listeria monocytogenes*. However, during one processing run the contamination level is such that resistant populations of *L. monocytogenes* emerge. In addition, the bacteriocin has eliminated other contaminants such as LAB, which may have held *L. monocytogenes* in check. Furthermore, the appearance and distribution of bacteriocin-resistant *L. monocytogenes* throughout a processing plant has now eliminated the bacteriocin's usefulness. Of course, this is a worst-case scenario; however, it is important to recognize some possible limitations or problems with bacteriocins in food preservation applications.

V. Assays for Bacteriocins in Foods

The method for nisin quantitation cited most often is the method of Tramer and Fowler (1964). Experimentally, nisin recovery from food systems has not been quantitative or predictable. Tramer and Fowler (1964) made suggestions for quantifying nisin in food substances. They extracted the nisin from any bond with food by boiling the food sample in 0.02 N HCl for 10 minutes, centrifuging the sample to remove the food, and testing the supernatant for residual nisin, using a plate agar diffusion method and back calculations to estimate that concentration in food. Tramer and Fowler (1964) incorporated 1% Tween 20 into the agar of the plate agar diffusion assay to enhance diffusion of the bacteriocin and used *Micrococcus luteus* as the indicator organism. The boiling step has been used in most assays for nisin in foods (Calderon et al., 1985; Chung et al., 1989; Jones, 1974; Rayman et al., 1981; Somers and Taylor, 1987). Also, this method of plate agar diffusion assay with *M. luteus* is used commonly and is the FAO/WHO method for nisin determination (Lipinska, 1977). Refrigerated preincubation of cultures in nisin quantitative assays was recommended by Mocquot and Lefebre (1956). Rogers and Montville (1991) used 1% Tween 20 in purified agar coupled with refrigerated preincubation and observed a trend of increased zone of inhibition in the plate agar diffusion assay.

Hirsch (1950) used *Streptococcus cremoris* or *S. agalactiae* as the indicator organism to quantitate nisin. *Lactobacillus lactis* and again *S. cremoris* were used as indicator organisms by Mocquot and Lefebre (1956). Tramer and Fowler (1964) used *Micrococcus luteus*. A screening procedure (Lewus et al., 1991) and side by side examinations of the respective sensitivities to nisin (Rogers and Montville, 1991) demonstrated that assays using *L. sake* as the indicator organism generated larger zones of inhibition (mm) than those with *M. luteus* as the indicator organism.

It is not uncommon for investigations using Tramer and Fowler's procedure to recover nisin at approximately 40% of the original amount added, similar to direct assay without boiling (Rogers, 1991; Somers and Taylor, 1987). Poor recovery of nisin from foods has been attributed to poor mixing, binding of nisin with food protein or other food components, and effects of food components on the sensitivity of the indicator organism used in the plate agar diffusion assay (Rogers, 1991; Somers and Taylor, 1981; Somers and Taylor, 1987). For example, Rayman et al. (1983) observed losses in nisin activity after heating nisin in pork slurries and postulated that heating increased either the tendency for nisin to bind to meat or the overall quantity of nisin bound.

All of the above described procedures are based on the bioassay method. Alternative methods for assaying bacteriocins, in particular for nisin, have recently been evaluated. Waites and Ogden (1987) described a procedure based on the observation that lysis of bacterial cells by nisin resulted in reduction of intercellular bacterial ATP and the extracellular appearance of

ATP in medium. A reproducible relationship between the amounts of ATP and nisin provided a means of estimating nisin concentration by bioluminescent assay of ATP. The assay was found to be useful for estimating nisin within the range of 0–120 IU/ml with an error of less than 5%. Although the method is more rapid than bioassay procedures, initial costs for ATP luminescence measuring instrumentation and enzyme reagents are high but are justifiable if nisin measurement becomes a routine quality control test in a food processing environment. An enzyme immunoassay for nisin was developed (Falahee et al., 1990) using polyclonal antisera raised against nisin in sheep. Nisin-specific antibody isolated by affinity chromotography was used to develop an ELISA protocol. The assay correlated well with the bioassay in cheese samples spiked with nisin. A detection limit of 1.9×10^{-2} IU was given and intentionally inactivated nisin did not give a positive reaction. The authors state that the method is no more rapid than the bioassay, but is less labor intensive and more suitable to automation.

Lactic acid bacteria produce as primary metabolites several low molecular weight compounds that exhibit antimicrobial activity toward certain microorganisms. These compounds include acetic acid, lactic acid, propionic acid, ethanol, diacetyl, and hydrogen peroxide. To distinguish between these types of compounds and bacteriocins, techniques have been developed to take advantage of certain fundamental differences between the two classes. The reader is referred to Chapter 2, this volume, and to Daeschel (1992) for discussions that address detecting bacteriocin activity and distinguishing it from that of other LAB inhibitors.

VI. Regulatory (United States) and Safety Considerations

Thus far, nisin is the only LAB bacteriocin permitted for use with foods in the United States (Federal Register, 1988). Currently there are petitions pending before the U.S. Food and Drug Administration requesting approval for broader use of nisin in foods (Duane Chase, private communication). However, nisin is allowed for food use in 47 other countries in a variety of products, primarily canned foods and dairy products (Delves-Broughton, 1990). The application of bacteriocinogenic starter cultures in fermented foods might be viewed as GRAS. Similarly, the addition of fermentation byproducts such as whey that may contain microbially produced bacteriocins could legally be permitted but may not be in the spirit of the law.

Why are there no toxicity studies among the recent multitude of reports that address newly characterized bacteriocins, bacteriocinogenic LAB, and their potential applications? A short-term answer may be that sufficient quantities of purified bacteriocins are not available to perform adequate testing. It is also possible that the precedent provided by the safe use of nisin for 30 years and the recognition that LAB are neither pathogenic or toxigenic suggests that all LAB bacteriocins are safe. This is a somewhat naive conclusion, though nevertheless pervasive. The literature supports this belief of implied safety in regard to LAB bacteriocins. The only investigations that directly address toxicity aspects are those pertaining to nisin (Frazer et al., 1962, Hara et al., 1962) and were conducted over 30 years ago. Bioactive proteins in small amounts can be strong potentiators in biological systems, for example, the action of botulinum and staphylococcal toxins. It is the obligation of the research community to address possible safety aspects prior to making any applied recommendations for food manufacture. Safety considerations should not only apply to purified bacteriocin preparations but to fermentation supernatants and bacteriocinogenic LAB. Although extensive toxicological investigations are expensive, some simple tests such as the Ames mutagenicity assay and cytotoxicity tests using tissue culture (Saito et al., 1979) can be initially considered.

Acknowledgments

Sincere appreciation is extended to A. M. Rogers of Kraft General Foods for her contributions to this chapter. Investigations by M. A. Daeschel and co-workers cited in this chapter were supported by grants from Diversitech, Inc., Kraft General Foods, Oregon Agricultural Research Foundation, Oregon Wine Advisory Board, Pickle Packers International, and the Western Dairy Foods Research Center.

References

Alifax, P. R., & Chevalier, R. (1962). Etude de la nisinase produite par *Streptococcus thermophilus*. *J. Dairy Res.* **29,** 233–240.

Aplin and Barrett Ltd. Trowbridge, Wiltshire, England. Technical Information Sheets: *The Food Preservative nisaplin*, Technical Data Ref. No. 1/88; and *Use of Nisaplin in processed cheese and processed cheese preparations*, Ref. No. 10/91.

Bell, R. G., and De Lacy, K. M. (1985). The effect of nisin-sodium chloride interactions on the outgrowth of *Bacillus licheniformis* spores. *J. Appl. Bact.* **59,** 127–132.

Bell, R. G., and De Lacy, K. M. (1986). Factors influencing the determination of nisin in meat products. *J. Food Tech.* **21,** 1–7.

Benkerroum, N., and Sandine, W. E. (1988). Inhibitory action of nisin against *Listeria monocytogenes*. *J. Dairy Science* **71,** 3237–3245.

Berry, E. D., Liewen, M. B., Mandingo, R. W., and Hutkins, R. W. (1990). Inhibition of *Listeria monocytogenes* by bacteriocin producing *Pediococcus* during the manufacture of fermented dry sausage. *J. Food Prot.* 53, 194–197.

Blackburn, P., Gusik, S., and Rubino, S. D. (1989). Intl. Patent Application No. WO 89/12399. Applied Microbiology, Inc. (12-28-89).

Bower, C. K., Watson, B. T., and Daeschel, M. A. (1992). Applications of bacteriocins in controlling bacterial spoilage and malolactic fermentation of Wine: Interactions between the bacteriocin nisin and components of red wines. Proceedings, 3rd Intl Symposium: Innovations in Wine Technology, May 25–27, Stuttgart, Germany.

Calderon, D., Collins-Thompson, D. L., and Usborne, W. R. (1985). Shelf life studies of vacuum packaged bacon treated with nisin. *J. Food Protect.* **48**, 330–333.

Campbell, L. L., Sniff, E. E., and O Brien, R. T. (1959). Subtilin and nisin as additives that lower the heat process requirements of canned food. *Food Tech.* **13**, 462–464.

Carlson, S., and Bauer, H. M. (1957). Nisin, ein antibakterieller wirkstoff aus *Streptococcus lactis* unter Berucksichtigung des resistenzproblems. *Arch. Hyg. und Bakt. Berl.* **141**, 445–60.

Chung, K. T., Dickson, J. S., and Crouse, J. D. (1989). Effects of nisin on growth of bacteria attached to meat. *Appl. Environ. Microbiol.* **55**, 1329–1333.

Collins-Thompson, D. L., Calderon, L. C., and Usborne, W. R. (1985). Nisin sensitivity of lactic acid bacteria isolated from cured and fermented meat products. *J. Food Prot.* **48**, 668–670.

Cowley, G. (1988). The microbe that ate Salmonella. *Newsweek*, August 22, p. 56.

Daba, H., Pandian, S., Gosselin, J. F., Simard, R. E., Huang, J., and Lacroix, C. (1991). Detection and activity of a bacteriocin produced by *Leuconostoc mesenteroides*. *Appl. Environ. Microbiol.* **57**, 3450–3455.

Daeschel, M. A. (1990). Application of Bacteriocins in Food Systems. In "Biotechnology and Food Safety" (Shain-Dow Kung, ed.), Butterworths, London.

Daeschel, M. (1992). Procedures to detect antimicrobial activities of microorganisms. In "Food Biopreservatives of Microbial Origin" (B. Ray and Daeschel, M. A. eds.). CRC Press, Boca Raton, Florida.

Daeschel, M. A., and Fleming, H. P. (1987). Achieving pure culture vegetable fermentations. In "Developments in Industrial Microbiology," **28**, pp. 141–148. Society for Industrial Microbiology, Arlington, Virginia.

Daeschel, M. A., and Klaenhammer, T. R. (1985). Association of a 13.6-megadalton plasmid in *Pediococcus pentosaceus* with bacteriocin activity. *Appl. Environ. Microbiol.* **50**, 1538–1541.

Daeschel, M. A., and Watson, B. T. (1991). Process for deacidifying wine. U. S. Patent, 5,059,431, October 22.

Daeschel, M. A., Andersson, R. E., and Fleming, H. P. (1987). Microbial ecology of fermenting plant materials. *FEMS Microbiol. Rev.* **46**, 357–367.

Daeschel, M., Jung, D. S., and Watson, B. T. (1991). Controlling wine malolactic fermentation with nisin and nisin-resistant strains of *Leuconostoc oenos*. *Appl. Environ. Microbiol.* **57**, 601–603.

Davey, G. P., and Pearce, L. E. (1980). The use of *Streptococcus cremoris* strains cured of diplococcin production as cheese starters. *N.Z. J. Dairy Sci. Technol.* **15**, 51–57.

Davey, G. P., and Richardson, B. C. (1981). Purification and some properties of diplococcin from *Streptococcus cremoris* 346. *Appl. Environ. Microbiol.* **41**, 84–89.

Degnan, A. J., Yousef, A. E., and Luchansky, J. B. (1992). Use of *Pediococcus acidilactici* to control *Listeria monocytogenes* in temperature-abused vacuum-packaged wieners. *J. Food Prot.* **55**, 98–103.

Delves-Broughton, J (1990). Nisin and its uses as preservative. *Food Tech.* **44**, 100–117.

Denny, C. B., Sharpe, L. E., and Bohrer, C. W. (1961). Effects of tylosin and nisin on canned food spoilage bacteria. *Appl. Microbiol.* **9**, 108–110.

Falahee, M. B., Adams, M. R., Dale, J. W., and Morris, B. A. (1990). An enzyme immunoassay for nisin. *Intl. J. Food Science and Tech.* **25**, 590–595.

Federal Register. (1988). Nisin Preparation; Affirmation of GRAS Status as a Direct Human Food Ingredient. 21 CFR Part 184, **Vol. 53**, 11247–11251.

Fleming, H. P., McFeeters, R. F., Daeschel, M. A., Humphries, E. G., and Thompson, R. L. (1988). Fermentation of cucumbers in anaerobic tanks. *J. Food Sci.* **53**, 127–133.

Foegeding, P. M., Thomas, A. B., Pilkington, D. H., and Klaenhammer, T. R. (1992). Enhanced control of *Listeria monocytogenes* by in situ-produced pediocin during dry fermented sausage production. *Appl. Environ. Micrbiol.* **58**, 884–890.

Forseth, B. R., Herman, R. E., and McKay, L. L. (1988). Cloning of the nisin resistant determinant and replication origin on 7.6 kilobase *Eco* R1 fragment of pNP40 from *Streptococcus lactis* subsp. *diacetylactis* DRC3. *Appl. Environ. Micrbiol.* **54**, 2136–2139.

Fowler, G. G. (1981). Nisin; Will It Be Used Here? *Food Eng.* **53**, 82–83.

Frazer, A. C., Sharratt, M., and Hickman, J. R. (1962). The biological effect of food additives. I.-Nisin *J. Sci Food Agric.* **13,** 32–42.
Galesloot, T. E. (1956). Lactic acid bacteria which destroy the antibiotic (nisin). *Neth. Milk and Dairy J.* **10,** 143–155.
Gasson, M. J. (1984). Transfer of sucrose fermenting ability, nisin resistance and nisin production into *Streptococcus lactis* 712. *FEMS Microbiol. Lett.* **21,** 7–10.
Gonzalez, C. F., and Kunka, B. S. (1987). Plasmid associated bacteriocin production and sucrose fermentation in *Pediococcus acidilactici. Appl. Environ. Microbiol.* **53,** 2534–2538.
Hara, S., Yakazu, K., Nakakawiji, K., Takenchi, T., Sata, M., Imai, Z., and Shibuya, T. (1962). An investigation of toxicity of nisin. *Tokyo Medical Journal.* **20,** 175–207.
Harding, C. D., and Shaw, B. G. (1990). Antimicrobial activity of *Leuconostoc gelidum* against closely related species and *Listeria monocytogenes. J. Appl. Bacteriol.* **69,** 648–654.
Harris, L. J., Daeschel, M. A., Stiles, M. E., and Klaenhammer, T. R. (1989). Antimicrobial activity of lactic acid bacteria against *Listeria monocytogenes, J. Food Prot.* **52,** 384–387.
Harris, L. J., Fleming, H. P., and Klaenhammer, T. R. (1991). Sensitivity and resistance of *Listeria monocytogenes* ATCC 19115, Scott A, and UAL 500 to nisin. *J. Food Prot.* **54,** 836–840.
Harris. L. J., Fleming, H. P., and Klaenhammer, T. R. (1992a). Characterization of two nisin-producing *Lactococcus lactis* subsp. *lactis* strains isolated from a commercial sauerkraut fermentation. *Appl. Environ. Microbiol.* **58,** 1477–1483.
Harris, L. J., Fleming, H. P., and Klaenhammer, T. R. (1992b). A novel paired starter culture system for sauerkraut consisting of a nisin-resistant *Leuconostoc mesenteroides* and a nisin-producing *Lactococcus lactis. Appl. Environ. Microbiol.* **58,** 1484–1489.
Heineman, B., and Williams, R. (1966). Inactivation of nisin by pancreatin. *J. Dairy Science* **49,** 312–314.
Henning, S., Metz, R., and Hammes, W. P. (1986a). Studies on the mode of action of nisin. *Int. J. Food Microbiol.* **31,** 21–134.
Henning, S., Metz, R., and Hammes, W. P. (1986b). New aspects for the application of nisin to food products based on its mode of action. *Intl. J. Food Microbiol.* **3,** 135–141.
Hirsch, A. (1950). The assay of the antibiotic nisin. *J. Gen. Microbiol.* **4,** 70–88.
Hirsch, A., Grinsted, E., Chapman, H. R., and Mattick, A. T. R. (1951). A note on the inhibition of an anaerobic sporeformer in Swiss-type cheese by a nisin producing *Streptococcus. J. Dairy Res.* **18,** 198–204.
Hoover, D. G., Dishart, K. J., and Hermes, M. A. (1989). Antagonistic effect of *Pediococcus* spp. against *Listeria monocytogenes. Food Biotech.* **3,** 183–196.
Hough, J. S., Briggs, D. E., Stevens, R., and Young, T. W. (1982). Microbial contamination in breweries. *In* "Malting and Brewing Science," Vol 2, 2nd ed, pp. 741–773. Chapman and Hall, London.
Hurst, A. (1972). Interactions of food starter cultures and food-borne pathogens: the antagonism between *Streptococcus lactis* and spore forming microbes. *J. Milk Food Technol.* **35,** 418–423.
Hurst, A. (1981). Nisin. *Advances Appl. Microbiol.* **27,** 85–123.
Jarvis, B. (1967). Resistance to nisin and production of nisin-inactivating enzymes from several *Bacillus* species. *J. Gen. Microbiol.* **47,** 33–48.
Jarvis, B. (1970). Enzymatic reduction of the C-terminal dehydralanyllysine of nisin. *Biochem. J.* **119,** 56.
Jarvis, B., and Farr, J. (1971). Partial purification, specificity and mechanism of action of the nisin-inactivating enzyme from *Bacillus cereus. Biochem. Biophys. Acta.* **227,** 232–240.
Jarvis, B., and Mahoney, R. R. (1969). Inactivation of nisin by alpha-Chymotrypsin. *J. Dairy Sci.* **52,** 1148–1450.
Jones, L. W. (1974). Effect of butterfat on inhibition of *Staphylococcus aureus* by nisin. *Can. J. Microbiol.* **20,** 1257–1260.
Jung, D. S., Bodyfelt, F. W., and Daeschel, M. A. (1992). Influence of fat and emulsifiers on the efficacy of nisin in inhibiting *Listeria monocytogenes* in fluid milk. *J. Dairy Sci,* **75,** 387–393.

Klaenhammer, T. R. (1988). Bacteriocins from lactic acid bacteria. *Biochimie* **70**, 337–349.
Klaenhammer, T. R., and Sanosky, R. B. (1984). Conjugal transfer from *Streptococcus lactis* ME2 of plasmids encoding phage resistance, nisin resistance and lactose fermenting ability: evidence for a high frequency conjugative plasmid responsible for abortive infection of virulent bacteriophage. *J. Gen. Microbio.* **131**, 1531–1541.
Kooy, J. S. (1952). Stammen van *Lactobacillus plantarum* die antibiotic van *Streptococcus lactis* onwekzaam maken. (strains of *Lactobacillus plantarum* which destroys the antibiotic made by *Streptococcus lactis*). *Neth. Milk and Dairy J.* **6**, 323–330.
Lewus, C. B., Kaiser, A., Montville, T. J. (1991). Inhibition of foodborne bacterial pathogens by bacteriocins from lactic acid bacteria isolated from meat. *Appl. Environ. Microbiol.* **57**, 1683–1688.
Lipinska, E. (1977). Nisin and its applications. In "Antibiotics and Antibiosis in Agriculture," (W. Woodbine, ed.), pp. 103–130. Butterworths, London.
Liu, W., and Hansen, J. N. (1990). Chemical and physical properties of nisin, a small protein antibiotic produced by *Lactococcus lactis*. *Appl. Environ. Micro.* **56**, 2551–2558.
McKay, L. L., and Baldwin, K. A. (1984). Conjugative 40-megadalton plasmid in *Streptococcus lactis* ssp. *diacetylactis* DRC3 is associated with resistance to nisin and bacteriophage. *Appl. Environ. Microbiol.* **47**, 68–74.
Mocquot, G., and Lefebre, E. (1956). A simple procedure to detect nisin in cheese. *J. Appl. Bacteriol.* **19**, 322–323.
Monticello, D. J., and O'Connor, D. (1990). Lysis of *Listeria monocytogenes* by nisin. In "Foodborne Listeriosis," (A. J. Miller, J. L. Smith, and G. A. Somkuti, eds.), pp. 81–83. Elsevier, Amsterdam.
Montville, T. J., Rogers, A. M., and Okereke, A. (1992). Differential sensitivity of *C. botulinum* to nisin. *J. Food Prot.* (in press).
Nielsen, J. W., Dickson, J. S., and Crouse, J. D. (1990). Use of a bacteriocin produced by *Pediococcus acidilactici* to inhibit *Listeria monocytogenes* associated with fresh meat. *Appl. Environ. Microbiol.* **56**, 2142–2145.
Oblinger, J. L. (1988). Bacteria associated with foodborne disease. Institute of Food Technologists scientific status summary. *Food Technology* **42**, no. 4.
Ogden, K. (1986). Nisin: a bacteriocin with a potential use in brewing. *J. Inst. Brew.* **92**, 379–383.
Ogden, K. (1987). Cleansing contaminated pitching yeast with nisin. *J. Inst. Brew.* **93**, 302–307.
Ogden, K., and Tubb, R. S. (1985). Inhibition of beer spoilage lactic acid bacteria by nisin. *J. Inst. Brew.* **91**, 390–392.
Ogden, K., and Waites, M. J. (1986). The action of nisin on beer spoilage bacteria. *J. Inst. Brew.* **92**, 463–467.
Ogden, K., Waites, M. J., and Hammond, J. R. M. (1988). Nisin and brewing. *J. Inst. Brew.* **94**, 23–38.
Okereke, A., and Montville, T. J. (1991a). Bacteriocin inhibition of *Clostridium botulinum* spores by lactic acid bacteria. *J. of Food Prot.* **54**, 349–356.
Okereke, A., and Montville, T. J. (1991b). Bacteriocin mediated inhibition of *Clostridium botulinum* spores by lactic acid bacteria at refrigeration and abuse temperatures. *Appl. Environ. Microbiol.* **57**, 3423–3428.
Oscroft, C. A., Banks, J. G., and McPhee, S. (1990). Inhibition of thermally-stressed *Bacillus* spores by combination of nisin, pH and organic acids. *Lebensm.-Wiss u.-Technol.* **23**, 538–544.
Oxford, A. E. (1944). Diplococcin, an antibacterial protein elaborated by certain milk streptococci. *Biochem. J.* **38**, 178–182.
Pucci, M. J., Vedamuthu, E. R., Kunka, B. S., and Vandenbergh, P. A. (1988). Inhibition of *Listeria monocytogenes* by using bacteriocin PA-1 produced by *Pediococcus acidilactici* PAC 1.0 *Appl. Environ. Microbiol.* **54**, 2349–2353.
Radler, F. (1990a). Possible use of nisin in winemaking. I. Action of nisin against lactic acid bacteria and wine yeasts in solid and liquid media. *Am. J. Enol. Vitic.* **41**, 1–6.

Radler, F. (1990b). Possible use of nisin in winemaking. II. Experiments to control lactic acid bacteria in winemaking. *Am. J. Enol. Vitic.* **41,** 7–11.

Ramseier, H. R. (1960). Die wirkung von nisin auf *Clostridium butyricum* prazm. *Arch. Microbiol.* **37,** 57–94.

Ray, B. (1992a). Nisin of *Lactococcus lactis* ssp. *lactis* as food biopreservative. *In* "Food Biopreservatives of Microbial Origin," (B. Ray and M. A. Daeschel, eds.). CRC Press, Boca Raton, Florida.

Ray, B. (1992b). Bacteriocins of starter culture bacteria as food biopreservatives: An overview. *In* "Food Biopreservatives of Microbial Origin," (B. Ray and M. A. Daeschel, Eds.). CRC Press, Boca Raton, Florida.

Rayman, M. K., Aris, B., and Hurst, A. (1981). Nisin; A possible alternative or adjunct to nitrite in the preservation of meats. *Appl. Environ. Microbiol.* **41,** 375–380.

Rayman, M. K., Malik, N., and Hurst, A. (1983). Failure of nisin to inhibit outgrowth of *Clostridium botulinum* in a model cured meat system. *Appl. Environ. Microbiol.* **46,** 1450–1452.

Rogers, A. M. (1991). Contribution of nisin to the inhibition of *Clostridium botulinum* in a Model Food System. Ph.D. dissertation, Rutgers University.

Rogers, A. M., and Montville, T. J. (1991). Improved diffusion assay for nisin quantification. *Food Biotechnol.* **5,** 161–168.

Saito, H., Watanabe, T., and Tomioka, H. (1979). Purification, properties, and cytotoxic effect of a bacteriocin from *Mycobacterium smegmatis*. *Antimicrob. Agents Chemo.* **15,** 504–509.

Schillinger, U. (1990). Bacteriocins of lactic acid bacteria. *In* "Biotechnology and Food Safety," (D. Bills and S. Kung, eds.), pp. 55–74. Butterworths, London.

Schillinger, U., and Lucke, F. C. (1989). Antibacterial activity of *Lactobacillus sake* isolated from meat. *Appl. Environ. Microbiol.* **55,** 1901–1906.

Scott, V. N., and Taylor, S. L. (1981a). Effect of nisin on outgrowth of *Clostridium botulinum* spores. *J. Food Sci.* **46,** 117–120.

Scott, V. N., and Taylor, S. L. (1981b). Temperature, pH, and spore load effects on the ability of nisin to prevent the outgrowth of *Clostridium botulinum* spores. *J. Food Sci.* **46,** 121–126.

Sobrino, O., Rodriguez, J. M., Moreira, W. L., Fernandez, M. F., Sanz, B., and Hernandez, P. E. (1991). Antibacterial activity of *Lactobacillus sake* isolated from dry fermented sausages. *Intl. J. Food Micro.* **13,** 1–10.

Somers, E. B., and Taylor, S. L. (1981). Further studies on the antibotulinal effectiveness of nisin in acidic media. *J. Food Sci.* **46,** 1972–1973.

Somers, E. B., and Taylor, S. L. (1987). Antibotulinal effectiveness of nisin in pasteurized cheese spreads. *J. Food Prot.* **50,** 842–848.

Spelhaug, S. R., and Harlander, S. K. (1989). Inhibition of foodborne bacterial pathogens from *Lactococcus lactis* and *Pediococcus pentosaceus*. *J. Food Prot.* **52,** 856–862.

Stevens, K. A., Sheldon, B. W., Klapes, N. A., and Klaenhammer, T. R. (1991). Nisin treatment for inactivation of *Salmonella* species and other Gram-negative bacteria. *Appl. Environ. Microbiol.* **57,** 3613–3615.

Taylor, S. L., and Somers, E. B. (1985). Evaluation of the antibotulinal effectiveness of nisin in bacon. *J. Food Protect.* **48,** 949–952.

Taylor, S. L., Somers, E. B., and Kreuger, L. A. (1985). Antibotulinal effectiveness of nisin-nitrate combinations in culture medium and chicken frankfurter emulsions. *J. Food Protect.* **48,** 234–239.

Todd, E. C. D. (1989). Preliminary estimates of foodborne disease in the United States. *J. Food Prot.* **52,** 595–601.

Tramer, J., and Fowler, G. G. (1964). Estimation of nisin in foods. *J. Sci. Food Agric.* **15,** 522–528.

Tubb, R. S., and Ogden, K. (1986). Use of nisin to prevent spoilage of alcoholic beverages. European Patent Application EP 186498 A2.

von Wright, A., Wessels, S., Tynkynen, S., and Saarela, M. (1990). Isolation of a replication region of a large lactococcal plasmid and use in cloning of a nisin resistance determinant. *Appl. Environ. Microbiol.* **56,** 2029–2035.

Waites, M. J., and Ogden, K. (1987). The estimation of nisin using ATP Bioluminometry. *J. Inst. Brew.* **93,** 30–32.

Whitehead, H. R. (1933). A substance inhibiting bacterial growth by certain strains of lactic streptococci. *Biochem. J.* **27,** 1793–1800.

Wiwobo, D. R., Eschenbruch, C. R., Davis, C. R., Fleet, G. H., and Lee, T. H. (1985). Occurrence and growth of lactic acid bacteria in wine: a Review. *Am. J. Enol. Vitic.* **36,** 302–313.

Yokotsuka, K., and Singleton, V. L. (1987). Interactive precipitation between graded peptides from gelatin and specific grape tannin fractions in wine-like model solutions. *Am. J. Enol. Vitic.* **38,** 199–206.

Yousef, A. E., Luchansky, J. B., Degan, A. J., and Doyle, M. P. (1991). Behavior of *Listeria monocytogenes* in wiener exudates in the presence of *Pediococcus acidilactici* H or Pediocin AcH during storage at 4 or 25°C. *Appl. Environ. Microbiol.* **57,** 1461–1467.

CHAPTER 5

The Molecular Biology of Nisin and Its Structural Analogues

J. NORMAN HANSEN

I. An Historical Perspective of Nisin

The discovery and development of antibiotics during the first half of the century has revolutionized the practice of medicine and created opportunities to study many important biological phenomena. Since the discovery of penicillin by Fleming in 1928, the pharmaceutical industry has searched the planet for new antibiotics. Only a small fraction of the thousands discovered have proved to be of practical therapeutic value. Many antibiotics have been useful laboratory tools; for example, puromycin was been used to elucidate protein biosynthesis, penicillin for cell wall biosynthesis, and oligomycin for oxidative phosphorylation. Most antibiotics, however, for a variety of reasons, have never been exploited.

Nisin activity was first observed in 1928 (Rogers and Whittier, 1928; Rogers, 1928), and it was studied as a discrete antimicrobial substance in 1944 (Mattick and Hirsch, 1944). Early attempts to find a practical application for nisin included testing its utility for treating mastitis in dairy herds (Taylor et al., 1949). Adverse reactions, apparently caused by impurities in the preparations, caused this work to be abandoned. From the 1940s through the 1970s, interest in nisin was maintained by a small number of laboratories, mainly Hurst and co-workers, who studied many aspects of the microbiology of nisin, including its mechanism of action and mode of biosynthesis (for a review, see Hurst, 1981), and Gross and co-workers, who studied the structure of nisin (Gross and Morell, 1971). Nisin also found a practical niche, where it proved to be a useful preservative in dairy products, and today holds the distinction of being the only "antibiotic" food additive that is permitted in the United States food supply.

Because the nisin story began long ago and unfolded slowly in few laboratories, its utility for the study of important biological problems was not widely appreciated until recently. An impediment to the recognition of nisin as representative of a unique biological system was its apparent similarity to other antibiotics. A description of nisin would include the fact that it is a bacterially produced peptide that contains an inordinate proportion of unusual amino acids. Peptide antibiotics that contain unusual amino acids are common, and most of them, such as valinomycin and Gramicidin S, are synthesized by nonribosomal mechanisms. As structural information about nisin became available, the presence of unusual amino acids led to the assumption that it was biosynthesized by a nonribosomal mechanism. However, experiments that tested this idea suggested a ribosomal mechanism was likely (Hurst, 1966). More than 20 years ago, Ingram proposed a scheme by which the unusual amino acids could be introduced by a novel sequence of posttranslational modifications of ordinary amino acids (Ingram, 1969; Ingram, 1970). Figure 1 shows the structure of mature nisin and Ingram's hypothetical scheme of posttranslational modifications, which include dehydration of serines and threonines to dehydro forms, followed by a Michael-type addition of cysteine sulfhydryl groups to give a thioether crosslink. The mature nisin structure accordingly contains dehydroalanine (Dha), dehydrobutyrine (Dhb), lanthionine (Ala-S-Ala), and β-methyllanthionine (Aba-S-Ala).

Although Ingram's hypothetical scheme was intriguing, the genetic methodology that could unambiguously prove it was not available at the time. Once cloning methods became available, it was possible to clone and sequence the nisin precursor gene. The sequence agreed perfectly with the hypothetical precursor predicted by Ingram's scheme, which is therefore accepted as the mechanism of posttranslational modification of the nisin precursor to mature nisin. Although this is convincing circumstantial evidence, it should be noted that the modification reactions have not yet been studied using biochemical methods.

The nisin precursor gene was first published by Buchman et al. (1988), and later by two other laboratories (Kaletta and Entian, 1989; Dodd et al., 1990). The sequence of the nisin gene and its immediately flanking regions are shown in Figure 2.

We now realize that nisin is but one example of a class of ribosomally synthesized peptide antibiotics that are formed by this mechanism. The first besides nisin to be discovered was subtilin (Jansen and Hirschmann, 1944), followed by several others in recent years, and include cinnamycin and duramycin (Gross, 1977), epidermin (Allgaier et al., 1986), and gallidermin (Kellner et al., 1988). The term "lantibiotic" has been coined as a name for this class (to connote a lanthionine-containing antibiotic), although its suitability is in dispute. Genes for several different lantibiotic precursors besides nisin have been cloned, including subtilin (Banerjee and Hansen, 1988),

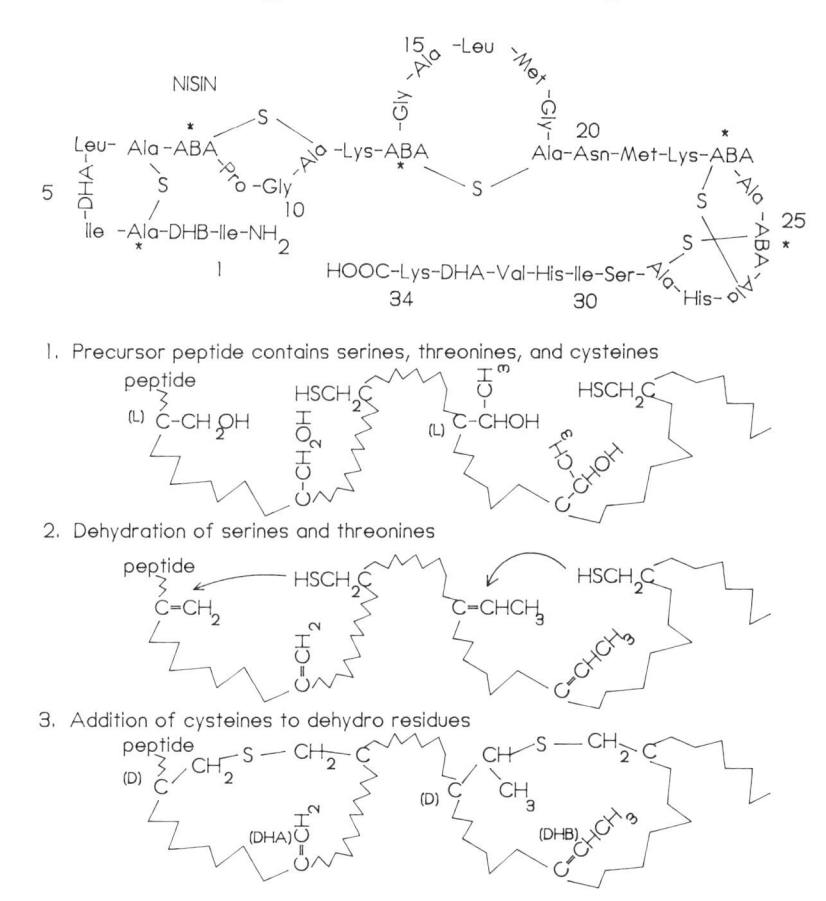

5. The Molecular Biology of Nisin and Its Structural Analogues

Figure 1 *(Top)* Structure of nisin as determined by Gross and co-workers (Gross and Morell, 1971). *(Bottom, 1–3)* Scheme for formation of unusual amino acids. Serine and threonine are dehydrated to dehydroalanine (Dha) and dehydrobutyrine (Dhb), respectively. Nucleophilic addition of the sulfhydryl group of cysteine to the dehydro residues proceeds with stereo-inversion. Lanthionine and β-methyllanthionine result from the formation of thioether cross-linkages. (Scheme is based on that proposed by Ingram, 1970.)

epidermin (Schnell et al., 1988), and Pep5 (Kaletta et al., 1989), and studies of the biosynthetic mechanism are under way in several laboratories. Although nisin languished for many years as one of Nature's curiosities, it has now become an important model biological system.

II. Significance of Posttranslationally Modified Peptides

The rapid evolution of genetic methodologies has greatly increased the scope of the biochemical experiments that can be performed with proteins and peptides. It has become respectable for protein chemists (now called protein engineers because they can design and modify proteins at will) to

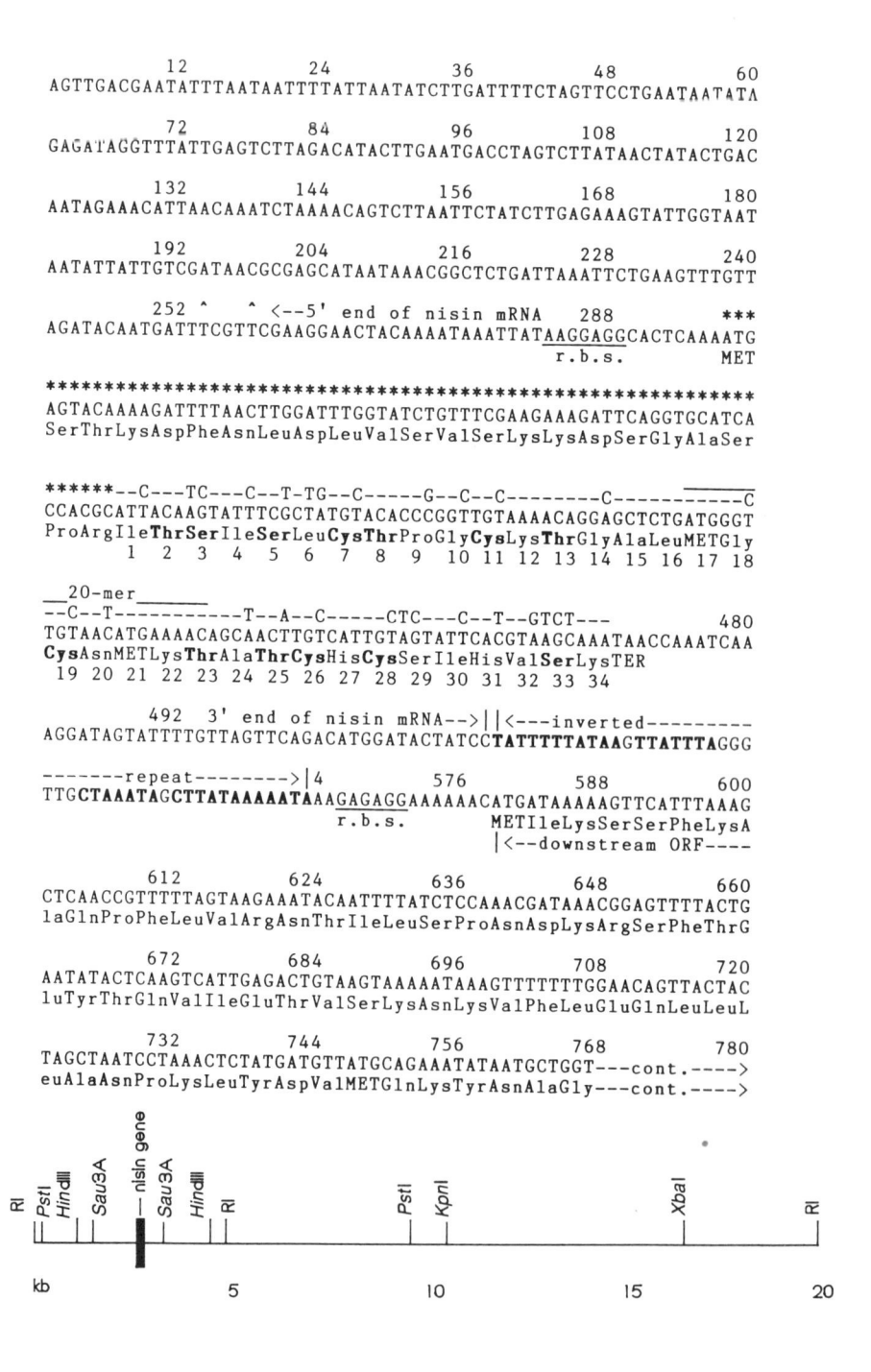

```
              12        24        36        48        60
AGTTGACGAATATTTAATAATTTTATTAATATCTTGATTTTCTAGTTCCTGAATAATATA
              72        84        96       108       120
GAGATAGGTTTATTGAGTCTTAGACATACTTGAATGACCTAGTCTTATAACTATACTGAC
             132       144       156       168       180
AATAGAAACATTAACAAATCTAAAACAGTCTTAATTCTATCTTGAGAAAGTATTGGTAAT
             192       204       216       228       240
AATATTATTGTCGATAACGCGAGCATAATAAACGGCTCTGATTAAATTCTGAAGTTTGTT
             252  ^   ^ <--5' end of nisin mRNA   288        ***
AGATACAATGATTTCGTTCGAAGGAACTACAAAATAAATTATAAGGAGGCACTCAAAATG
                                           r.b.s.             MET
*************************************************************
AGTACAAAGATTTTAACTTGGATTTGGTATCTGTTTCGAAGAAAGATTCAGGTGCATCA
SerThrLysAspPheAsnLeuAspLeuValSerValSerLysLysAspSerGlyAlaSer

******--C---TC---C--T-TG--C-----G--C--C--------C-----------C
CCACGCATTACAAGTATTTCGCTATGTACACCCGGTTGTAAAACAGGAGCTCTGATGGGT
ProArgIleThrSerIleSerLeuCysThrProGlyCysLysThrGlyAlaLeuMETGly
 1   2   3   4   5   6   7   8   9  10  11 12 13 14  15 16  17 18
   20-mer
--C--T----------T--A--C-----CTC---C--T--GTCT---          480
TGTAACATGAAAACAGCAACTTGTCATTGTAGTATTCACGTAAGCAAATAACCAAATCAA
CysAsnMETLysThrAlaThrCysHisCysSerIleHisValSerLysTER
 19  20  21  22  23  24  25  26  27  28  29  30  31  32  33 34

             492  3' end of nisin mRNA-->||<---inverted---------
AGGATAGTATTTTGTTAGTTCAGACATGGATACTATCCTATTTTTATAAGTTATTTAGGG

-------repeat-------->|4      576       588       600
TTGCTAAATAGCTTATAAAAATAAAGAGAGGAAAAAACATGATAAAAAGTTCATTTAAAG
                         r.b.s.         METIleLysSerSerPheLysA
                                        |<--downstream ORF----
             612       624       636       648       660
CTCAACCGTTTTTAGTAAGAAATACAATTTTATCTCCAAACGATAAACGGAGTTTTACTG
laGlnProPheLeuValArgAsnThrIleLeuSerProAsnAspLysArgSerPheThrG
             672       684       696       708       720
AATATACTCAAGTCATTGAGACTGTAAGTAAAAATAAAGTTTTTTGGAACAGTTACTAC
luTyrThrGlnValIleGluThrValSerLysAsnLysValPheLeuGluGlnLeuLeuL
             732       744       756       768       780
TAGCTAATCCTAAACTCTATGATGTTATGCAGAAATATAATGCTGGT---cont.---->
euAlaAsnProLysLeuTyrAspValMETGlnLysTyrAsnAlaGly---cont.---->
```

study approaches to the design and construction of new enzymes and other proteins with novel and useful properties. A tacit assumption is that these new proteins will be constructed solely from the common 20 amino acids that are defined by the genetic code. This would ultimately limit the scope of the kinds of physical and chemical properties that these designer proteins could have. However, the existence of nisin and the several other ribosomally synthesized lantibiotics shows us that some biological systems can introduce new amino acids in a controlled manner into what would otherwise be ordinary peptides. If we could somehow arrange to direct the posttranslational modification system to make these kinds of changes in other peptides or proteins, we would suddenly have the capability of constructing proteins with chemical and physical properties that are unattainable with the ordinary amino acids. Whether the nisin biosynthesis system could be appropriately adapted to achieve this depends entirely on the details of the recognition–modification mechanism, and whether the signals and structures that cause the modifications to occur can be inserted into new peptides without destroying their function.

Another obvious advantage of the lantibiotics is their structures are unambiguously dictated by genes. Unlike most antibiotics synthesized by multienzyme metabolic pathways, the fact that lantibiotics are gene encoded and synthesized ribosomally means it is possible to make specific structural alterations in the mature lantibiotic by mutating the precursor gene. One can thus expect to be able to make improved versions of the known lantibiotics, carry out structure–function studies with relative ease, and develop approaches to the rational design and construction of novel lantibiotics that could have properties and spectra of action that are unknown in Nature.

III. Lantibiotics Could be Adapted to Multiple Purposes

Our knowledge about how lantibiotics are biosynthesized and how they work is limited; thus one can still be optimistic about what is possible to achieve with them. One of the most optimistic possibilities is linked to the processing signals in the lantibiotic precursor peptide, which reside com-

---

Figure 2 Sequence of the nisin gene and its flanking regions as determined by Buchman et al. (1988). The sequence and the translation of the peptide precursor gene is shown, in which the leader region is overscored with asterisks and the structural region is numbered as in Figure 1. Amino acids that undergo modification are in bold letters and correspond to the locations of the unusual amino acids in mature nisin (Figure 1). The 5' and 3' terminals of the nisin transcript were determined by S1 mapping (Buchman et al., 1988). The restriction map shows major restriction sites in relation to the location of the gene sequence. The EcoRI site at position 0 kb on the restriction map is an artificial site created by the cloning process. The 5-kb *Eco*RI fragment is a truncated form of a 9.5-kb genomic fragment, and the site of truncation is upstream from the nisin structural gene.

pletely within the leader region. Any peptide or protein that is fused to the leader will undergo processing, including dehydration of serines and threonines and formation of thioether cross-linkages. One could accordingly aspire to place electrophilic dehydro residues in the active site of an enzyme or the recognition site of a binding protein such as an antibody. The enzyme could thus catalyze a novel reaction or the binding protein could become covalently attached to its target, thus increasing its affinity and effectiveness. Equally intriguing is the possibility of introducing enhanced thermal stability into a protein by incorporating thioether cross-bridges. Nisin can be autoclaved at pH 2 without undergoing inactivation (Hall, 1966; Hurst, 1981). This remarkable thermal stability is attributed in part to the five thioether cross-bridges that may confer enhanced conformational stability. It is widely recognized that a major limitation in the use of enzymes for industrial processes is their thermal instability. Discovery of a thermally stable enzyme that allows an industrial process to be carried out at an elevated temperature in comparison to a less stable enzyme can make the difference between a process that is economically unfeasible and one that is highly profitable. If the remarkable thermal stability of nisin is indeed the result of thioether cross-bridges, introduction of these into industrial enzymes could have a large economic impact, as well as hasten the day when much industrial manufacturing will be based on biological processes. Such processes hold promise of being more efficient, less energy intensive, and less damaging to the environment.

IV. A Dilemma Posed by Nisin Resistance

A question that always arises with respect to bacteria that produce antibiotics effective against other bacteria concerns the mechanism by which the producer bacterium protects itself against its own antibiotic. It has been shown that *Lactococcus lactis* is sensitive to nisin and acquires good nisin resistance only when it is producing nisin (Hurst and Kruse, 1972). This implies that a nisin resistance factor is coproduced with nisin itself.

The question of resistance is central to the rationale of using the nisinlike antibiotics as part of a strategy for making new antibiotics. The pharmaceutical industry is in a constant race against antibiotic resistance among microbial populations that are exposed to antibiotics during the course of therapeutic treatment. An important approach in drug design is to make structural changes in a natural antibiotic to improve its spectrum of action and overcome resistance to the natural form. Thousands of derivatives have been made of penicillin alone, with the result that penicillinlike antibiotics are still extremely useful (Hedge and Spratt, 1985). The major barrier in this approach is that the structural changes are made by arduous organic chemical synthesis, which is difficult, time consuming, and expensive. Moreover, manufacture of these modified antibiotics requires expen-

5. The Molecular Biology of Nisin and Its Structural Analogues

sive scaling-up of size of what is often an exotic laboratory synthesis. Treatments requiring such antibiotics are accordingly expensive. In contrast, a ribosomally synthesized antibiotic such as nisin can be structurally modified by mutating the gene and letting the genetically engineered cell do the organic chemistry required to construct the structural analogue. Subsequent step-up for manufacture of the analogue would be a simple step-up of a generic fermentation process. Techniques for mutagenesis have become simple enough that one can envision generating thousands (or millions) of nisin mutants and screening for those that possess desired characteristics, allowing one to obtain a large number of useful nisin variants in a relatively short time.

Ribosomally synthesized antibiotics that can evolve by mutation and selection for function are obviously useful to the producer organism. They should be able to adapt rapidly to the needs of the host as well as changes among the competitor population. A problem is created, however, by the need for the producer organism to remain resistant to its own antibiotic. If the antibiotic gene mutated to give a dramatically altered form that has a different spectrum of action, a modified target specificity, or an altered mechanism of action, the producer resistance factor might not be able to cope with these changes and the producer cell would die. This implies that evolution would have to occur in a lock-step fashion, in which mutations in the antibiotic would be followed by selection of a new resistance factor that was optimized to protect against the mutant antibiotic. It is not obvious how the cell can achieve this, but it is hoped that the elucidation of the mechanism of self-resistance will shed light on it.

Adding to the resistance dilemma, it has been observed that a cell that produces one lantibiotic is only partially resistant to other lantibiotics. For example, subtilin-producing *Bacillus subtilis* is not resistant to nisin, nor is nisin-producing *Lactococcus lactis* resistant to subtilin. If this turns out to be generally true for lantibiotic producers, it would suggest that the respective resistance factors are relatively specific. It seems unlikely that the mechanisms by which these organisms achieve resistance are completely different. If it is to be assumed that the "purpose" of these lantibiotics is biological warfare among bacteria that occupy overlapping biological niches, then the specificity of resistance factors is important. If a single resistance factor could confer resistance to all lantibiotics, lantibiotics would cease to be useful weapons among competing lantibiotic producers. The fact that there seems to be little cross-resistance suggests that lantibiotic producers managed not to fall into this evolutionary trap. What remains to be explained is how the specificity is achieved and how resistance within an organism can evolve in parallel with lantibiotic activity. Inasmuch as we have no idea about the nature of the resistance factor or its mechanism of action, we are a long way from explaining how this parallel evolution might occur.

Equally confusing is the relationship between the resistance exhibited by *Lactococcus lactis* against the nisin that it is producing, and the resistance that some nonnisin-producing organisms have against nisin. The best-understood example is the resistance by *Bacillus cereus* to nisin, which has been attributed to a nisin reductase (Jarvis, 1967; Jarvis and Farr, 1971), in which inactivation apparently occurs by reduction of one or more critical dehydro residues in nisin. Another example of nisin resistance is that reported by Froseth and McKay (1991), who cloned a gene that encoded a nisin resistance factor from *Lactococcus lactis* subsp. *lactis* biovar *diaceylactis* DRC3, which is a nonproducer of nisin. The gene encodes a 318-residue protein that did not show homologies to other known proteins. It has not yet been possible to establish the relationship, if any, between this gene and the resistance factor of a nisin producer.

V. The Molecular Biology of Nisin Biosynthesis Is of Unknown Complexity

All that can be said with certainty about nisin biosynthesis is that it is synthesized as a precursor consisting of a 23-residue leader region and a 34-residue structural region (Buchman et al., 1988). The leader region is excised, the structural region undergoes posttranslational modifications, and mature nisin is secreted outside the cell. Neither the sequence of these events, nor the location in the cell where they occur, nor the mechanism by which they occur, is known. The one gene that is indisputably indispensable is the nisin gene itself. All the posttranslational events could conceivably be auto-catalyzed by the precursor peptide, or be performed by proteins that are not directly related to nisin production. For example, secretion and removal of the leader could be performed by the secretion system employed for other secreted proteins.

At the other extreme is a situation in which nisin is synthesized by a dedicated system that contains a full complement of proteins that orchestrate a complex biosynthesis pathway. In this scenario, it is possible that the precursor peptide contains signals that target the precursor to a unique processing-secretion apparatus consisting of binding and transport proteins, a leader peptidase, and enzymes that catalyze dehydration and cross-linking. It is easy to imagine that such a complex system might require a dozen or more genes that may be organized as one or more operons regulated in a highly coordinated manner.

VI. Cloning of the Genes for the Nisin and Subtilin Precursor Peptides

Despite the fact that nisin has been known for over sixty years, and that a ribosomal mechanism for its biosynthesis was proposed over twenty years ago (Ingram, 1970), it was not until the gene that encodes the nisin precursor peptide was cloned that the nisin system could be studied using techniques of molecular genetics. The most gratifying discovery related to the precursor gene was that the elegant and painstaking work on the bizarre and difficult structures of nisin and subtilin performed by Erhard Gross and co-workers for many years (Gross, 1975) turned out to be correct in every detail. It also confirmed Ingram's hypothesis about the posttranslational mechanism of nisin biosynthesis. Without the structural information provided by Gross, and the insight provided by Ingram on how to infer the original translation product from the structure of the mature antibiotic, it would not have been possible to construct the hybridization probes used to identify clones that contained the nisin and subtilin precursor genes. The nisin gene that was cloned by Buchman et al. (1988) and the subtilin gene that was cloned by Banerjee and Hansen (1988) both encoded serines, threonines, and cysteines at exactly the positions predicted by Ingram's scheme, simultaneously proving that structures determined by Gross and the scheme proposed by Ingram were correct. Cloning these genes, however, provided a great deal more than this. The gene sequences also revealed that the precursor peptides of nisin and subtilin are synthesized with a leader region that is removed during maturation. By the use of the gene sequences as hybridization probes, it was possible to study the expression of the genes into transcripts, to determine their time of synthesis, and to measure their half-lives (Banerjee and Hansen, 1988; Buchman et al., 1988). The genes also provided access to the region of the chromosome in which they are located, and it immediately became possible to search for other genes located nearby that are required by the lantibiotic biosynthesis pathway. The organization of these genes can now be studied, and a main purpose of this chapter is to discuss what has been learned up to the present.

VII. Evolutionary and Functional Relationships between Nisin and Subtilin Implied by Comparison of Their Structural Genes

A powerful way to gain insight about the functional role of structural features in a biological molecule is by comparison with other molecules of similar function. In this we are fortunate that both nisin and subtilin, which are close structural homologues, are available. Comparison of the nisin and subtilin structures (Figure 3) shows that the amino acids are different at twelve positions, the rings resulting from thioether cross-linkages are the same size and in the same positions, and they both have two Dha residues and one Dhb residue. The Dha residues are in the same relative positions, but the Dhb residue is in a different position.

On one hand, the structural similarities suggest that nisin and subtilin have evolved from a common ancestor; on the other hand, the structural differences suggest they have been evolving separately for a long time

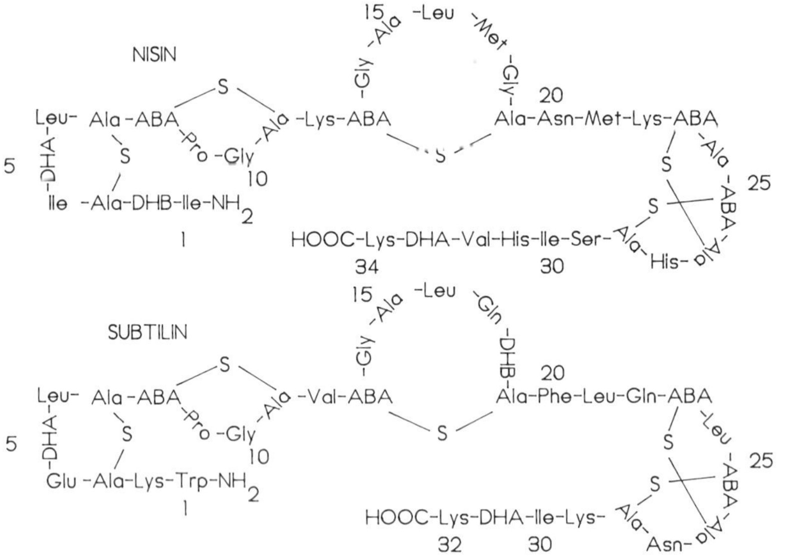

Figure 3 *(Top)* Comparison of the mature forms of nisin and subtilin. *(Bottom)* Homologies between the amino acid sequences of the nisin and subtilin precursor peptides prior to posttranslational modification. Dashes indicate identical amino acids; gaps are inserted to improve homologies.

```
                Amino Acid Homologies
        (Leader Region)        (Structural Region)
MSTK DFNLDLVSVSKKDSGASPR  ITSISLCTPGCKTGALMGCNMKTATCHCSIHVSK  Nisin
--KFD--D--V-K---Q--KIT-Q  WK-E-------V----QT-FLQ-L--N-K-     Subtilin
```

(Buchman et al., 1988). While gazing at the sequences of the two precursor peptides, it is important to remember that evolutionarily conserved features include the signals required to orchestrate the maturation process as well as provide the functional properties of the mature lantibiotic. As far as is known, the only peptide in the cell that undergoes the posttranslational modifications is the lantibiotic precursor itself. The processing system ignores the thousands of other proteins in the cell. One must conclude that the precursor peptide contains highly specific signals that instruct the processing machinery to carry out the modifications. These signals could reside in only the leader region, only the structural region, or they could be distributed throughout the entire peptide. Locating and identifying these signals will be important to our understanding of the processing mechanism. These signals must accomplish the export of the lantibiotic outside the cell, either prior to, during, or after the other processing events have occurred. Most proteins are exported by cells contain signals that recognize a transport apparatus that carries out the translocation process. In prokaryotes, the best-understood mechanism is the Sec system, in which the protein to be exported contains a leader with a characteristic hydrophobic signal region that is recognized by the SecB protein in the cytoplasm; this complex is in turn recognized by SecA on the membrane. Other Sec proteins, such as

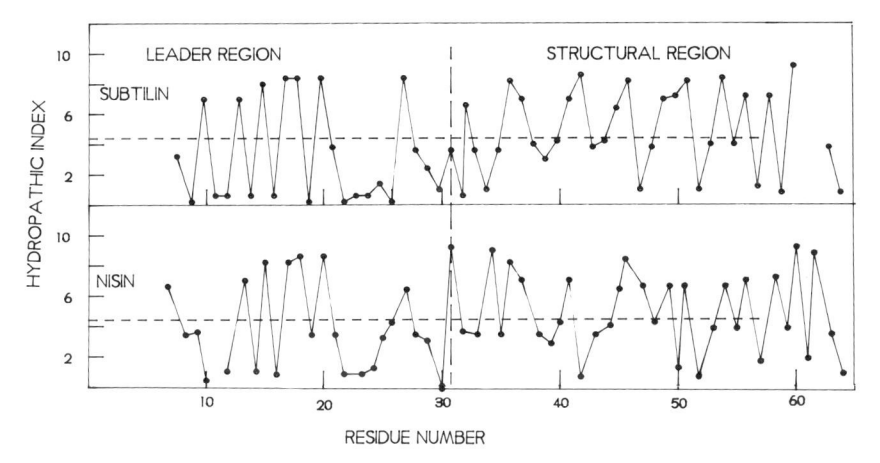

5. The Molecular Biology of Nisin and Its Structural Analogues 103

Figure 4 Comparison of the hydropathic profiles of subtilin and nisin. The greatest similarities occur in the leader regions. (Hydropathic profiles calculated by Buchman et al., 1988.)

SecE, SecY, SecD, and SecF participate in further stages of binding and translocation, and a leader peptidase cleaves off the signal-containing leader region (Bassford et al., 1991). Another mechanism of translocation is found in the hemolysin protein of *Escherichia coli*, in which the C-terminal end of the hemolysin protein is recognized by the HlyD protein in the membrane, and the translocation is achieved by another protein called HlyB. There is no cleavage of a signal region, and the hymolysin protein is translocated without modification (Holland et al., 1990). Nisin and subtilin could be secreted by one of these mechanisms, or by yet a third mechanism. The hydropathic profiles of the leader regions of the nisin and subtilin precursor peptides (Figure 4) are devoid of a hydrophobic stretch that is characteristic of the Sec-mediated export system, suggesting that the Sec system is not involved. On the other hand, the leader region of the precursor is cleaved off, which is different from the hemolysin system. This is despite the fact that a gene that encodes a protein with strong homology to the HlyB translocation protein is located just upstream from the subtilin structural gene (Chung et al., 1992). Although it is merely circumstantial evidence, the presence of the HlyB-like gene in the immediate vicinity of the subtilin gene suggests that it may encode a protein that is involved with subtilin translocation across the membrane.

VIII. Expression of the Genes for Nisin and Subtilin and Characterization of Their Transcripts

It is hypothesized that nisin synthesis requires the participation of several genes, in addition to the nisin gene itself. The early work of Hurst established the principle that nisin production occurred mainly, although not exclusively, during the later stages of growth (late log phase and continuing into stationary phase) (Hurst and Dring, 1968). Evidence was also presented that a precursor form of nisin was stored in the exterior cell envelope prior to undergoing one or more final stages of maturation (Hurst and Peterson, 1971). It is interesting to compare these results with the kinetics of expression of the nisin structural gene during the growth cycle of *Lactococcus lactis* ATCC 6633 as observed by Buchman et al. (1988). RNA was isolated during all growth stages, from early log phase into stationary phase, and examined for nisin transcripts by Northern analysis. After normalizing for total RNA content, it was determined that the nisin gene was expressed at about the same extent at all stages tested. Hurst's observation that active nisin was not observed until late growth stages was confirmed, despite the fact that nisin gene transcripts were observed for several hours before the onset of detectable nisin activity (Buchman et al., 1988). It is most unlikely that the delay between the onset of nisin gene expression and appearance of nisin activity can be attributed to the time required for maturation of the nisin precursor. Either the transcript is not translated at early times, or the translation product is unstable, or there are one or more components required for maturation that do not appear until late growth stages. The third possibility seems the most likely, and would require at least one gene that is subject to differential regulatory control with respect to the nisin gene itself. It would further imply that an incompletely processed precursor form would need to be stored somewhere to await the appearance of the needed maturation component(s), which is consistent with the observation of a nisin precursor in the cell envelope (Hurst and Peterson, 1971). It is also consistent with observations reported by Weil et al. (1990), who found that a precursor form of the lantibiotic called Pep5 could be isolated from *Staphylococcus epidermidis* that had undergone dehydrations to give dehydro residues but no cross-linkages with the cysteine residues. The fact that such a precursor could be isolated indicates that there is a subsequent rate-limiting step that causes it to build up to significant levels. Confirmation of these ideas will have to await the identification and study of all genes that participate in nisin biosynthesis.

Banerjee and Hansen (1988) performed similar experiments on the expression of the subtilin gene and found sharp contrast to what was observed with nisin. In medium A, which is the classic high-sucrose medium (Feeney et al., 1948) that promotes high subtilin production, the subtilin gene was not significantly expressed until well into stationary phase, where it appeared at levels that were about 200-fold greater than during log phase. Subtilin activity appeared simultaneously with the onset of the expression of the subtilin gene. It was also observed that the subtilin transcript was extremely long lived for a prokaryotic mRNA, having a half-life of about 45 minutes instead of a more usual 2–3 minutes (Banerjee and Hansen, 1988). The stability of the nisin transcript was much less, with a half-life of about 7 minutes (Buchman et al., 1988). The reasons for these differences and their implications for any differences in the biosynthetic mechanism are far from clear. On one hand, it could mean that profound changes in the mechanism of maturation have been wrought by evolutionary time. On the other hand, it could merely be different regulation imposed by the fact that *Bacillus subtilis* is a spore-forming bacterium, whereas *Lactococcus lactis* is not. *Bacillus subtilis* typically expresses antibiotics as secondary metabolites during the early stages of sporulation. It may turn out that an understanding of the regulation of subtilin biosynthesis will have important implications for the mechanisms of gene regulation in bacilli, particularly in relation to sporulation, and that there may be many contrasts between the mechanism of regulation employed by spore-forming Gram-positive bacteria and non-spore-forming Gram-positive bacteria. These differences may be reflected in the way these organisms regulate lantibiotic biosynthesis.

IX. The Ability to Produce Subtilin Can Be Transferred among Strains of *Bacillus subtilis*

The similarity among lantibiotics is the result of either convergent or divergent evolution. If we consider only nisin and subtilin, their structural and functional similarities are great enough to suggest they evolved from a common ancestor (Buchman et al., 1988). The ability to produce an antibiotic is a trait that is frequently transmitted between bacteria. Since most antibiotics are synthesized by multistep pathways that require several proteins, genes for antibiotic production are often found clustered within some kind of genetic element such as a plasmid or transposon (Hopwood et al., 1985). Such clustering of genes facilitates the transfer of antibiotic production between bacterial species, and also makes it much easier to identify and characterize them. Once a new strain has acquired the ability to produce the antibiotic, the antibiotic becomes subject to the selective pressures that exist within the new host. If it is presumed that the new host will occupy a different ecological niche than the one from which the antibiotic was derived, the fact that the lantibiotics are gene encoded makes them particularly able to adapt to the needs of their new host. This is because they have access to the same mechanism of mutation and selection that is available to all proteins. The exquisite perfection of enzymes in their ability to recognize and catalyze substrates is testimony to the power of this process. Lantibiotics should therefore be able to adapt to the host's needs to a similar level of perfection, to become particularly effective against organisms that occupy the same ecological niche as the new host. That this occurs is supported by Hurst's observation that *Streptococcus cremoris*, with nutritional and growth condition requirements that are similar to *Lactococcus lactis*, is extremely sensitive to nisin (Hurst and Collins-Thompson, 1979). This can be explained by nisin having evolved, by a process of mutation and natural selection, to become optimized against the competitors of the nisin-producing host.

Even though the original event that resulted in transmission of the capability to make an antibiotic may have involved a genetic element that contained all of the necessary genes required for antibiotic production, there is no guarantee that the organization of the genetic element will be retained over evolutionary time. If genes for an antibiotic biosynthesis pathway entered the cell encoded in a plasmid, some of the genes might become transferred from the plasmid to the chromosome, or the plasmid might become integrated into the chromosome and subsequently randomized. Such randomization could occur with a transposon as well.

One way to study the organization of the genes is to attempt to transfer genes for lantibiotic production from a producer strain to a nonproducer strain, to convert the latter into a lantibiotic producer. If this can be done, one can be confident that the genetic material from the producer strain provides whatever the nonproducer lacks. One can then characterize this genetic material. One cannot be certain, however, that this DNA contains all of the genes required for antibiotic production. For example, a majority of *Lactococcus lactis* strains that have been isolated from the environment are nisin producers, which suggests that nisin production is a natural trait for this organism. A nonproducing strain could accordingly be a mutant that is defective in a single protein or regulatory element. Such a mutant could be reverted by a small piece of DNA, giving the impression that nisin production requires few genes. The same can be said if a nonproducer is converted to a producer by transformation with a plasmid or transposon, since the recipient strain could possess some crucial genes in its chromosome that are complemented by those in the plasmid or transposon.

Having stated these caveats, transfer of lantibiotic production between strains is still an important experiment to perform. This laboratory has achieved the transfer of subtilin production between *Bacillus subtilis* strains, and several laboratories have achieved the transfer of nisin between species and strains of *Lactococcus*. Since nisin and subtilin are so closely related, it is presumed that information about one will provide useful insight about the other.

It is notable that subtilin production is not widespread among strains of *Bacillus subtilis*. In particular, subtilin is not produced by the Type strain (otherwise known as the Marburg strain or strain 168), nor does the Type strain contain the subtilin gene (Banerjee and Hansen, 1988). Indeed, it appears that *Bacillus subtilis* ATCC 6633 may be the only strain in which subtilin has been observed, and that other subtilin-producing strains are merely the ATCC 6633 strain renamed in other collections. In any event, subtilin production is not typical of *Bacillus subtilis* in the way that nisin production is typical of *Lactococcus lactis*. Although this does not eliminate the possibility that nonsubtilin-producing strains of *Bacillus subtilis* possess a partial complement of genes required for subtilin production, it makes it less likely.

Of all the experiments that we could perform in the transfer of subtilin genes, the most valuable is the conversion of *Bacillus subtilis* 168 to a subtilin producer because there is an enormous amount of genetic information known about strain 168, whereas subtilin-producing *Bacillus subtilis* ATCC 6633 is completely uncharacterized. Moreover, a great variety of techniques for recombinant DNA and genetic manipulations are available for strain 168, and their suitability for strain ATCC 6633 is unknown.

The transfer of subtilin production to strain 168 was achieved by first placing a chloramphenicol-selectable chloramphenicol acetyltransferase (CAT) gene immediately downstream from the subtilin structural gene in the chromosome of *Bacillus subtilis* ATCC 6633. High molecular weight DNA was isolated and used to transform competent *Bacillus subtilis* 168 cells, and chloramphenicol-resistant colonies were selected. A large majority of these transformants showed antibiotic production using a halo assay, and one of them, called LH45, was shown to produce authentic subtilin based on its chromatographic behavior, amino acid composition, N-terminal amino acid sequence, and its biological activity (Liu and Hansen, 1991). It was further established that all the genes that were required to confer subtilin production were contained within a 60-kb *Mlu*I restriction fragment, and that about 40 kb of the central portion of this fragment was unique to, and derived from, strain ATCC 6633 (Liu and Hansen, 1991). A restriction map of the *Mlu*I fragment that shows identified restriction sites is shown in Figure 5. This is an important result for several reasons. Although one cannot be certain that this piece of DNA contains all the genes required for subtilin production, it shows that subtilin-related genes are highly clustered and have not become randomized around the chromosome. It also shows that recognition signals in the subtilin precursor peptide and the processing proteins are fully functional in a new host. This implies that the processing system is reasonably flexible and bodes well for our future attempts to use the system to process heterologous proteins.

This result also allows us to begin to test hypotheses about the organization of the genes within this fragment that are involved with subtilin production. It is notable that a sequence that corresponds to a strong ρ-inde-

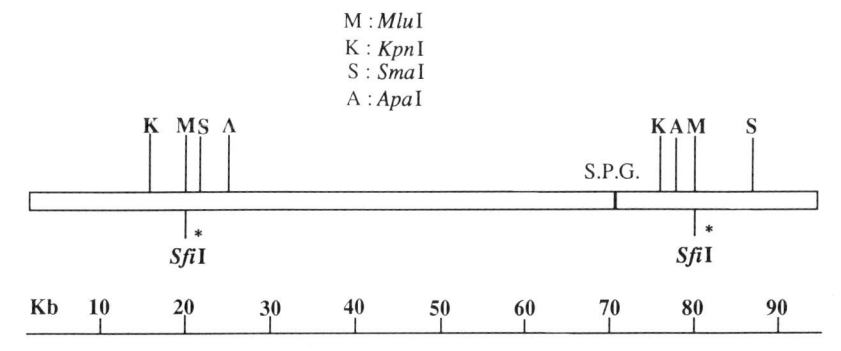

Figure 5 Locations of rarely cutting restriction in the vicinity of the subtilin precursor gene (S.P.G.). The 60-kb *Mlu*I fragment contains all the genes required to convert *Bacillus subtilis* 168 to a subtilin producer (see text). (Data from Liu and Hansen, 1991.)

pendent terminator is located immediately downstream from the subtilin gene, and that a plasmid with a CAT gene has been inserted about 300 nucleotides below the terminator. Transcriptional read-through of the subtilin gene into the downstream genes therefore seems unlikely. The fact that LH45 produces subtilin as efficiently as strain ATCC 6633 (Liu and Hansen, 1991) shows that insertion of the plasmid and CAT gene does not interfere with subtilin production. This suggests that the subtilin gene is at the 3' end of an operon and that downstream genes may not be required for subtilin production. However, this has not yet been experimentally confirmed by showing that deletion of the downstream genes does not interfere with subtilin biosynthesis.

X. The Ability to Produce Nisin Can Be Transferred between Strains of *Lactococcus lactis*

Strains of *Lactococcus lactis* generally contain several plasmids, some of which are quite large. It has been demonstrated that traits of nisin production, nisin resistance, sucrose fermentation, and phage resistance tend to be genetically linked in this organism (Klaenhammer and Sanozky, 1985; Broadbent and Kondo, 1991). These traits can be cotransferred in conjugation experiments (Fuchs et al., 1975; Gasson, 1984; Klaenhammer and Sanozky, 1985), which has variously been attributed to their location on a plasmid or transposon. The nisin structural gene itself is the only component whose location can be unambiguously identified by nucleic acid hybridization experiments. Donkersloot and Thompson (1990) used a nisin gene probe to show that nisin was chromosomally located in *Lactococcus lactis* K1. Steen et al. (1991) showed that the nisin gene hybridized to restriction fragments of *Lactococcus lactis* ATCC 11454 that were too large to be in plasmids, and therefore concluded a chromosomal location for the nisin gene. On the other hand, Kaletta and Entian (1989) reported a plasmid location of the nisin gene in *Lactococcus lactis* 6F3. Buchman et al. (1988) discovered an open reading frame upstream from the nisin gene that encoded a protein with strong homology to known transposases from *E. coli* insertion elements, suggesting that the nisin gene was part of a transposable element such as a transposon. Dodd et al. (1990), found an IS904 insertion element upstream from the nisin gene, and proposed that it also may play a role in mediating the transfer of nisin production between strains. It has been reported in a symposium that nisin production can be transferred among strains of *Lactococcus lactis* on a conjugative transposon with a molecular size of about 70 kb (Gasson, 1990). It is tempting to conclude that this large transposon contains all the genes required for nisin production along with those of other traits generally linked with nisin production (nisin resistance, sucrose fermentation, phage resistance), and that the IS904 element is the border of this transposon. Other observations suggest it may not be that simple. For one thing, these traits have been reported to be in plasmids by various laboratories studying various strains (Fuchs et al., 1974; Kaletta and Entian, 1989). In addition, primer extension studies of nisin gene transcripts described below indicate that the promoter that drives expression of the nisin gene lies several kilobases upstream from it, which is far beyond the putative transposon border. This means either that this is not the border of the transposon or expression of the nisin gene is driven from a fortuitous chromosomal promoter upstream from the site of transposon insertion. One possibility is that the site of insertion into the chromosome is highly specific, so that insertion next to an appropriate regulatory element occurs consistently. None of these explanations can accommodate the disparate observations of the location of nisin-related genes made by different laboratories over the years. Unless some of those observations turn out to be in error, a scheme for gene organization that can encompass all experimental observations has yet to be proposed.

XI. What is Known about the Organization of Genes Associated with Nisin Biosynthesis

The ultimate goal is to identify all the genes that are involved in the biosynthetic pathway of nisin from the time it is synthesized as a primary translation product to the time it is secreted as a mature active antibiotic. Once identification is complete, we will know the primary structures of all the proteins that are required and can begin the process of determining their functional roles. Much can be inferred by comparing these sequences with the protein database, but their functions must ultimately be confirmed by genetic and biochemical analysis. Once the genes are defined, the roles of regulatory elements can be studied. The difficulty of this task depends on the complexity of organization of the genes involved. It will be relatively easy if all the genes are organized in a single operon. It will be more difficult if multiple operons are involved, and it could be extremely difficult if the operons are scattered and regulated as a regulon.

The place to begin the analysis is with the nisin gene itself. Buchman et al. (1988) searched the sequences that flanked the nisin gene for regulatory elements. They established the 5' and 3' ends of the 267-nucleotide nisin gene transcript using S1-mapping. The 5' end was "ragged," suggesting that the nisin transcript is a processing product of a larger mRNA rather than the synthesis of a monocistronic mRNA. No obvious promoter sequence could be found. There was a stem-loop structure at the 3' end of the transcript that lacked the stretch of Us typical of ρ-independent terminator.

There was also a long open reading frame immediately downstream that appeared to be expressed as read-through from the nisin gene. They concluded that the nisin gene appeared to be expressed as part of a polycistronic mRNA.

Steen et al. (1991) extended these observations, and used primer extension analysis of mRNA to show that the nisin gene was expressed off a promoter that was at least 4 kb or more upstream from the nisin gene. They also sequenced the open reading frame downstream from the nisin gene to reveal an 851-residue protein with a C-terminal membrane anchor. Secondary structural analysis indicated that protein had many helices, most of which were amphipathic. They concluded that the protein was associated with, and anchored to, the cytoplasmic side of the membrane. A search for homologies in the protein data base supported the idea of it being a membrane-associated protein, but did not otherwise provide insight into the role that this protein might play in nisin biosynthesis. A good ρ-independent terminator was found downstream from this large open reading frame, suggesting a transcription stop site and a potential end of an operon. Shortly downstream from the terminator, two canonical promoters were found, immediately followed by an open reading frame, suggesting the beginning of a new operon. Whether this new operon contains genes required for nisin biosynthesis has not been explored. Figure 6 shows the organization of these genes.

The conclusions one can draw from these results are limited. The most definite conclusions are that the nisin gene and the downstream 851-residue open reading frame are probably in the same operon, and that the operon terminates below the 851-residue open reading frame. The role of any downstream operons cannot be assessed. It has also been established that the nisin gene is expressed from a promoter that is several kilobases upstream. However, the possibility is discussed above that the IS904 insertion element immediately above the nisin gene may represent the border of a conjugative transposon. If it does, the promoter that is driving the expression of the nisin gene lies outside the transposon and would appear to have a different purpose besides regulating the expression of the nisin gene. Clearing up these ambiguities will require a great amount of work.

XII. What is Known about the Organization of Genes Associated with Subtilin Biosynthesis

Now that information about genes for subtilin biosynthesis is becoming available, it appears that there are significant differences between the organization of genes for nisin biosynthesis and those for subtilin biosynthesis. This was apparent with the subtilin structural gene itself. The subtilin gene has an excellent ρ-independent terminator, and there is no indication that

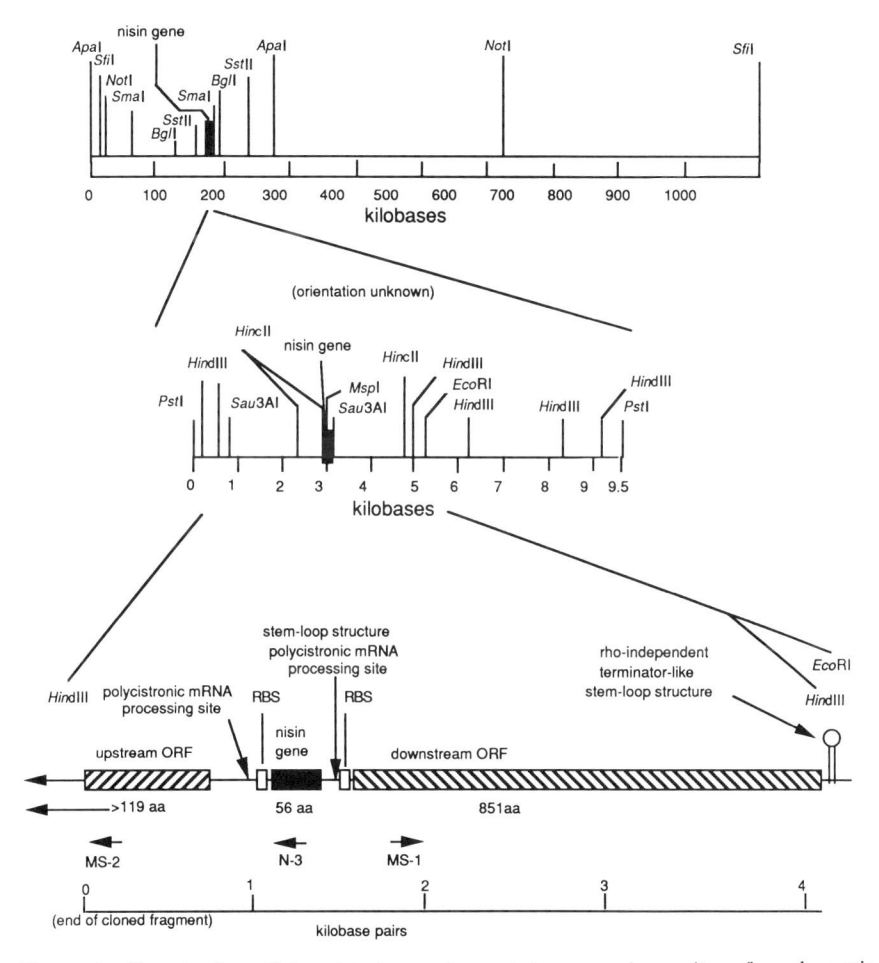

5. The Molecular Biology of Nisin and Its Structural Analogues   111

Figure 6 Organization of the nisin locus. A restriction map shows sites of rarely cutting enzymes within a *NotI* fragment that contains the nisin structural gene *(top)*. *(Middle)* The region in the immediate vicinity of the nisin gene. *(Bottom)* A diagram of the proposed polycistronic mRNA, showing the ORFs around the nisin locus. The direction of transcription of the nisin gene is from left to right. (Data from Steen et al., 1991.)

read-through from the subtilin gene to downstream open reading frames occurs. Indeed, a CAT gene with its own terminator can be placed after the subtilin terminator with no detrimental effect on subtilin production (Liu and Hansen, 1991). The subtilin gene accordingly seems to be the 3' terminal gene of an operon, whereas the nisin gene is at the penultimate position in its operon. However, the participation of genes downstream from the subtilin gene cannot yet be ruled out. There are also differences in the regions upstream from the nisin and subtilin genes. Whereas nisin has sequences corresponding to a transposase and IS904 insertion sequence that may be the border of a transposon, there is nothing like this upstream

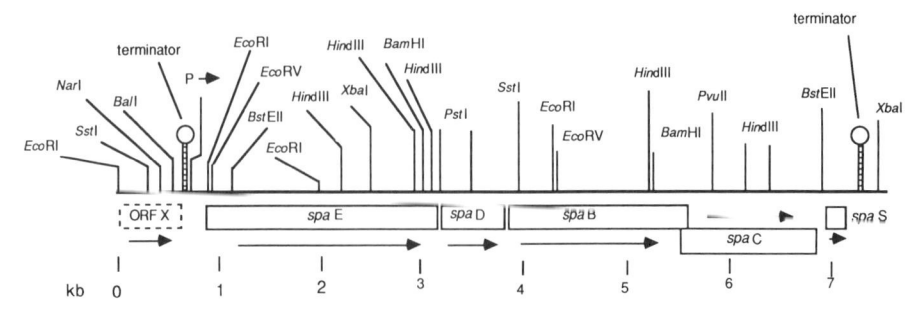

Figure 7 Organization of open reading frames within the subtilin (*spa*) operon. The subtilin structural gene is *spa*S, and the operon contains four other open reading frames, called *spa*E, *spa*D, *spa*B, and *spa*C. The promoter of the operon is designated P, and the direction of transcription by an arrow. The terminator of the operon lies immediately downstream from *spa*S. ORF X is a partial open reading frame of unknown function that lies upstream from the *spa* operon. A putative terminator is located between ORF X and the promoter of the *spa* operon. (Data from Chung and Hansen, 1992.)

from the subtilin gene. There are, however, several open reading frames; one of them, *spa*B, has strong homology to the *E. coli* HlyB protein (Chung et al., 1992). As discussed above, the HlyB protein participates in the translocation of the hemolysin protein. This protein also has strong homology to a wide variety of other translocating proteins, particularly the multidrug resistance proteins in the human, mouse, rabbit, and other mammals (Gerlach et al., 1986; Ames, 1986). It seems likely that *spa*B similarly has a translocation function and may participate in secretion of subtilin. Primer extension analysis of mRNA transcripts using an oligonucleotide sequence from within the subtilin gene has established that the subtilin gene is expressed as part of a polycistronic operon that includes these upstream reading frames (Chung et al., 1992). This indicates that earlier speculation that the subtilin gene may be encoded as a monocistronic operon with its own promoter (Banerjee and Hansen, 1988) is incorrect. A map showing these open reading frames and known restriction sites in the vicinity of the subtilin gene is shown in Figure 7.

XIII. Strategies and Systems to Express the Structural Genes for Nisin and Other Lantibiotics

The probability that nisin and subtilin, as well as other lantibiotics, all evolved from a common ancestor raises questions and creates opportunities. One question is the extent to which the various processing steps and the enzymes that catalyze them have been conserved. There are two kinds of conservation that may have occurred. One kind concerns the proteins that

5. The Molecular Biology of Nisin and Its Structural Analogues

carry out the processing steps, the particular kinds of steps that are performed, and the biochemical mechanisms by which they are achieved. Another kind of conservation that occurs in the lantibiotics is the recognition signals that appear in the precursor peptide that are recognized by the processing proteins. Even though the amino acid sequences in the precursor peptide have undergone extensive evolutionary changes, the structures and conformations they adopt may be highly conserved and remain relatively unchanged with respect to how they are recognized by the processing proteins. If such conservation has occurred, one could expect that the precursor peptide of a lantibiotic that is produced by one kind of cell would be recognized and accurately processed by the processing system of another cell that produced a different lantibiotic. For example, if the nisin precursor peptide were expressed in subtilin-producing *Bacillus subtilis* ATCC 6633, the processing enzymes in ATCC 6633 might recognize the nisin precursor and convert it to active nisin. Conversely, if a cell were to express the subtilin precursor in nisin-producing *Lactococcus lactis*, the processing enzymes in *Lactococcus lactis* might convert the subtilin precursor to active subtilin. If such heterologous expression were to work, it would provide strong confirmation of their evolutionary relatedness. Such heterologous expression would permit a great deal of flexibility in the design of lantibiotic expression hosts. Many of the lantibiotic producers have fastidious growth requirements and are accordingly expensive and troublesome to cultivate. *Lactococcus lactis* prefers to grow on complex media such as milk. Moreover, most lantibiotic producers have not been genetically well characterized, and genetic engineering methodologies are generally not well developed. It is therefore not a simple task to optimize lantibiotic expression so that its production can be performed economically in the natural hosts.

There are relatively few organisms that have been groomed for industrial production. It is unfortunate that *E. coli* is a Gram-negative organism that tends to produce low levels of secretion products, and those it does produce may be contaminated with endotoxins that must be removed. *E. coli* is therefore of limited usefulness. Derivatives of *Bacillus subtilis* 168 have been widely exploited for industrial production of biological materials. This organism is second only to *E. coli* in the amount of genetic information that is available for it, and the number of genetic manipulations that can be performed with it. As a Gram-positive organism, it tends to secrete products and it produces no endotoxins. It is certainly fortuitous that subtilin, which is one of the most important lantibiotics, is produced by *Bacillus subtilis*, and that *Bacillus subtilis* 168 has been successfully converted to a subtilin producer (Liu and Hansen, 1991). It is thus clear that this widely used industrial organism is capable of making lantibiotics. Many of the lantibiotics may prove to be of practical use if they can be manufactured economically; it would therefore be expedient if they could be biosynthesized by *Bacillus subtilis* 168, since the process for production of one could be employed to produce the others. To do so would require the incorporation of genetic material from the various lantibiotic producers into *Bacillus subtilis* 168. We would like to know whether the entire genetic machinery for the entire processing pathway will be necessary, or whether only the structural gene will be required. That is, if one wished to produce nisin in subtilin-producing *Bacillus subtilis*, would it be sufficient to express the nisin precursor sequence and let the subtilin-processing enzymes carry out the modifications, or would it be necessary to include all of the nisin-processing enzymes as well? If it is the latter, it will be more difficult, but if it is possible to make all of the lantibiotics in the same optimized organism, the most effective use of manufacturing methods will be realized.

XIV. Processing of Chimeric Precursor Peptides

Heterologous expression of nisin and subtilin has been attempted. An initial attempt to produce active nisin by expressing the nisin structural gene in *Bacillus subtilis* ATCC 6633 was only partially successful. One experiment placed the nisin structural gene in a plasmid downstream from a constitutive promoter. This plasmid could not be cloned into ATCC 6633 without undergoing rearrangement, and all the rearranged forms examined failed to produce nisin gene transcripts when examined by Northern analysis (Hawkins, 1990), even if the nisin gene was present. It was as if forms of the plasmid that expressed the nisin gene were lethal, and the only forms that survived were those that had undergone rearrangement in such a way that the nisin gene was not expressed. Since ATCC 6633 is only slightly resistant to nisin, a possible explanation of this result is that expression of copious quantities of the nisin precursor indeed led to formation of mature nisin, but it was lethal to the cell. There are other possible explanations and it is not yet possible to know which is correct.

Another experiment was attempted, in which the subtilin leader peptide sequence was fused in-frame to the nisin mature region sequence (Hawkins, 1990). The rationale was that the subtilin leader region should be optimal for recognition by the subtilin processing enzymes, which might then proceed to process the nisin structural region of the precursor to the mature form. If processing occurred correctly, the product should be extracellular mature active nisin. An extracellular peptide with chemical and physical properties of nisin was recovered. It was subjected to amino acid composition analysis, N-terminal sequence analysis, and biological activity analysis. The amino acid composition and N-terminal sequence analysis were the same as natural nisin. Moreover, the elution profile from the amino acid analyzer of nisin produced as the chimera was the same as for natural nisin. The only way that this could have occurred is if the chimeric subtilin-nisin precursor peptide had undergone processing to dehydrate the serines and threonines and form cross-linkages with the cysteine residues, and if the subtilin leader region had been cleaved off. If any of these events had not occurred, the amino acid composition and the analyzer elution profile of chimeric nisin could not have been the same as that of natural nisin. However, despite the identical chemical composition, the nisinlike peptide produced as the chimera was not biologically active (Hawkins, 1990). This result was rationalized by assuming that all aspects of processing had gone correctly, except the thioether cross-linkages that formed were not the correct ones. This conclusion was supported by an additional observation that the chimerically produced nisin consisted exclusively of a molecular size that corresponded to a dimeric form of the peptide (Hawkins, 1990), whereas natural nisin normally is mainly monomeric with a mixture of multimeric forms (Liu and Hansen, 1990). This suggests that intermolecular thioether linkages may have formed in addition to the normal intramolecular linkages. If these interpretations are correct, it suggests that conformational interactions between the leader and the structural region of the nisin precursor peptide participate in guiding the formation of the correct thioether cross-linkages; however, these conformational effects are not required to carry out the dehydrations of serines and threonines (Hawkins, 1990). It must be emphasized that these interpretations are still quite tentative and must be subjected to further tests. The hypothesis being tested is that normal active nisin is produced when the complete nisin precursor peptide is expressed in a strain of *Bacillus subtilis* that possesses a subtilin processing system, but when a chimeric peptide that consists of a subtilin leader region and a nisin structural region is expressed, processing occurs but incorrect thioether cross-linkages are formed. If this hypothesis is supported by further experiments, it will establish that there is a great deal of homology between the nisin and subtilin processing systems, but the folding and conformational signals between the leader regions and the structural regions are not functionally identical. It would also suggest that thioether cross-linkage formation is a carefully directed process that is strongly influenced by the conformation of the precursor peptide.

XV. Production of Natural and Engineered Nisin Analogues in *Bacillus subtilis*

The results of these experiments will have strong bearing on the feasibility of expressing a variety of lantibiotics in an industrial-grade organism such as *Bacillus subtilis* 168. If nisin produced in *Bacillus subtilis* is toxic to the host, a solution may be to put a nisin-resistance gene, such as the one that has been characterized by Froseth and McKay (1991), into the cell before attempting to express the nisin precursor. If this proves successful, and one can indeed create nisin-producing strains of *Bacillus subtilis* 168, it will have a considerable effect on the economics of nisin production and should alter the role that nisin may play as a food preservative and therapeutic agent. As the cost of nisin drops, its use in a variety of new applications is likely to be explored. Success in these applications will open the door to nisin and subtilin structural analogues and lantibiotics in general. If these can be produced inexpensively on a large scale, they may have a large impact on human health and nutrition.

XVI. Structural and Functional Analysis of Lantibiotic Analogues: The Dehydro Residues Provide a Window through Which the Chemical State of Nisin and Subtilin Can Be Observed

To achieve the goal of rationally designed improved forms of lantibiotics, one must first have a firm understanding of the mechanism by which they exert their antimicrobial effects, and the role that various structural features such as the dehydro residues and thioether cross-linkages play in the chemical and physical properties of the molecule. For example, one must understand exactly how nisin interacts with the cell to inhibit it if one desires to make structural changes that improve its performance. Similarly, one must know what determines the chemical stability if one desires to make changes to improve stability. There are at present two distinct hypotheses of the mechanism of nisin action, which are not necessarily mutually exclusive. The hydropathic profile of nisin shows it to be a relatively hydrophobic basic peptide, and it has been proposed that it may behave as a cationic detergent (Ruhr and Sahl, 1985; Sahl, 1985). The ability of nisin and other lantibiotics to associate with membranes, cause them to become conductive, and collapse voltage, pH, and ion gradients has been well documented (Ruhr and Sahl, 1985; Kordel et al., 1988; Schuller et al., 1989). An attractive mechanism accordingly is that nisin creates membrane pores or in some other way destabilizes the integrity of the membrane so that intracellular components leak out. The loss of membrane potentials and gradients accordingly would also destroy the ability of the cell to generate energy through electron transport. A deficiency of this cationic detergent mechanism is that it provides no obvious role for the dehydro residues and the thioether rings, which are highly conserved features among the lantibiotics. Although it is possible that these features do not participate in the antimicrobial mechanism of lantibiotics, it is difficult to rationalize why they are so highly conserved if they do not. That they do play a role is supported by observations that the dehydro residues are required for antimicrobial activity (Liu and Hansen, 1990; Hansen et al., 1991).

An alternative mechanism is that the dehydro residues act as Michael acceptors to react with nucleophilic groups such as sulfhydryl groups in the target cell. It has been shown that nisin inactivates sulfhydryl groups in the membranes of germinated bacterial spores, so that they will no longer react with iodoacetate (Morris et al., 1984). It has also been shown that unmodified membrane sulfhydryl groups are necessary to permit spore outgrowth (Morris et al., 1978; Morris and Hansen, 1981; Buchman and Hansen, 1987). One way these two mechanisms can be reconciled is if the cationic–detergent properties of nisin direct it to the cell membrane; it may then react with critical sulfhydryl group in the membrane. The physical presence of nisin in the membrane, as well as covalent modification of membrane proteins, may disrupt the structural integrity of the membrane and its ability to participate in energy transductions. Membrane leakage and collapse of gradients would then lead to the death of the cell.

Yet another possibility is that the nisin molecule may be able to function by two distinct mechanisms. While the studies that implicate nisin action with membrane sulfhydryl groups modification were carried out on bacterial spores, the studies that implicated nisin in the collapse of membrane integrity were carried out on log-phase cells. The mechanism by which nisin interacts with cells in these disparate stages of growth could be different. It is to be expected that the availability of structural analogues of nisin will make studies of the mechanism of nisin action simpler and more straightforward. One can control the detergentlike properties and the potential for reaction with sulfhydryl groups by making appropriate structural modifications in hydrophobic and dehydro residues. One can also assess the role of the thioether cross-linkages by removing them or moving them around the molecule. By expressing nisin in heterologous hosts that grow on minimal defined media, one will be in a better position to make highly radioactive nisin by growing nisin-producing cells on appropriate radioactive precursors. The labeled nisin can then be used to monitor covalent and noncovalent incorporation of nisin into cells.

The dehydro residues in lantibiotics are both a frustration and a blessing. On one hand, they constitute an unknown quantity that confers unusual properties on the molecule, making its behavior unpredictable. On the other hand, they are one of the reasons that lantibiotics are so interesting and have much potential as a biological system. They also provide a unique and useful window through which one can peer at the interior workings of the molecule. NMR spectroscopy is particularly useful because the dehydro residues have unique vinyl protons that give resonances well separated from the complex region of the spectrum caused by the hydrocarbon and peptide groups (Liu and Hansen, 1990). Even better, all three dehydro residues (one Dhb and two Dha residues) are in sufficiently different enough chemical environments that they give well-separated and readily identifiable resonances. The NMR spectrum of nisin is shown in Figure 8.

One can therefore readily assess relationships between antimicrobial activity and any or all of the dehydro residues. It has been shown that treatment of nisin with mercaptans both inactivates its activity and destroys the dehydro residues, presumably by addition across the double bond (Liu and Hansen, 1990). Subtilin is appreciably less stable than nisin and undergoes spontaneous inactivation when in aqueous solution. The time course of inactivation follows the same kinetics as the loss of the Dha_5 resonance in the subtilin NMR spectrum (Hansen et al., 1991). Nisin also has a de-

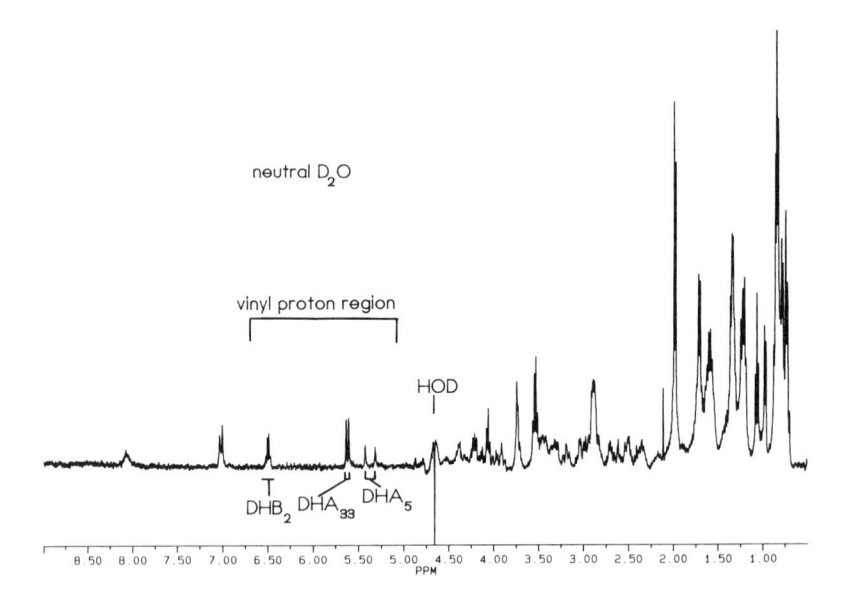

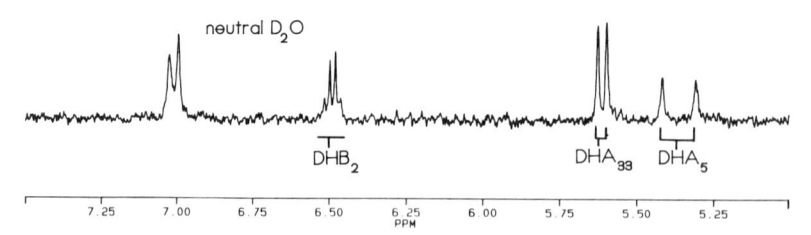

Figure 8 Proton NMR spectra (400 MHz). Complete proton NMR spectrum of nisin in neutral D$_2$O *(top)*. Expanded-scale partial spectrum in the region where the unique resonances of the vinyl protons of Dha and Dhb appear *(bottom)*. (Data from Liu and Hansen, 1990.)

hydroalanine at position 5, but it does not disappear under the conditions that inactivate subtilin; nor does nisin become inactive under these conditions. It thus appears that this residue is critical for subtilin activity, and it would accordingly seem that the difference between the stability of nisin and subtilin is attributable to differences in the chemical environment around the Dha$_5$ residue. This is a question that can be studied with appropriate structural mutants of nisin and subtilin. This work is in progress.

XVII. Conclusions and Future Prospects

Nisin and other lantibiotics are still in the infancy of their development. The dramatic increase in the number of laboratories studying them creates confidence that the fundamentals of their molecular biology, mechanism of biosynthesis, and mechanism of action will be elucidated within a few years.

When this information is available, it will be possible to assess the scope of biological problems and practical applications that can be addressed by exploiting the unusual and unique characteristics of lantibiotics. Much depends on the precise details of the biosynthetic pathway of lantibiotics, particularly the way that the processing signals in the lantibiotic precursor peptides are recognized by the posttranslational processing machinery, and how the subsequent steps of modification are orchestrated. It if turns out that the processing system is sufficiently flexible to easily make structural analogues of lantibiotics, and if dehydro and lanthionine residues can be introduced into heterologous proteins in a controlled manner, it would be difficult to overestimate the impact that lantibiotics may have on many areas of medicine and biotechnology. It would also be surprising if lantibiotics constitute the only bacterial system in which such extensive posttranslational modifications occur. As knowledge of how to screen for lantibiotic production has become established, the number of known lantibiotic-producing organisms has grown rapidly. There are probably many other lantibiotics that have yet to be discovered. If the study of lantibiotics realizes its potential, it will provide an incentive to search for other kinds of posttranslational modification systems. If they can be found, the lessons learned by studying lantibiotics can be used to exploit these alternate systems.

Acknowledgments

This work was supported by NIH Grant AI24454, the National Dairy Promotion and Research Board, and Applied Microbiology, Inc., New York, New York. I wish to thank Andre Hurst for reading this manuscript and providing critical comments.

References

Allgaier, H., Jung, G., Werner, R. C., Schmelder, U., and Zahner, H. (1986). *Eur. J. Biochem.* **160**, 9–22.
Ames, G. F. L. (1986). *Cell* **47**, 323–324.
Banerjee, S., and Hansen, J. N. (1988). *J. Biol. Chem.* **263**, 9508–9514.
Bassford, P., Beckwith, J., Ito, K., Kumamato, C., Mizushima, S., Oliver, D., Randall, L., Silhavy, T., Tai, P. C., and Wickner, B. (1991). *Cell* **65**, 367–368.
Broadbent, J. R., and Kondo, J. K. (1991). *Appl. Environ. Microbiol.* **57**, 517–524.
Buchman, G. W., and Hansen, J. N. (1987). *Appl. Environ. Microbiol.* **53**, 79–82.
Buchman, G. W., Banerjee, S., and Hansen, J. N. (1988). *J. Biol. Chem.* **263**, 16260–16266.
Chung, Y. J., Steen, M. T., and Hansen, J. N. (1992). *J. Bacteriol.* **174**, 1417–1422.
Chung, Y. J., and Hansen, J. N. (1992). *J. Bacteriol.* **174**, 6699–6702.
Dodd, H. M., Horn, N., and Gasson, M. J. (1990). *J. Gen. Microbiol.* **136**, 555–566.
Donkersloot, J. A., and Thompson, J. (1990). *J. Bacteriol.* **172**, 4122–4126.
Feeney, R. E., Garibaldi, J. A., and Humphreys, E. M. (1948). *Arch. Biochem. Biophys.* **17**, 435–445.
Froseth, B. R., and McKay, L. L. (1991). *Appl. Environ. Microbiol.* **57**, 804–811.
Fuchs, P. G., Zajdel, J., and Dobrzanski, W. T. (1975). *J. Gen. Microbiol.* **88**, 189–192.
Gasson, M. J. (1984). *FEMS Microbiol. Lett.* **21**, 7–10.

Gasson, M. J. (1990). Third International Conference on Streptococcal Genetics. Minneapolis, Minnesota.
Gerlach, J. H., Endicott, J. A., Juranka, P. F., Henderson, G., Sarangi, R., Deuchers, K. L., and Ling, V. (1986). *Nature (London)* **324**, 485–489.
Gross, E. (1975). In "Peptides: Chemistry, Structure, and Biology" (R. Walter and J. Meienhofer, eds.), pp. 31–42. Ann Arbor Sci., Ann Arbor, Michigan.
Gross, E. (1977). In "Protein Cross-Linking" (M. Friedman, ed.), pp. 131–153. Plenum, New York.
Gross, E., and Morell, J. L. (1971). *J. Am. Chem. Soc.* **93**, 4634–4635.
Hall, R. H. (1966). *Process. Biochem.* **1**, 461–464.
Hansen, J. N., Chung, Y. J., Liu, W., and Steen, M. J. (1991). In "Nisin and Novel Lantibiotics" (G. Jung and H. G. Sahl, eds.). ESCOM Science Publishers, Leiden, Germany.
Hawkins, G. (1990). Investigation of the site and mode of action of the small protein antibiotic subtilin and development and characterization of an expression system for the small protein antibiotic nisin in *Bacillus subtilis*. University of Maryland, College Park, Maryland.
Hedge, P. J., and Spratt, B. G. (1985). *Nature (London)* **318**, 478–480.
Holland, I. B., Kenny, B., and Blight, M. (1990). *Biochimie* **72**, 131–141.
Hopwood, D. A., Malpartida, F., Kieser, H. M., Ikeda, H., Duncan, J., Fujii, I., Rudd, B. A. M., and Floss, H. G. (1985). *Nature (London)* **314**, 642–644.
Hurst, A. (1966). *J. Gen. Microbiol.* **44**, 209–220.
Hurst, A. (1981). In "Advances in Applied Microbiology" (D. Perlman and A. I. Laskin, eds.), pp. 85–123. Academic Press, Inc., New York.
Hurst, A., and Collins-Thompson, D. (1979). *Adv. Microb. Ecol.* **3**, 79–133.
Hurst, A., and Dring, G. J. (1968). *J. Gen. Microbiol.* **50**, 383–390.
Hurst, A., and Kruse, H. (1972). *Antimicrobial Agents and Chemotherapy* **1**, 277–279.
Hurst, A., and Peterson, G. M. (1971). *Can. J. Microbiol.* **17**, 1379–1384.
Ingram, L. (1969). *Biochem. Biophys. Acta* **184**, 216–219.
Ingram, L. (1970). *Biochem. Biophys. Acta* **224**, 263–265.
Jansen, E. F., and Hirschmann, D. J. (1944). *Arch. Biochem.* **4**, 297–309.
Jarvis, B. (1967). *J. Gen. Microbiol.* **47**, 33–48.
Jarvis, B., and Farr, J. (1971). *Biochem. Biophys. Acta* **227**, 232–240.
Kaletta, C., and Entian, K. D. (1989). *J. Bacteriol.* **171**, 1597–1601.
Kaletta, C., Entian, K. D., Kellner, R., Jung, G., Reis, M., and Sahl, H. G. (1989). *Arch. Microbiol.* **152**, 16–19.
Kellner, R., Jung, G., Horner, T., Zahner, H., Schnell, N., Entian, K. D., and Gotz, F. (1988). *Eur. J. Biochem.* **177**, 53–59.
Klaenhammer, T. R., and Sanozky, R. B. (1985). *J. Gen. Microbiol.* **131**, 1531–1541.
Kordel, M., Benz, R., and Sahl, H. G. (1988). *J. Bacteriol.* **170**, 84–88.
Liu, W., and Hansen, J. N. (1990). *Appl. Environ. Microbiol.* **56**, 2551–2558.
Liu, W., and Hansen, J. N. (1991). *J. Bacteriol.* **173**, 7387–7390.
Mattick, A. T. R., and Hirsch, A. (1944). *Nature (London)* **154**, 551–552.
Morris, S. L., and Hansen, J. N. (1981). *J. Bacteriol.* **148**, 465–471.
Morris, S. L., Levin, R. A., Wright-Wilson, C., and Hansen, J. N. (1978). In "Spores VII" (G. Chambliss and J. C. Vary, eds.), pp. 85–89. American Society for Microbiology, Washington, D.C.
Morris, S. L., Walsh, R. C., and Hansen, J. N. (1984). *J. Biol. Chem.* **259**, 13590–13594.
Rogers, L. A. (1928). *J. Bacteriol.* **16**, 321–325.
Rogers, L. A., and Whittier, E. O. (1928). *J. Bacteriol.* **16**, 211–229.
Ruhr, E., and Sahl, H. G. (1985). *Antimicrob. Agents Chemother.* **27**, 841–845.
Sahl, H. G. (1985). *Microbiol. Sci.* **2**, 212–217.
Schnell, N., Entian, K. D., Schneider, U., Gotz, F., Zahner, H., Kellner, R., and Jung, G. (1988). *Nature (London)* **333**, 276–278.
Schuller, F., Benz, R., and Sahl, H. G. (1989). *European J. Biochem.* **182**, 181–186.
Steen, M., Chung, Y. J., and Hansen, J. N. (1991). *Appl. Environ. Microbiol.* **57**, 1181–1188.
Taylor, J. I., Hirsch, A., and Mattick, A. T. R. (1949). *The Veterinary Record* **61**, 197–198.
Weil, H. P., Beck-Sickinger, A. G., Metzger, J., Stevanovic, S., Jung, G., Josten, M., and Sahl, H. G. (1990). *Eur. J. Biochem.* **194**, 217–223.

CHAPTER 6

Nonnisin Bacteriocins in Lactococci: Biochemistry, Genetics, and Mode of Action

JAN KOK
HELGE HOLO
MARCO J. VAN BELKUM
ALFRED J. HAANDRIKMAN
INGOLF F. NES

I. Summary

In recent years we have seen a rapid increase in our knowledge of the structure, genetics, and mode of action of a number of bacteriocins that are produced by certain strains of lactococci. A number of the new bacteriocins have been purified to homogeneity, allowing their amino acid sequence analysis. The genetic information for both the production of and immunity against a number of lactococcins has been analyzed at the nucleotide level and has revealed that these bacteriocins are processed at their amino-terminal end. In addition to the bacteriocin structural and immunity genes, two genes have been identified that are essential for lactococcin production and, on the basis of protein homology studies, the products of these genes tentatively form a dedicated secretion system for lactococcin.

The effect of purified lactococcin A on whole lactococcal cells and vesicles indicates that the bacteriocin increases the permeability of the cytoplasmic membrane of sensitive lactococci in a voltage-independent way. The specificity of lactococcin A for lactococci seems to stem from the fact that the bacteriocin recognizes a *Lactococcus*-specific membrane receptor protein.

II. Introduction

Antagonistic activity in lactic acid bacteria (LAB) has usually been associated with the production of lactic acid and the concomitant drop in pH. A number of other low molecular weight compounds produced by LAB, such as hydrogen peroxide, organic acids, alcohols, aldehydes, and ketones may also possess antimicrobial activity. In addition to these low molecular weight metabolites, members of all species of LAB have been implicated in the production of bacteriocins.

Bacteriocins are potent antimicrobial substances that are produced by a great number of different bacterial species. By definition, bacteriocins are proteins, or substances of proteinaceous nature, with bactericidal activity against bacteria closely related to the producing organism. Although bacteriocin production by LAB has been known for years, relatively limited information on the structure, genetics, and mode of action of their bacteriocins was available until recently. One exception is the extensively studied LAB bacteriocin nisin, a 3353-Da pentacyclic lanthionin-containing polypeptide produced by certain strains of *Lactococcus*. Nisin is reviewed by Hansen in Chapter 5, this volume.

A number of nonnisin (nonlantibiotic) bacteriocins have been identified in lactococci. For many of these no follow-up studies have been published since the review on LAB bacteriocins by Klaenhammer (1988). The recently developed methodology of genetic manipulation of lactococci (for reviews, see de Vos, 1987; Kok, 1991) has been used to dissect the genetics of lactococcal bacteriocins and has provided us with a wealth of new information covered here. Moreover, rapid progress has been made in the biochemical field and a number of bacteriocins have been purified to homogeneity. In the case of lactococcin A described here, this has allowed a detailed study of its mode of action. In this chapter we limit ourselves to the description of the nonnisin bacteriocins produced by lactococci. We describe, in the order given, the three main groups of nonnisin bacteriocins, namely diplococcin, lactostrepcins, and the lactococcins and will discuss their biochemistry, genetics, and mode of action. Since the lactococcins have been studied in greatest detail, emphasis will inevitably be on this group of bacteriocins.

III. Nomenclature

With the description and isolation of an ever increasing number of "new" bacteriocins from LAB, the nomenclature of bacteriocins of LAB is becoming more confusing. The publication of a book solely dedicated to the bacteriocins of LAB seems an excellent opportunity to establish a consensus for a logical and useful nomenclature. We conform to the proposal of Tagg et al. (1976) and use the name lactococcin for those lactococcal bacteriocins that consist of solely nonmodified amino acids. The abbreviation for the protein would be LcnB (for lactococcin B) and *lcn*B for the corresponding structural gene. In this respect, it is advisable to try to prevent the use of an abbreviation that has already been used in literature. For example, *lcn*A, the structural gene for lactococcin A, has also been used for the completely different bacteriocin, leucocin A-UAL187 (Hastings et al., 1991). When the primary amino acid sequence for a lactococcal bacteriocin is not known, a temporary designation should be made until a definitive name can be given according to the suggested nomenclature.

IV. Diplococcin

A. Identification and Purification of Diplococcin

As early as 1933 an antimicrobial agent was described that was produced by *Lactococcus lactis* subsp. *cremoris* (Whitehead, 1933). This inhibitory substance was termed diplococcin, as the producing bacteria exhibited a diplococcal arrangement. Diplococcin was partly purified, shown to be of proteinaceous nature, and affected only lactococci (Oxford, 1944). More than three decades later, an extensive screening for antimicrobial activity among a total of 150 *L. lactis* subsp. *cremoris* strains yielded 11 strains that produced antimicrobial substances with the characteristics of diplococcin (Davey and Pearce, 1980). The inhibitory spectrum of the producers was restricted to lactococci. One of the strains under study, *L. lactis* subsp. *cremoris* 346, produced diplococcin throughout the exponential growth phase and the bacteriocin was purified from the supernatant of stationary phase cells of this strain (Davey and Richardson, 1981). The procedure employed included ammonium sulfate precipitation (60% saturation) and cation-exchange column chromatography on carboxymethyl cellulose. Approximately 1000-fold purification was obtained. Purified diplococcin was rather unstable and, therefore, a number of characteristics of the bacteriocin were examined with a partially purified preparation (the redissolved ammonium sulfate precipitate). The partially purified diplococcin was stable at 100°C at pH 5.0 for at least 1 hour, while the purified bacteriocin lost 75% of its activity under these conditions. Both diplococcin preparations were sensitive to trypsin, pronase, and α-chymotrypsin, indicative of the proteinaceous nature of the molecule. Gel filtration of partially purified diplococcin using Sephadex G-100, combined with amino acid composition analysis of purified diplococcin, suggested that diplococcin has a size of approximately 5300 Da. No modified amino acids except one ornithine residue per diplococcin molecule were detected by the amino acid composition analysis.

B. Effects of Diplococcin on Bacterial Cells

The effect of purified diplococcin from *L. lactis* subsp. *cremoris* 346 on a number of both sensitive and immune strains of *L. lactis* subsp. *cremoris* was studied (Davey, 1981). Diplococcin at a concentration of 4 arbitrary units (AU) per milliliter was efficiently adsorbed within 15 minutes after addition to exponentially growing as well as stationary phase cells of all bacteria tested, and to heat-killed cells. A rapid reduction of 6 orders of magnitude in the viable count of exponentially growing *L. lactis* subsp. *cremoris* 448 cells was observed within 3 hours at 30°C. Under the same conditions, stationary phase cells of the same strain showed only a 2 log reduction in viability. Lysis of the cultures was not observed by optical density measurements. DNA and RNA synthesis completely ceased within 2 minutes after the addition of 8 AU/ml of diplococcin to cells of strain 448 at 30°C, possibly causing the immediate but incomplete cessation of protein synthesis as a secondary effect.

The effect of diplococcin on the permeability of the bacterial cytoplasmic membrane, a possible primary target of the bacteriocin (see Section VI,D), has not been examined. Future studies of this kind may shed more light on how diplococcin acts on susceptible cells.

C. Plasmids and Diplococcin Production

Plasmids are commonly found in lactococci and most strains carry several different plasmid species (Davies and Gasson, 1981; McKay, 1983). In the early eighties several groups assessed the involvement of plasmid DNA in bacteriocin production in this group of organisms. Plasmid curing as well as conjugation strategies have been employed to obtain either an indication or definite proof that specific plasmids specified bactericidal activity. In a first attempt to link diplococcin production (Dip$^+$) to plasmid DNA in *L. lactis* subsp. *cremoris*, Davey and Pearce (1982) cured three Dip$^+$ strains of the ability to produce the bacteriocin by growth at a high sublethal temperature. The Dip$^-$ variants appeared more frequently (0.15–0.19%) than expected for a chromosomal mutation and no Dip$^+$ revertants were observed during subculturing of the Dip$^-$ isolates. Dip$^+$ was transferred by a conjugationlike process to *L. lactis* subsp. *cremoris* C13, and from a transconjugant of this strain to *L. lactis* subsp. *lactis* (Davey and Richardson, 1981). Examination of the plasmid profiles of the strains obtained from the curing and conjugation experiments was thwarted by the multiplasmid nature of the strains used, allowing no correlation of the Dip$^+$ phenotype with any particular plasmid. In a later report, Davey (1984) used plasmid-cured derivatives of wild-type strains of *L. lactis* subsp. *cremoris* (Dip$^-$) as recipients and strain 346 as the diplococcin-producing donor in filter mating experiments. Dip$^+$ transconjugants were isolated with a frequency of up to 10^{-1} transconjugants per donor. All the Dip⁺ transconjugants tested contained a 54-MDa (81-kb) plasmid and had also acquired immunity (Imm⁺) to diplococcin. Spontaneous Dip⁻, Imm⁻ colonies could be isolated from the transconjugants with a frequency of 0.1–0.2% with concurrent loss of the 81-kb plasmic (Davey, 1984). These experiments strongly suggest that diplococcin production and immunity are encoded by a conjugative plasmid of 81 kb in *L. lactis* subsp. *cremoris* 346. However, the final proof for plasmid linkage of diplococcin production has to be provided through isolation and sequencing of the structural gene for diplococcin.

V. Lactostrepcins

A. Identification and Definition of Lactostrepcin

Since the late seventies a group of lactococcal antagonists active at acidic pH have been under study at the Medical Academy in Warsaw, Poland (Kozak et al., 1978; Bardowski et al., 1979; Dobrzanski et al., 1982; Zajdel and Dobrzanski, 1983; Zajdel et al., 1985). The initial screening for bacteriocin activity was carried out with 67 nonnisin-producing lactococci (Kozak et al., 1978). After growth of the various strains and removal of the cells, the culture media were adjusted to pH 4.6 and examined for antimicrobial activity. Of the strains tested, 44 produced antagonistic activity against at least one of the three indicator strains used in the test. The producers could be divided into two groups, group I killing only one of the indicator strains and group II killing all three indicator strains used. The antagonists were bactericidal, as less than 10% of the indicator cells survived, whereas no decrease of optical density was observed. The antagonistic activities were destroyed by treatment with various proteinases, indicating that they were proteinaceous in nature. They were resistant to heating at 100°C for 10 minutes and fully active within the pH range of 4.2–5.0. By raising the pH in the bacteriocin-containing supernatant, the activity was gradually lost. At pH 8 the bacteriocins from most of the strains (43 out of 45) were totally inactivated. The activities were restored, however, when the pH in the supernatants was readjusted to 4.6. The acidic pH optimum for activity was the main property that differentiated this group of bacteriocins from the other lactococcal bacteriocins, and the name "lactostrepcin" (Las) was proposed (Kozak et al., 1977; 1978).

B. Purification and Mode of Action of Lactostrepcin 5

A detailed study was conducted on lactostrepcin 5 (Las 5) produced by *L. lactis* subsp. *cremoris* 202 (Zajdel and Dobrzanski, 1983). Las 5 was not included in the initial screening by Kozak et al. (1978) and was different from the previously described lactostrepcins in its ability to inactivate three *Bacillus cereus* strains. In addition, Las 5 inactivated approximately 60% of the lactococcal strains tested, 2 out of 6 *Leuconostoc* strains, and 3 out of 8 *Lactobacillus helveticus* strains. Las 5 was purified from a boiled, cell-free culture supernatant of strain 202 by binding to bentonite and extraction of the bentonite-adsorbed bacteriocin with a 10% aqueous pyridine solution. The extract was concentrated by ammonium sulfate precipitation (Zajdel et al., 1985). This protocol resulted in only 170-fold purification, suggesting that the bacteriocin was only partly pure. In the final concentration step by ammonium sulfate the bacteriocin was found in the floating lipidlike material. This has also been observed during purification of bacteriocins from other LAB, and suggested that Las 5 is hydrophobic and associated with other cell-produced or growth medium components (Mørtvedt et al., 1991; Klaenhammer, 1988).

Las 5 exerted a strong and rapid bactericidal effect on susceptible bacteria as had previously been shown for other lactostrepcins (Kozak et al., 1978; Zajdel and Dobrzánski, 1983). Treatment of the indicator cells with proteinases prior to exposure to Las 5 resulted in reduced killing, suggesting that the proteinase had destroyed a proteinaceous Las 5 receptor (Zajdel and Dobrzanski, 1983). When Las 5 was incubated with the indicator prior to the proteinase treatment it expressed full bactericidal effect. This finding suggested that either the putative receptor/Las 5 complex was not available for the proteinase or that the receptor had already carried out its possible function of mediating the bacteriocin to the susceptible cell at the time the protease was added. The idea of the presence of a target or specific receptor on the surface of a sensitive bacterium was also substantiated by the observation that protoplasts prepared from Las 5-sensitive cells were not susceptible to the bacteriocin. Las 5 inhibited uridine uptake and caused leakage of K^+ ions and ATP from the cells. Las 5 treatment also inhibited DNA, RNA, and protein synthesis, which may well be secondary effects caused by transport inhibition of precursor required for macromolecular synthesis, energy depletion, and/or leakage of low molecular weight components required for various metabolic activities in the bacterium. In contrast to nisin, Las 5 was also active against energy-depleted cells of the susceptible strains. Lactococcin A has also been shown to act on deenergized cells (see Section VI,D).

C. Plasmid Curing and Loss of Las 5 Production

The initial description of Las 5 (Zajdel and Dobrzanski, 1983) also involved a plasmid analysis of the producer strain *L. lactis* subsp. *cremoris* 202. This strain possessed four to five different plasmids. In order to link the Las 5 trait to a particular plasmid, plasmid curing agents were used to obtain Las 5 negative mutants. Several of such mutants were isolated but a comparison of their plasmid profiles with strain 202 was not conclusive because of the complexity of the profiles and the irreproducibility of the plasmid isolation procedure (Dobrzanski et al., 1982; Zajdel and Dobrzanski, 1983).

VI. Lactococcins

We now give a résumé of the data obtained by four different research groups on a set of lactococcal bacteriocins that have been named lactococcins. According to nucleotide sequencing data, these groups, involved in analysis of bacteriocin production in different strains, namely in the *L. lactis* subsp. *cremoris* strains 9B4 (Geis et al., 1983; Neve et al., 1984; van Belkum et al., 1989; 1991a; 1992) and LMG2130 (Holo et al., 1991), and in *L. lactis* subsp. *lactis* biovar. *diacetylactis* WM4 (Scherwitz et al., 1983; Scherwitz-Harmon and McKay, 1987; Stoddard et al., 1992) have been working on similar systems. The bacteriocin plasmids p9B4-6 and pNP2 of strains 9B4 and WM4, respectively, and the 55-kb Bac plasmid of strain LMG2130 all seem to carry the genes for the production of at least two of the three different lactococcins A, B, and M. As their results are complementary, integration of the genetic and biochemical data gathered by the four groups gives us a clear-cut picture of lactococcin production and will be given below.

A. Genetics of Lactococcins

1. Plasmid Involvement in Lactococcin Production

In an extensive screening of bacteriocin production in 280 lactococcal strains, 16 were shown to produce a bacteriocinlike substance (Geis et al, 1983). The activities were found in the culture supernatants and could be precipitated with ammonium sulfate. A variety of bacteria were tested for their susceptibility to the ammonium sulfate concentrated bactericidal activities. Only close relatives and some other Gram-positive bacteria were sensitive, while none of the Gram-negative bacteria tested were affected. Based on these studies and their sensitivity to trypsin and heat treatment, the bacteriocins were divided into eight different groups. No bacteriocin inhibited the growth of its producer or any of the other bacteriocin producers in its group. When a bacteriocin producer lost its ability to make bacteriocin it became sensitive to its "own" bacteriocin and to the other bacteriocins in the same group. As discussed by Geis et al. (1983), the bacteriocins in group II and III resemble diplococcin. The bacteriocins in group VI, all produced by *L. lactis* subsp. *lactis* strains, resembled nisin. In fact, Kaletta and Entian (1989) cloned the nisin structural gene from *L. lactis* subsp. *lactis* 6F3, one of the group VI strains analyzed by Geis et al. (1983).

Of the 16 bacteriocin producers identified by Geis et al. (1983), 13 were tested for their ability to transfer this trait by conjugation to the plasmid-free recipient *L. lactis* subsp. *lactis* biovar. *diacetylactis* Bu2-60 (Neve et al., 1984). Four of the donor strains produced Bac$^+$ progeny after 18 hours of mating on filters. In transconjugants of 3 of the 4 donors, single plasmids of 39.6 MDa (60 kb) and 75 MDa (113 kb) had been transferred. In some cases, an additional small plasmid was present. By incubation of the transconjugants for 24 hours and more at sublethal temperatures, Bac$^-$ variants were isolated and all of these had lost their large plasmid. Taken together, these results indicated that in *L. lactis* subsp. *cremoris* strains 9B4 and 4G6 bacteriocin production was specified by 60-kb plasmids. These two strains may be closely related, and the two Bac plasmids may be identical, as their plasmid profiles apparently are the same (Geis et al., 1983). The type I bacteriocin activity of *L. lactis* subsp. *cremoris* 9B4 has been analyzed extensively over the last 5 years. The notion that bacteriocins of one type may actually be a heterogeneous group (Geis et al., 1983) was clearly confirmed by the genetic analysis of the 60-kb Bac plasmid p9B4-6 of this strain: it carried the genes for three different bacteriocins that have been named lactococcin A, B, and M (see below).

Upon the transfer of lactose-fermenting ability (Lac$^+$) from the donor *L. lactis* subsp. *lactis* biovar. *diacetylactis* strain WM4 to a plasmid-free Lac$^-$ *L. lactis* recipient, Scherwitz et al. (1983) noted that two types of transconjugants appeared on the lactose indicator agar. Some of the Lac$^+$ transconjugant colonies were surrounded by a clear zone of growth inhibition of the confluent lawn of Lac$^-$ background. The inhibitory substance was heat resistant, sensitive to proteolysis, and distinguishable from nisin. Using a highly sensitive indicator strain and a plasmid isolation protocol that enabled the detection of large plasmids, the authors identified an 88-MDa plasmid [pNP2, later estimated at 131.1 kb (Scherwitz-Harmon and McKay, 1987)] that specified bacteriocin production. This bacteriocin plasmid was self-transmissible (Tra$^+$) and facilitated the transfer of the resident Lac plasmid (Scherwitz et al., 1983; Scherwitz-Harmon and McKay, 1987), explaining the simultaneous transfer of the Lac and Bac traits in some of the transconjugants.

2. Cloning of Lactococcin Determinants

A *Bcl*I digest of plasmid PNP2 (131.1 kb; Bac$^+$, Tra$^+$) of *L. lactis* subsp. *lactis* biovar. *diacetylactis* WM4 was ligated into PGB301 and used to transform the plasmid-free strain *L. lactis* subsp. *lactis* LM0230 (Scherwitz-Harmon and McKay, 1987). A Bac$^+$ transformant with three PNP2-derived inserts in PGB301 was isolated. Two of the fragments were essential for bacteriocin production and it was concluded [and later confirmed by DNA sequence analysis (Stoddard et al., 1992)] that a *Bcl*I site was present at a critical position in the genetic determinant for the WM4 bacteriocin. This work represents the first attempt to trace the genetic determinant of a lactococcal bacteriocin by recombinant DNA technology. The size of the

DNA involved in bacteriocin production was considerably reduced but still encompassed 18.4 kb of the original 131.1 kb of DNA of pNP2.

Because the original bacteriocin producer *L. lactis* subsp. *cremoris* 9B4 carried multiple plasmids, van Belkum et al. (1989) decided to use *L. lactis* subsp. *lactis* biovar. *diacetylactis* Bu2-61, a transconjugant carrying only the bacteriocin plasmid p9B4-6 of strain 9B4 (Neve et al., 1984), in an attempt to locate the bacteriocin determinant. Strain Bu2-61 served as the source of p9B4-6 DNA and approximately 70% of the 60-kb plasmid was cloned in various subfragments into *Escherichia coli*. The authors used cloning vectors derived from the broad host-range lactococcal replicon pWV01 (Kok et al., 1984) and since no antagonistic activity was observed in *E. coli*, the recombinant plasmids were subsequently transferred to the plasmid-free *L. lactis* subsp. *lactis* strain IL1403. Two fragments of p9B4-6 were identified that specified bacteriocin activity, one with low and one with high antagonistic activity (see Figure 1). Both fragments were relatively large and by deletion analysis could be reduced to 1.8 kb, specifying the low, and 1.3 kb specifying the high antagonistic activity. In a later study, van Belkum et al. (1992) identified a third bacteriocin determinant on p9B4-6 (see below). The three bacteriocins had different specificities, as a clone carrying one of the bacteriocin determinants inhibited the growth of cells carrying either of the other two. Not only had the genetic information for the production of three bacteriocins been cloned but, concomitantly, their respective immunity genes had also been cloned. The mutual inhibition of the three bacteriocin clones showed that the immunity genes were specific for their respective bacteriocin.

Lactococcin A from *L. lactis* subsp. *cremoris* LMG2130 was purified to homogeneity and its complete amino acid sequence was determined (Holo et al., 1991; see below). A 26-mer oligodeoxynucleotide probe derived from the amino acid sequence hybridized with a 55-kb plasmid present in strain LMG2130. A derivative of this strain that was cured for lactococcin A production and immunity and lacked the 55-kb plasmid did not react with the probe. Subsequently, a 4-kb hybridizing *Hind*III fragment was cloned in pUC18, partially sequenced, and was shown to carry the lactococcin A structural gene. In addition, in this case no bacteriocin production was observed in *E. coli*. Therefore, the various pUC18 derivatives carrying the structural gene for lactococcin A were fused to pIL253 at their unique *Eco*RI sites and transferred to *L. lactis*. Plasmid pIL253 (Simon and Chopin, 1988) is a derivative of the broad host-range Gram-positive MLS plasmid pAMβ1, and functions as the replicon in *Lactococcus*. With the two constructs obtained in this way, pON2 and pON7, lactococcin A production, albeit at considerably reduced levels, was detected in *L. lactis* subsp. *lactis* IL1403. Neither of the two plasmids caused detectable bacteriocin production in *L. lactis* subsp. *cremoris* BC101 or in any other strain of *L. lactis*. Both plasmids did, however, confer immunity to the host cells, whether in IL1403 or BC101. The significance of these results will be discussed below.

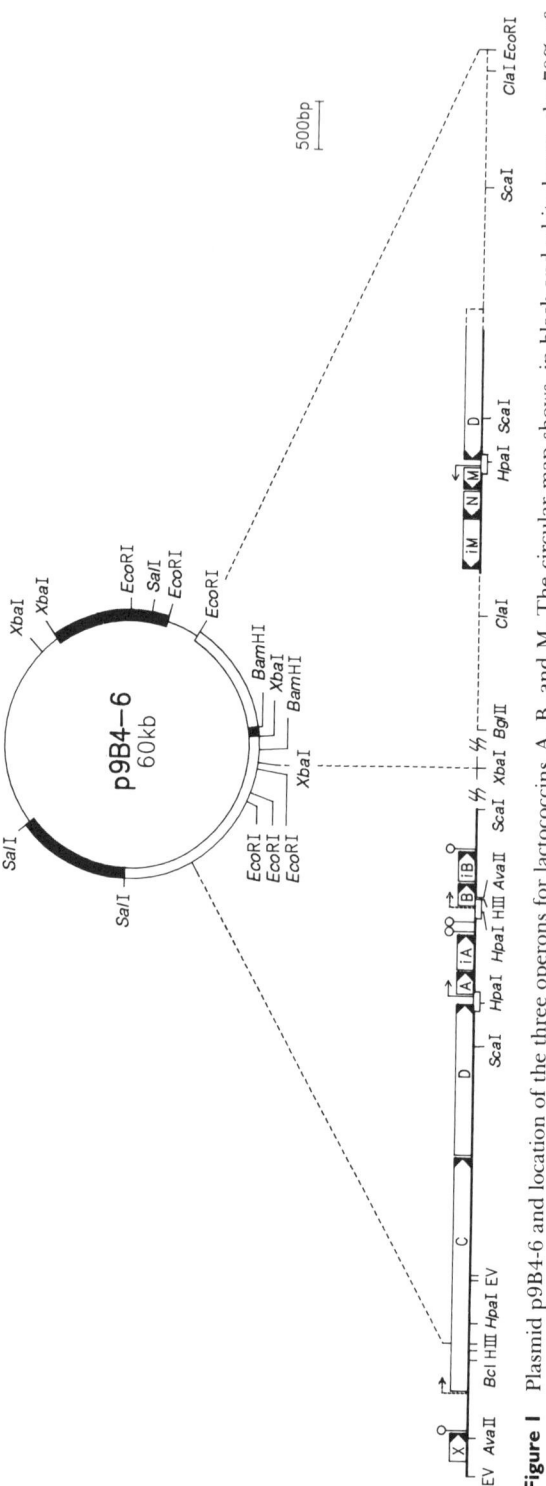

Figure 1 Plasmid p9B4-6 and location of the three operons for lactococcins A, B, and M. The circular map shows, in black and white boxes, the 70% of p9B4-6 that have been cloned in *E. coli*. The white boxes represent the 7.9-kb *Eco*RI/*Bam*HI and the 15-kb *Xba*I/*Sal*I fragments that conferred bacteriocin production and immunity to *L. lactis* (van Belkum et al. 1989). Below the circle a linear map is given of the three lactococcin operons on p9B4-6. It represents the most probable reconstruction based on the nucleotide sequence data of p9B4-6 (van Belkum et al., 1991a; 1992), of the 55-kb bacteriocin plasmid of strain LMG2130 (Holo et al. 1991), and of pNP2 (Stoddard et al. 1992), and the following argumentation: limited restriction enzyme analysis data of three plasmids show similarities in the region in question; the nucleotide sequence of the region encompassing the 3' terminal 75% of *lcn*D and the two terminatorlike structures behind *lci*A has been determined independently by the three groups and found to be identical; the nucleotide sequence specifying the two nucleotide binding areas of *lcn*C, determined on p9B4-6, was identical to the equivalent DNA sequence in pNP2. The base line represents the DNA sequence and is shown with restriction enzyme cutting sites, allowing easy access to the original papers. Where nucleotide sequence data are not available, the base line is broken. The direction of transcription of the genes is indicated by the arrowheads. The rectangles below the baseline indicate the promoter-containing regions conserved upstream of all three lactococcin structural genes. The right/left turn arrows indicate promoters. If the transcription site has not been determined, the arrow is stippled. Lollipops are presumed terminator structures. x: ORF$_x$; C, D, A, iA, iB, M, N, iM: *lcn*C, *lcn*D, *lcn*A, *lci*A, *lcn*B, *lci*B, *lcn*M, *lcn*N, *lci*M, respectively.

3. Nucleotide Sequence and Mutation Analysis of Three Different Lactococcin Determinants

Van Belkum et al. (1991a) published the nucleotide sequence of two different bacteriocin determinants from plasmid p9B4-6. The low-activity determinant resided on a 1.8-kb DNA fragment that contained three small open reading frames (ORFs) of 69, 77, and 154 codons, whereas the fragment specifying high bacteriocin activity carried two ORFs of 75 and 98 codons (see Figure 1). By detailed mutation analyses on both DNA fragments, the ORFs of 69 and 77 codons of the low-activity determinant and the ORF of 75 codons of the high-activity determinant were implicated in the production of the two different bacteriocins. In the same way, the ORFs of 154 and 98 codons were shown to be involved in immunity against the respective bacteriocins (see below). As the bacteriocins had not been purified, the authors were, strictly speaking, not able to incontrovertibly show that the structural genes for bacteriocin production had been sequenced, and no genetic nomenclature was proposed for the various ORFs.

Holo et al. (1991) sequenced a 1.2-kb *Hind*III-*Rsa*I fragment that reacted with a DNA probe designed from the N-terminus of purified lactococcin A. The published sequence appeared to be identical to that of the high-activity determinant of p9B4-6. Comparison of the amino acid and nucleotide sequences identified the first ORF on the DNA fragment as the structural gene for lactococcin A and the gene was termed *lcn*A. This nomenclature was adopted by van Belkum et al. (1992) and these authors proposed to use the term *lcn* for the various genes involved in lactococcin production and, in analogy to the practice in the colicin field, to reserve *lci* for the immunity genes. The high-activity determinant encoding lactococcin A, thus consists of the structural gene, *lcn*A, and *lci*A, the gene conferring immunity to lactococcin A. The low-activity determinant contains two genes, *lcn*M and *lcn*N, that are involved in the production of a bacteriocin termed lactococcin M. The immunity against lactococcin M is conferred by *lci*M. Figure 1 gives an overview of the genetic structure of the two lactococcin determinants.

Nucleotide sequences analysis of a 5.2-kb DNA fragment derived from the bacteriocin plasmid pNP2 of *L. lactis* subsp. *lactis* biovar. *diacetylactis* WM4 revealed that it specified lactococcin A (Stoddard et al., 1992). The sequences of the three lactococcin A-specifying regions, independently obtained from different lactococcal strains, are identical for as many sequences as are available. Van Belkum et al. (1991a) noted that the 5' ends of the structural genes of lactococcin A and M and the region immediately upstream thereof were identical. This upstream region contained the putative promoter for both lactococcin determinants. Removal of the promoter region and the 5' end of the structural gene *lcn*A resulted in loss of bacteriocin production and, simultaneously, in loss of bacteriocin immunity. Apparently, the lactococcin A production and immunity genes are transcribed from one promoter immediately upstream of *lcn*A. Replacement of the

DNA fragment carrying this promoter and the 5' part of *lcn*A by a piece of DNA carrying a strong lactococcal promoter restored lactococcin A immunity only. Similar results were obtained when analyzing the putative transcription initiation region of the *lcn*M, *lcn*N, and *lci*M genes. These data showed that the lactococcin A structural and immunity genes, as well as those for lactococcin M, are located in operon structures. The start site of transcription of each operon was determined by primer extension. Although the sequences in the promoter regions were identical, the messenger of the *lcn*A operon was initiated at a position two bases downstream of the transcription start site used for the *lcn*M operon (van Belkum et al., 1991a).

Immediately following the putative transcription terminator of the *lcn*A operon a sequence of 187 nucleotides is present that is almost identical to the sequences encompassing the promoters and 5' sequences of *lcn*A and *lcn*M. This observation suggested that another bacteriocin determinant was located downstream of the *lcn*A operon (van Belkum et al., 1991a). Nucleotide sequence analysis confirmed this idea and identified the operon for lactococcin B immunity (van Belkum et al., 1992). In addition, in this case the operon consisted of a structural gene, *lcn*B, and the lactococci B immunity gene, *lci*B (See Figure 1). As the upstream regions of the three lactococcin operons are nearly the same and the -35 and -10 sequences are identical, the *lcn*B operon is, most probably, transcribed from the same promoter as those identified upstream of the other two. Unpublished sequence data (H. Holo, Ø. Nilssen, and I. F. Nes) indicate that the *lcn*A operon of the *L. lactis* subsp. *cremoris* strain LMG2130 is followed by *lcn*B. As growth of strain LMG2130 is not inhibited by a strain producing lactococcin M, it is likely that the *lci*M gene is also present in this strain (H. Holo and M. van Belkum, unpublished data).

By comparing the restriction enzyme map of the region downstream of the *lcn*A operon of strain WM4 with the known nucleotide sequence of the *lcn*B operon, it seems that this strain also contains the lactococcin B structural and immunity genes (Stoddard et al., 1992).

Holo et al. (1991) pointed out that a palindromic structure consisting of a 19-nucleotide inverted repeat could represent a binding site for a regulatory protein(s) affecting *lcn*A transcription. The palindromic structure overlaps the promotor sequence, a configuration also seen upstream of some colicin genes of *E. coli* (Akutsu et al., 1989; Ebina et al., 1983). In these cases, the inverted repeats have been shown to be SOS repressor boxes and binding sites for the *lex*A repressor (Ebina et al., 1982; 1983; Van den Elzen et al., 1982). At the moment, there are no data available indicative of a similar regulatory mechanism of *lcn*A expression.

4. Lactococcin Immunity and Resistance

The three lactococcin operons each contain one or, in the case of lactococcin M, two genes involved in the production of lactococcin, and one gene (*lci*) specifying immunity against the particular lactococcin. When cells carrying the immunity gene for one of the lactococcins are used as indicators in an overlay test, no halo is formed around a colony of cells producing that lactococcin. Holo et al (1991) observed that when a 1.8 kb DNA fragment containing both lactococcin A and its immunity gene was cloned (pON7) into other lactococci they became immune to the bacteriocin while the bacteriocin itself was not expressed. Nissen-Meyer et al. (1992) purified and characterized the ImmA protein which was shown to be a protein containing 98 amino acid residues as suggested by the DNA sequence analysis of its gene. The immunity protein appeared to be a major cell protein component and one cell may contain (to an order of magnitude) 10^5 molecules. Exposing lactococcin A sensitive cells to an excess of the immunity protein did not affect the lactococcin A-induced killing, indicating that the immunity protein does not protect cells by simply binding to lactococcin A nor to externally exposed domains on the cell surface. Immunoblot analysis suggested that the ImmA protein is associated with the cell membrane since the membrane fraction was enriched in ImmA protein as compared to the soluble cytoplasmic fraction (Nissen-Meyer et al., 1992). Van Belkum et al. (1989) noted that *L. lactis* spontaneously and rather frequently acquired a slight degree of resistance to the lactococcins produced by an *L. lactis* strain carrying the bacteriocin plasmid p9B4-6. It was, therefore, important to confirm that the immunity observed in the strains carrying *lci* was really attributable to the immunity gene. This was done with cells containing the *lcn*M operon. These cells produce and are immune to lactococcin M. To rule out the possibility that the cells had acquired a chromosomal mutation that would render them resistant to their own (low) lactococcin M activity, the plasmid carrying *lcn*M and *lci*M was cured from the strain. Loss of the plasmid was accompanied by the loss of both lactococcin M production and immunity, implying that *lci*M indeed confers immunity to lactococcin M. Similarly, Holo et al. (1991) found that LMG2130 became sensitive to *lcn*A when the 55-kb Bac plasmid was lost in curing experiments.

Van Belkum et al. (1991a, 1992) made deletion or frame shift mutations in either of the three immunity genes they had identified and at the same time left the corresponding bacteriocin structural gene intact. The constructs which produced lactococcin in *E. coli* were subsequently transferred to *L. lactis*. Unexpectedly, *L. lactis* transformants were obtained with the various constructs, although the colonies initially were small and the cells grew poorly. That the cells produced lactococcin was visible by halo formation on a lawn of indicator cells. They gradually started to grow more rapidly with a concomitant increase in the level of lactococcin production (M. van Belkum, unpublished data). If lactococcin [30 to 50% (vol/vol) of a supernatant of a lactococcin-producing culture] was added to the plates on which the initial transformants were to be selected, no transformants developed. Apparently, the initial concentration of lactococcin produced by the transformants was insufficient to completely inhibit the growth of the transformed cells. Moreover, these results show that cells can overcome the lethal action of lactococcin at a stage when little of the bacteriocin is produced, probably by allowing for the selection of cells carrying a chromosomal mutation rendering them resistant to the bacteriocin. Several possibilities leading to lactococcin resistance can be envisaged at this point, one of which, in the case of resistance against lactococcin A, is the loss by mutation of the putative lactococcin A receptor (see below). In this chapter, we restrict the use of the term immunity to lactococcin-insensitivity caused by the product of the *lci* genes. All other forms of insensitivity to lactococcin are termed lactococcin resistance.

B. The Lactococcin A Secretion Machinery

1. In Addition to the Structural Gene, Two Other Genes Are Essential for Lactococcin Production

As noted by both the Norwegian and the Dutch group working on lactococcin, the 3' end of a putative ORF is present in the regions of identity located upstream of the *lcn*M operon (van Belkum et al., 1992) and that of lactococcin A (van Belkum et al., 1992; Holo et al., 1991). Stoddard et al. (1992) sequenced 4.5 kb upstream of the *lcn*A operon of *L. lactis* subsp. *lactis* biovar. *diacetylactis* WM4 and showed that, indeed, a complete ORF (*lcn*D) with an AUG start codon and a likely ribosome binding site is present at this position. The ORF contains 474 codons and could encode a protein of 52.5 kDa. Another ORF (*lcn*C) with the capacity to encode a protein of 78.8 kDa is located immediately upstream of *lcn*D (see Figure 1). Transposon (Tn5) mutagenesis was carried out to elucidate the role of *lcn*C and *lcn*D in lactococcin production. The mutagenesis was performed in *E. coli*, and plasmids containing single Tn5 insertions into either of the ORFs were introduced in *L. lactis*. The transformants were screened for halo formation on an indicator strain. Insertions in *lcn*C and *lcn*D resulted in the loss of halo-forming ability and both genes are, apparently, involved in some step of lactococcin production. Protein homology search revealed that the presumed products of *lcn*C and *lcn*D share significant similarity with proteins that form secretion systems for extracellular proteins produced by various Gram-negative bacteria whose secretion does not depend upon the general signal peptide-dependent export pathway (Pugsley, 1988; Randall et al., 1987). Hemolysin and colicin V of *E. coli*, cyclolysin of *Bordatella pertussis*, the *Pasteurella haemolytica* leukotoxin, and metalloproteases B and C of *Erwinia chrysanthemi* are examples of secreted proteins that lack the typical N-terminal signal peptide and are dependent on the products of at least two other genes for their externalization (Wagner et al., 1983; Glaser et al. 1988; Strathdee and Lo, 1989; Létoffé et al. 1990; Delepelaire and Wandersman, 1990; Gilson et al., 1987; 1990). Most or all of these secretion genes are genetically closely linked to the corresponding structural genes of the secreted proteins and their products form a dedicated export apparatus. Three proteins are necessary for export of hemolysin (the products of *hly*B, *hly*D, and *tol*C; Wagner et al., 1983; Delepelaire and Wandersman, 1990). The proteins essential for the secretion of protease B of *E. chrysanthemi* are PrtD, PrtE, and PrtF (Létoffé et al., 1990). One of the proteins invariably found in the signal peptide-independent export systems is an ATP-dependent membrane translocator of the hemolysin B subfamily (See Table 1 in Blight and Holland, 1990). This subfamily, exemplified by HlyB and PrtD, is a large class of membrane proteins involved in the secretion of a wide variety of molecules (allocrites) and is found in bacteria (Higgins et al., 1986), yeast (McGrath and Varshavsky, 1989), protozoans (Foote et al., 1989), and mammals (Riordan et al., 1989; Endicott and Ling, 1989). PrtD and HlyB are attached to the inner membrane with their hydrophobic N-termini. This domain, with weak primary sequence similarity but striking structural similarities, is present in all the HlyB homologues (Blight and Holland, 1990). The C-terminal parts of HlyB and PrtD are located in the cytoplasm and constitute highly conserved ATP binding domains of approximately 200 amino acids (Higgins et al., 1986). The presumed product of the first ORF in the *lcn*A regulon, *lcn*C, shares similarity with the ATP-dependent translocators. The similarity is most pronounced in the C-terminal stretch of 200 amino acids of *lcn*C, which contains the conserved ATP binding domain. LcnC and HlyB show 19.7% identity and 35.3% similarity at the amino acid level. An analysis of LcnC by the method of Rao and Argos (1986) predicts three hydrophobic domains in the N-terminus of the protein. The second protein of the protease B secretion apparatus, PrtE, is also located in the inner membrane (Delepelaire and Wandersman, 1991). The extreme N-terminus of PrtE is located in the cytoplasm. It has one N-terminal transmembrane domain and a large C-terminal domain exposed to the periplasm. HlyD has 23% identity with PrtE and contains an N-terminal hydrophobic region that might be a transmembrane domain. Hydrophobicity analysis of the predicted product of *lcn*D indicates that it is largely hydrophilic with exception of the N-terminal 43 amino acids. This region does not contain a typical signal peptide found in most transported proteins. In this respect it resembles HlyD and PrtE. The third protein in the hemolysin secretion system is TolC, a minor outer membrane protein of *E. coli* (Wandersman and Delepelaire, 1990). Its homologue in *E. chrysanthemi* is PrtF. TolC and PrtF are hydrophilic proteins and contain typical N-terminal signal peptides. PrtF was shown to be located in the outer membrane (Delepelaire and Wandersman, 1991).

The protein homology comparisons suggest that the step blocked in both insertion mutants is that of secretion of the lactococcin. As a consequence of the above, the following working model of lactococcin A production and secretion emerges (Figure 2): lactococcin A is first produced as a 75-amino-acid precursor (see Section VI,C,2). The 21 N-terminal amino acid residues are removed at some stage during the export of the peptide to the external medium. This secretion involves at least two proteins, an ATP-binding protein, LcnC (possibly providing the energy for the secretion/maturation), and LcnD, which might be attached to the membrane by

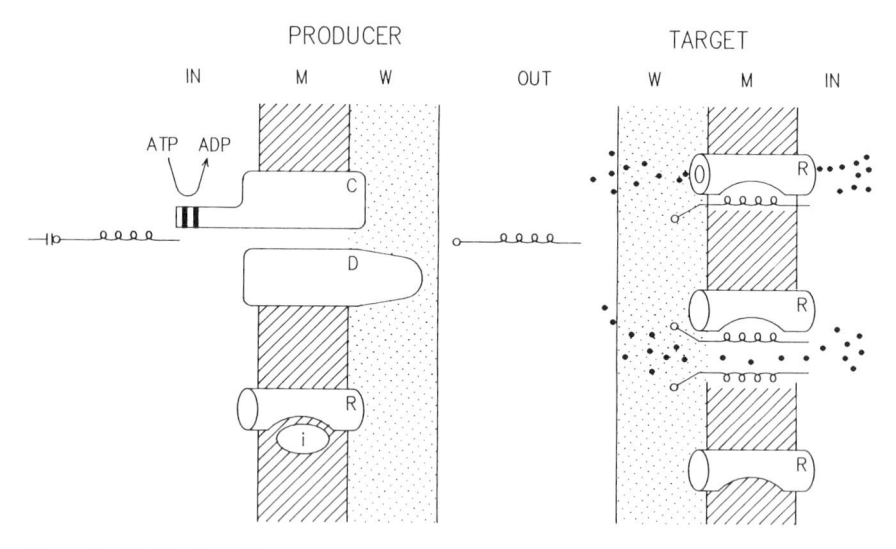

Figure 2 Production of lactococcin A and effect on sensitive target cells. The producer cell is shown on the left. LcnC (C) is considered to be associated with the cytoplasmic membrane (M) with its N-terminus, while the C-terminus, containing the ATP binding area (black boxes), is facing the cytoplasm (IN). LcnD (D) is thought to be attached to the cytoplasmic membrane with its hydrophobic N-terminus. The looped line is pre-lactococcin A, with its N-terminal extension containing the glycine doublet (the two small vertical lines). Secreted lactococcin A (OUT) is devoid of the N-terminal extension. On the right, two mechanisms of lactococcin A permeabilization of the cytoplasmic membrane of the sensitive target cell are envisaged: either lactococcin A interacts with a membrane protein(s), allowing the free diffusion of ions and amino acids (*top*), or lactococcin A itself forms pores (*middle*). In either case, the action is mediated by an *L. lactis*-specific receptor protein. In the producer cell this receptor protein is "protected" by the lactococcin A-specific immunity protein (i). W: Cell wall, R: Receptor.

its N-terminal hydrophobic domain. In the Gram-negative examples mentioned above, the allocrites must cross two membranes. Two inner membrane proteins, HlyB and HlyD, and the minor outer membrane protein TolC are needed for the secretion of hemolysin by *E. coli*, while PrtD/PrtE and PrtF are the inner and outer membrane components of the protease B secretion apparatus of *E. chrysanthemi*, respectively. The possibility that lactococcin A translocation across only a single membrane structure requires, apart from the ATP-translocator, only one additional (cytoplasmic membrane) protein (LcnD) is at the moment under investigation. Although speculative, the model shown in Figure 2 provides an excellent basis for further analysis of what might be an example of a "signal-sequence-independent" secretory pathway in a Gram-positive microorganism.

2. *The* L. lactis *IL1403 Chromosomal Lactococcin Secretion Function*

From the above it is clear that for secretion of lactococcin the plasmid-encoded secretion functions are essential. The contiguity of the lactococcin

A and B operons with *lcn*C and *lcn*D suggests that both bacteriocins use *lcn*C and *lcn*D for externalization. Preliminary sequence data (M. van Belkum et al., unpublished data) indicate that immediately upstream of *lcn*M an exact copy of *lcn*D is present (see Figure 1). The possibility that an *lcn*C homologue is specified by the DNA upstream of *lcn*D is at present under investigation (see below). In retrospect, it is surprising that lactococcin production and secretion was found with plasmids that carried only the lactococcin structural and immunity genes but did not encode the postulated secretion functions. Holo et al. (1991) were able to detect only (reduced) production of lactococcin A in *L. lactis* strain IL1403. This strain was also used throughout the work of van Belkum et al. (1989; 1991a; 1992). In all other strains tested no lactococcin production was detectable (Holo et al., 1991; M. van Belkum et al., unpublished data). This observation can now be explained by assuming that the plasmid-free strain IL1403 carries genes for an analogous secretion system on the chromosome. When using a DNA fragment encompassing most of *lcn*C, *lcn*D, *lcn*A and *lci*A as a probe in Southern hybridizations, a signal with chromosomal DNA of strain IL1403 is indeed obtained (J. Kok and A. J. Haandrikman, unpublished data). This putative chromosome-located secretion machinery is capable of externalization of all three lactococcins (van Belkum et al., 1989; 1992). The reduction in the halo size of strain IL1403 carrying a plasmid with only a lactococcin structural gene, as compared to the situation where the secretion genes are located on a plasmid together with the structural gene (van Belkum et al., 1989; 1991a), could be explained in two ways. The lower copy number of the chromosome-located secretion genes (probably only one) could result in a lower secretion efficiency, as has been observed for the secretion of protease B by *E. coli* (Létoffé et al., 1990). Alternatively, the lactococcins are not the natural substrates of the secretion system encoded by the chromosome of strain IL1403 and partial complementation would thus explain the reduced secretion efficiency. Heterologous complementation has been observed in the Gram-negative examples of this kind of secretion machinery: the secretion functions of α-hemolysin can specifically promote the secretion of leukotoxin (Strathdee and Lo, 1989) and cyclolysin (Masure et al., 1990). The α-hemolysin secretion functions could partially replace the secretion functions of protease B of *E. chrysanthemi* but not vice versa (Delepelaire and Wandersman, 1990).

C. Biochemistry of Lactococcin A

1. Production and Purification of Lactococcin A

Both *L. lactis* subsp. *cremoris* strains 9B4 and LMG2130 produce lactococcin A throughout the exponential growth phase. The activity is quite stable in the first 24 hours of stationary phase. The growth characteristics of the two producer strains are different. While strain 9B4 follows normal growth into stationary phase, strain LMG2130 grows slowly into stationary phase and stops growing at an $OD_{600\,nm}$ of approximately 1.5 in lactose-containing M17 medium. The amounts of lactococcin A produced by the two strains are comparable. The production of lactococcin appears to be affected by the composition of the media. Geis et al. (1983) observed that growth in lactic broth increased the activity 27-fold when compared with M17 broth. Lactococcin activity of LMG2130 was increased only twofold when lactic broth was used (H. Holo and I. F. Nes, unpublished data). The enhanced activity observed by Geis et al. (1983) can possibly be explained by an effect on the bacteriocin titer rather than an effect on the production itself (H. Holo and I. F. Nes, unpublished data). The initial work of Geis et al. (1983) showed that *L. lactis* subsp. *cremoris* 9B4 prevented growth not only of other lactococci but also of some clostridia. Van Belkum et al. (1991a; 1992) showed that plasmid p9B4-6 of this strain encodes three different bacteriocins and this also seems to be the case with the 55-kb bacteriocin plasmic in strain LMG2130 (H. Holo and I. F. Nes, unpublished data). Both the crude extract of LMG2130 and purified lactococcin A inhibit growth of lactococci. Of over 120 different lactococci tested, only one strain was insensitive to lactococcin, while all other Gram-positive bacteria tested were insensitive to lactococcin A. Although all the lactococci tested were sensitive to lactococcin A, the minimal inhibitory concentrations of growth (50% growth inhibition) varied 2500-fold from the most sensitive to the least sensitive strain. The most susceptible lactococcal strains were inhibited by 7 pmol lactococcin A in lactic broth, which corresponds to about 400 molecules of lactococcin A per colony forming unit (CFU) (Holo et al., 1991).

A three-step procedure was developed for the purification of lactococcin A to homogeneity (Holo et al., 1991). The first step involved concentration of the activity from the supernatant of an early stationary phase culture. The easiest and quantitatively best results were obtained with ammonium sulfate precipitation (280 grams of ammonium sulfate per liter). The 23-fold concentrated bacteriocin activity was then bound to a cation-exchange resin (CM-Sepharose, Pharmacia, Sweden) and step-eluted from the column with $0.3M$ NaCl. This fraction was subjected to reverse phase chromatography using a phenyl-Superose (Pharmacia, Sweden) column. Pure lactococcin A eluted at about 40% ethanol. This procedure resulted in 2300-fold purification of lactococcin A, and approximately 0.5 mg bacteriocin was obtained from one liter of culture. A scaled-up protocol for lactococcin A purification yielded approximately 6–7 mg of lactococcin A from 20 liters of cells.

The bacteriocin activity from strain LMG2130 is stable in growth broth and can be kept in broth at 4°C for several days without any significant loss of bactericidal activity. When lactococcin A was purified, its stability decreased with increasing purity. It was crucial to store pure lactococcin A in a solvent less polar than water, since otherwise a large portion of the biological activity was lost. This may partly be due to low solubility of the bacteriocin in aqueous solutions, resulting in precipitation. In order to keep lactococcin A in a soluble and active form, 50–70% ethanol was used combined with mild heating prior to use. Purified lactococcin A at a concentration of up to 1.0 mg/ml could be kept in the soluble form at −20°C when a mixture of 50% propanol with 0.1% trifluoro acetic acid was used as the solvent (H. Holo et al., unpublished data).

During the purification of lactococcin A no bacteriocin activities other than lactococcin A were detected although the strain used, LMG2130, should be able to produce two other lactococcins, namely lactococcin B and M (see above). This could possibly be explained by the large production of lactococcin A, overshadowing activity of the other two. Lactococcin B has been purified from a recombinant strain containing only the *lcn*B gene [*L. lactis* subsp. *lactis* IL1403 (pMB580)] by using the procedure developed for lactococcin A. The recombinant strain produces only 50 BU (bacteriocin units)/ml broth whereas *L. lactis* subsp. *cremoris* LMG2130 produces approximately 20,000 BU/ml under the same conditions. Approximately 1000 BU of lactococcin B were purified from one liter of culture. No protein was detectable in the purified lactococcin B sample and, therefore, no biochemical analysis was permitted (H. Holo et al., unpublished data).

2. Physical Characterization and Properties of Lactococcin A

Purified lactococcin A was used to determine the amino acid sequence of the molecule (Holo et al., 1991). The active extracellular bacteriocin consists of 54 amino acids and is shown in Figure 3. By comparing the amino acid sequence with the DNA sequencing data from the structural gene, it was clear that lactococcin A was synthesized as a precursor of 75 amino acids with an N-terminal presequence of 21 amino acids (Figure 3). The N-terminal extension and the processing site do not follow the rules of von Heijne (1986) for genuine signal peptides, which suggests that the extension in pre-lactococcin A is not a signal sequence in its classical sense (Holo et al., 1991). Processing takes place behind a glycine doublet (Figure

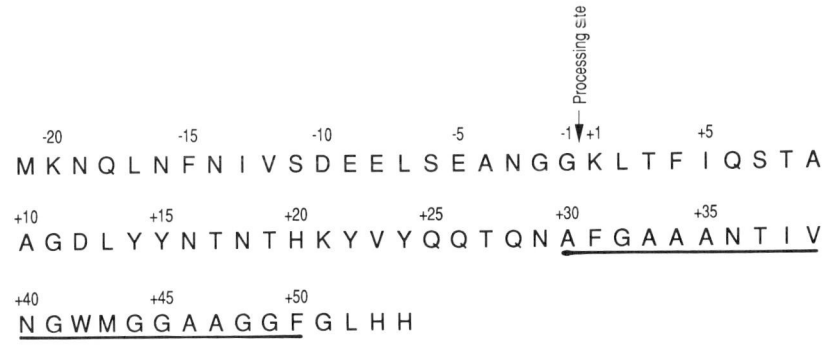

```
        -20         -15         -10         -5       -1 ↓ +1      +5
        M K N Q L N F N I V S D E E L S E A N G G K L T F I Q S T A

        +10         +15         +20         +25         +30         +35
        A G D L Y Y N T N T H K Y V V Y Q Q T Q N A F G A A A N T I V

        +40         +45         +50
        N G W M G G A A G G F G L H H
```

Figure 3 The amino acid sequence of the pre-lactococcin A molecule. The processing site is indicated by the arrow between the amino acid residues −1 and +1. The proposed hydrophobic membrane embedded segment of LcnA is underlined.

3). The structural genes of two other lactococcins, lactococcin B and M, have been sequenced (see Section VI,A,3). The predicted primary translation products of these genes are entirely different from each other and from pre-lactococcin A except for a high similarity in the N-termini. Although neither lactococcin B nor lactococcin M have been purified to homogeneity, and no amino acid sequencing data are available for the two bacteriocins, the similarity strongly suggests that lactococcin B and M are also produced as precursors with N-terminal extensions of 21 amino acids ending with two glycines (van Belkum et al., 1991a; 1992). Recently, the nucleotide sequences of the structural genes of bacteriocins of three other species of LAB have been published. Pediocin PA-1 of *Pediococcus acidilactici* PAC 1.0 (Henderson et al. 1992; Marugg et al., 1992), lactacin F of *Lactobacillus acidophilus* 11088 (Muriana and Klaenhammer, 1991), and leucocin A-UAL187 of *Leuconostoc gelidum* UAL-187 (Hastings et al., 1991) are all produced as precursors. The first two contain N-terminal extensions of 18 amino acids whereas the latter was reported to have a 24-amino-acid extension (see Figure 4). All three are processed behind a glycine doublet, leading to the speculation that a general processing mechanism for the maturation of small hydrophobic bacteriocins may exist in LAB (van Belkum et al., 1992). The possibility that the similar N-terminal extensions in the lactococcin precursors have a function in recognition by the secretion apparatus is under study.

Lactococcin A is particularly rich in alanine, glycine, and threonine (8, 8, and 6 residues, respectively; see Figure 3). It contains only one negatively charged residue (Asp.12) and two positively charged residues (Lys.1 and Lys. 21) and the estimated charge at pH 7 is 1.26. It is a cationic peptide with a pI of 9.21. The extinction coefficient is calculated to be $1.081 \times 10^4/M \cdot cm$ at 280 nm wavelength at pH 6.5. The absorption of a 0.1%

```
Lactococcin A        M K N Q L N F N I V S D E E L S E A N G G ↓ K L T
Lactococcin B        M K N Q L N F N I V S D E E L A E V N G G   S L Q Y
Lactococcin M        M K N Q L N F E I L S D E E L Q G I N G G   I R G T
Leucocin A-UAL 187   M M N M K P T E S Y E Q L D N S A L E Q V V G G   K Y Y G
Lactacin F           M K - - - Q F N Y L S H K D L A V V V G G   R N N W
Pediocin PA-1        M K - - - K I E K L T E K E M A N I I G G   K Y Y G
                     * *               .  . . . . . .           * *
```

Figure 4 Comparison of the N-terminal amino acid presequences of a number of small bacteriocins of lactic acid bacteria. Identical amino acids are indicated by an asterisk, functionally similar residues are highlighted by a dot below the sequences. The arrow shows the position where lactococcin A, leucocin A-UAL 187, lactacin F, and pediocin PA-1 are processed (Holo et al., 1991; Hastings et al. 1991; Muriana and Klaenhammer, 1991; Marugg et al. 1992). On the basis of the homology, lactococcin B and M are thought also to be processed behind the glycine doublet (van Belkum et al., 1992) The amino acid sequence of leucocin A-UAL 187 is shown in order to give maximum fit with the other peptides. On the basis of the comparison it may be argued that the third methionine in the sequence of leucocin A-UAL 187 is the actual start of the protein.

solution (1 mg/ml) is 1.87 at 280 nm. The hydrophobicity of lactococcin A is evident from both its behavior in reverse phase column chromatography (elution at 40% ethanol) and from its low solubility in water. The maximum solubility of pure lactococcin A in 50% ethanol (in 5mM phosphate buffer, pH 7.0) was approximately 30 μg/ml at −20°C, while at room temperature it was approximately ten times higher.

Although the Edman degradation gave no indication of amino acid modifications in lactococcin A, modified amino acids may still be present. To establish whether modifications are present in lactococcin A, its exact molecular weight was determined by mass spectroscopy. Two independent mass spectroscopic analyses revealed a monomeric unit of lactococcin A of 5783 ± 4 (H. Holo and I. F. Nes, unpublished data). The molecular weight calculated from the sequencing data is 5778, which strongly suggests that no posttranslational modification had taken place on the lactococcin A molecule. A statistical analysis of the hydropathic character of lactococcin A showed an overall average hydropathic value (GRAVY score) of −2.45, which did not classify lactococcin A as a fully membrane-embedded peptide (Kyte and Doolittle, 1982).

D. Mode of Action of Lactococcin A

Lactococcin A is a small 54-amino-acid hydrophobic peptide that specifically inhibits the growth of lactococcal strains (Holo et al., 1991). As such, a possible target for its action could be the cytoplasmic membrane of sensitive cells and it would, thereby, resemble several other peptide antibiotics that permeabilize membranes (Gao et al., 1991; Galvez et al., 1991; Kordel and Sahl, 1986; Kordel et al., 1988; Schaller et al., 1989). Lactococcin A has been shown to mediate influx of propidium iodide and 4′,6′-diamidino-2 phenylindole 2.HCl (molecular weight 668.4 and 368.3, respectively) into susceptible bacteria. Lactococcal cells treated with lactococcin A became leaky for ultraviolet absorbing material, while susceptible cells gently pretreated with lysozyme released lactate dehydrogenase within a few minutes of lactococcin A treatment (H. Holo and I. F. Nes, unpublished data).

Van Belkum et al. (1991b) investigated the possibility of membrane permeabilization in detail and studied the action of purified lactococcin A on whole cells and membrane vesicles of sensitive and immune lactococcal strains. At a lactococcin A concentration that did not affect immune cells, the peptide rapidly dissipated the membrane potential of glucose-energized sensitive cells. This observation suggested that lactococcin A increased the permeability of the cytoplasmic membrane, an assumption that was corroborated by amino acid uptake experiments in whole cells and cytoplasmic membrane vesicles of *L. lactis*. Energized cells of *L. lactis* rapidly lost preaccumulated amino-isobutyric acid (AIB), a nonmetabolizable alanine analogue that is taken up in symport with one proton and, thus, is driven by the proton motive force (Konings et al., 1989). Preincubation of sensitive cells with lactococcin A completely blocked AIB uptake. To show that the efflux of substrates by lactococcin A was a direct consequence of increase in the permeability of the cytoplasmic membrane, and not only of the absence of a proton motive force, uptake of L-glutamate was studied. This amino acid is taken up by *L. lactis* in a phosphate bond-linked unidirectional process that is not driven by the proton motive force (Poolman et al., 1987). Efflux of accumulated glutamate was observed when sensitive cells were treated with lactococcin A, even if the proton motive force in these cells was collapsed prior to the addition of the peptide (Figure 5). This indicated that lactococcin A indeed affects the permeability of the cytoplasmic membrane of *L. lactis* and that it can do so in a voltage-independent way. In the control experiments with immune cells of *L. lactis* no effects of lactococcin A (at the same concentration as that used for the sensitive cells) on membrane permeability were observed. The addition of high concentrations of lactococcin A did lead to the dissipation of membrane potential of immune cells. Apparently, *lci*A renders *L. lactis* cells immune to only a certain concentration of lactococcin A above which the *lci*A product is titrated out (H. Holo et al., unpublished data), a phenomenon that is known as immunity breakdown (Pugsley, 1984). The effect of lactococcin A on the cytoplasmic membrane was directly analyzed by using cytoplasmic membrane vesicles derived from

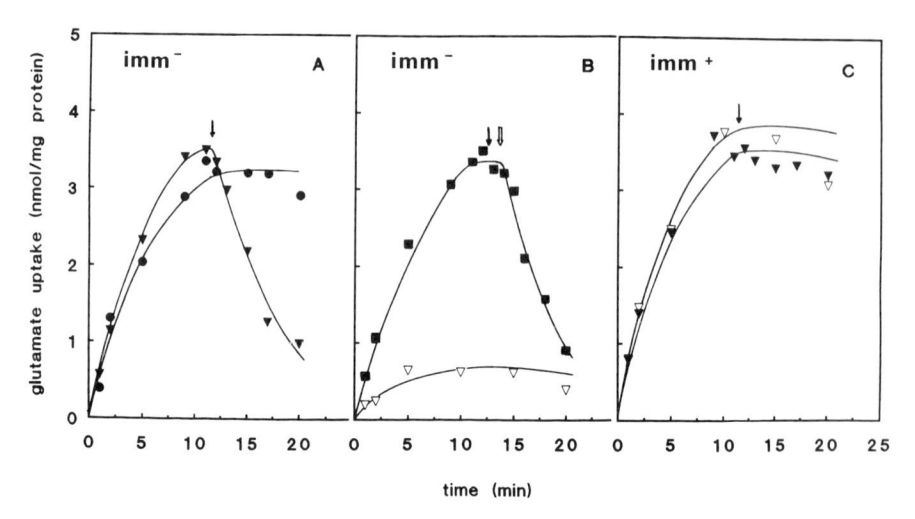

Figure 5 Effects of lactococcin A on L-glutamate uptake in energized cells of *L. lactis* IL1403 (A and B) and *L. lactis* IL1403pMB563, a plasmid expressing the immunity protein (C). In separate experiments valinomycin (1μM) plus nigericin (0.5μM) (●) or lactococcin A (0.029 μg/mg of protein) (▼) were added to the cell suspension of strain IL1403 or strain IL1403(pMB563) at the times indicated by the arrows. Uptake was also monitored in an experiment (B) in which first valinomycin (1μM) plus nigericin (0.5μM) and subsequently lactococcin A (0.029 μg/mg of protein) were added to the cell suspension (indicated with the closed and the open arrow, respectively) (■). Uptake in cells of IL1403 and IL1403(pMB653) preincubated for 3 minutes with lactococcin A (0.29 μg/mg of protein) is indicated (▽). The assays were started by the addition of 1.75μM ^{14}C-labeled glutamate. Imm$^+$, Imm$^-$: immune and not immune, respectively.

both sensitive and immune cells of L. *lactis*. Leucine uptake in these vesicles can be driven by an artificially imposed proton motive force (Konings et al., 1989), and was inhibited by lactococcin A. Surprisingly, lactococcin A was only fully effective when it was preincubated with the vesicles for at least 5 minutes before the uptake experiment was started. Lactococcin A action on the cytoplasmic membrane vesicles was also studied in the absence of a proton motive force by performing leucine counterflow experiments (Driessen et al., 1988). Vesicles loaded with a high concentration of unlabeled leucine and diluted in a buffer with a low concentration of [^{14}C]leucine are exposed to an outwardly directed leucine concentration gradient. This gradient gives rise to a leucine carrier-mediated exchange of nonradioactive leucine in the vesicles for radioactive leucine in the medium. The exchange was severely inhibited when the vesicles were preincubated with lactococcin A, showing that lactococcin A can permeabilize lactococcal cytoplasmic membrane vesicles in the absence of a proton motive force, and thus acts voltage-independently. Electron microscopic examination of whole cells treated with lactococcin A showed that no apparent lysis or other morphological alterations occurred, even after prolonged incubation with the bacteriocin. The integrity of the cytoplasmic membrane vesicles was apparently not affected by the addition of lactococcin A and no aggregation of vesicles was observed. These observations and the results of the substrate efflux experiments led to the speculation that lactococcin A might permeabilize the cytoplasmic membrane of sensitive cells by the formation of pores (van Belkum et al., 1991b). Leucine uptake by cytoplasmic membrane vesicles of *Bacillus subtilis*, *Clostridium acetobutyricum*, and *Escherichia coli* was not significantly inhibited by lactococcin A, indicating that the cytoplasmic membrane was involved in the species-specific response to lactococcin A. Neither preincubation with nor addition of high concentrations of lactococcin A to liposomes derived from phospholipids of a sensitive *L. lactis* strain affected an artificially imposed membrane potential in these liposomes. Both observations suggest that lactococcin A exerts its effect by interacting with a membrane receptor specific for lactococci. Cytoplasmic membrane vesicles derived from a lactococcal strain carrying the lactococcin A immunity gene were not or were only slightly affected by lactococcin A at concentrations of the peptide that did affect the sensitive vesicles, indicating that the immunity protein is attached to the cytoplasmic membrane. These findings were also supported by data obtained by Nissen-Meyer et al. (1992). The immunity protein could operate either by blocking the membrane receptor for lactococcin A or by preventing the (putative) pore formation in the cytoplasmic membrane (van Belkum et al., 1991b).

It is both tempting and instructive to speculate on how lactococcin A could increase the permeability of the cytoplasmatic membrane. By plotting the hydrophilicity along the lactococcin A sequence according to the method of Kyte and Doolittle (1982) it was noted that the 21-amino-acid sequence between positions Ala.30 and Phe.50 (Figure 3) could possibly form a membrane-spanning helix (Klein et al., 1985). The hydrophobicity of the hypothetical transmembrane spanning segment is 0.52 as compared to the average hydrophobicity of 0.18 for the lactococcin A molecule as a whole. A large number of pore-forming toxins are known to create channels through a "barrel-stave" mechanism (Ojcius and Young, 1991). This formation requires that the toxins aggregate like barrel staves surrounding a central water-filled pore. The helical wheel representation of the segment between Ala.30 to Phe.50 according to Schiffer and Edmundson (1967) would agree with a barrel model of pore formation in the cytoplasmic membrane. For example, a helical wheel based on the 16 amino acids between the residues in positions +35 and +50 has a hydrophobic moment of 0.26. These computer predictions suggest that lactococcin A could be anchored to the cytoplasmic membrane of sensitive bacteria by the hypothetical transmembrane helical segment, thereby inducing leakage by pore formation. The data and speculations summarized above can be integrated into a working model for lactococcin A action and immunity that is presented in Figure 2.

VII. Conclusions and Future Prospects

With the advent of recombinant DNA technology for lactococci, our knowledge of the bacteriocins produced by these industrially important microorganisms is expanding rapidly. The genes for three lactococcins and their respective immunity genes have been cloned and sequenced. The genetic analysis has revealed the existence of what might ultimately turn out to be a secretion system dedicated to the externalization of these bacteriocins. One of the lactococcins, lactococcin A, has been purified to homogeneity. Comparison of the DNA and protein sequences has shown that active, secreted lactococcin A is formed by proteolytic cleavage of 21 amino acids from the N-terminus of a 75-amino-acid precursor. The fact that the N-termini of the primary translation products of the other two lactococcins, and those of a number of small hydrophobic bacteriocins of other species of LAB, are very similar suggests that a similar processing mechanism may underlie the maturation of these bacteriocins. The exciting possibility exists that N-terminal processing is a part of the secretion mechanism of the bacteriocins and it will be interesting to see whether bacteriocins can be processed and secreted by heterologous (putative) maturation/secretion systems.

The availability of pure lactococcin A allowed a detailed analysis of its mode of action. These studies have shown that the bacteriocin acts by increasing the permeability of the cytoplasmic membrane of sensitive cells. Depolarization of the cytoplasmic membrane and efflux of essential compounds from the cells can explain the growth inhibition and ultimate death of lactococcal cells exposed to lactococcin A. A producer cell escapes the deleterious effect of lactococcin by producing a membrane-associated immunity protein, LciA. The lactococcins have a narrow inhibition spectrum and lactococcin A seems to inhibit lactococci only.

The availability of the nucleotide sequences of the structural genes for a number of lactococcins offers the possibility of changing amino acids in the molecules by several mutagenesis techniques. Alternatively, the knowledge of the amino acid sequences of lactococcins allows the synthesis of selected parts of the peptides or even, by virtue of their small sizes, the whole lactococcin molecules. Such lactococcin derivatives, created in either way, would enable the study of a number of important features of the lactococcins. In the near future we may expect such mutations to shed more light on the secretion and maturation process, the amino acid (sequences) important for specific binding to the susceptible cell, and those that take part in the permeation of these cells. Mutations in the putative C-terminal transmembrane anchor will be used to test the "barrel-stave" model of pore formation. The way the immunity proteins interfere with lactococcin action is fully unclear and unraveling this mystery will be an important future goal. These studies may ultimately lead to knowledge-based alterations in the molecule, with the aim of increasing or modifying the inhibitory spectrum of the bacteriocin.

The availability of a protocol for the isolation of large quantities of lactococcin A is important, as it will enable to determine the molecular structure by NMR studies. This would allow the detailed analysis of structure/function relationships of lactococcin A, the next challenging step in the study of this fundamentally and industrially interesting molecule.

In the near future we will undoubtedly see isolation and characterization of more lactococcal bacteriocins as a consequence of the worldwide increased interest in these molecules. Only recently, two new reports of bacteriocin production by lactococci have been published. In a study of Dufour et al. (1991) a new bacteriocin termed lactococcin was purified from *L. lactis* subsp. *lactis* ADRI 85LO30 to apparent homogeneity as judged by urea-SDS polyacrylamide gel electrophoresis. Unfortunately, no amino acid sequence was obtained by Edman degradation of the purified bacteriocin. The authors suggested this may be due to modification of the amino acids in the polypeptide. According to our suggested definition of lactococcin (Section III) the bacteriocin of Dufour et al. (1991) should be given a more neutral and temporary designation until its structure is determined. Analysis of the bacteriocin-producing strain ADRI 85LO30 revealed the presence of several plasmids. Six nonbacteriocin-producing (Bac$^-$) and non immune (Imm$^-$) variants were obtained by plasmid curing. All Bac$^-$ Imm$^-$ isolates had lost plasmids POS4 (32 kb) and POS5 (70 kb). A DNA region of about 10 kb from POS5 seemed to be required for wild-type expression of both bacteriocin activity and immunity (Dufour et al., 1991). Additional studies are required to establish whether this "lactococcin" contains only regular amino acids or if it is a nisin-type bacteriocin.

A recent publication reported the finding of a bacteriocin from *L. lactis* subsp. *diacetylactis* that was active only against lactococci, including a nisin producer (Kojic et al., 1991). The bacteriocin was susceptible to proteinases but resistant to boiling. No molecular properties of the bacteriocin were published.

The structure of our chapter on the study of lactococcin production during the last 5 years was chosen deliberately to stress that some caution is needed in future research in this field. As we have seen, all bacteriocin plasmids analyzed thus far are (self)transmissible by conjugation. These plasmids, although differing quite considerably in size, all seem to carry quite similar bacteriocin structural and immunity genes. The availability of DNA fragments and oligonucleotide primers of the three lactococcin operons studied so far should allow the rapid assessment (by PCR technology) of whether a "new" bacteriocin determinant is yet another example of the *lcn*A/*lcn*B/*lcn*M system that has apparently been widely disseminated in the lactococci.

Acknowledgments

We thank Gerard Venema for critical reading of the manuscript, Tjakko Abee for helpful discussions, and Henk Mulder for preparing the figures. We are endebted to our colleagues for submitting papers prior to publication.

References

Akutsu, A., Masaki, H., and Ohta, T. (1989). Molecular structure and immunity specificity of colicin E6, an evolutionary intermediate between E-group colicins and cloacin DF13. *J. Bacteriol.* **171,** 6430–6436.

Bardowski, J., Kozak, W., and Dobrzanski, W. T. (1979). Further characterization of lactostrecins—acid bacteriocins of lactic streptococci. *Acta Microbiol. Pol.* **28,** 93–99.

Blight, M. A., and Holland, I. B. (1990). Structure and function of hemolysin B, P-glycoprotein and other members of a novel family of membrane translocators. *Mol. Microbiol.* **4,** 873–880.

Davey, G. P. (1981). Mode of action of diplococcin, a bacteriocin from *Streptococcus cremoris* 346. *New Zealand J. Dairy Sci. Technol.* **16,** 187–190.

Davey, G. P. (1984). Plasmid associated with diplococcin production in *Streptococcus cremoris*. *Appl. Environ. Microbiol.* **48,** 895–896.

Davey, G. P., and Pearce, L. E. (1980). The use of *Streptococcus cremoris* strains cured of diplococcin production as cheese starters. *New Zealand J. Dairy Sci. Technol.* **15,** 51–57.

Davey, G. P., and Pearce, L. E. (1982). Production of diplococcin by *Streptococcus cremoris* and its transfer to nonproducing group N streptococci. *In* "Microbiology-1982" (D. Schlessinger, ed.), pp. 221–224. American Society for Microbiology, Washington, D.C.

Davey, G. P., and Richardson, B. C. (1981). Purification and some properties of diplococcin from *Streptococcus cremoris* 346. *Appl. Environ. Microbiol.* **41,** 84–89.

Davies, F. L., and Gasson, M. J. (1981). Reviews of the progress of dairy science: genetics of lactic acid bacteria. *J. Dairy Res.* **48,** 363–376.

Delepelaire, P., and Wandersman, C. (1990). Protein secretion in Gram-negative bacteria. *J. Biol. Chem.* **265**, 17118–17125.

Delepelaire, P., and Wandersman, C. (1991). Characterization, localization and transmembrane organization of the three proteins PrtD, PrtE and PrtF necessary for protease secretion by the Gram-negative bacterium *Erwinia chrysanthemi*. *Mol. Microbiol.* **5**, 2427–2434.

de Vos, W. M. (1987). Gene cloning and expression in lactic streptococci. *FEMS Microbiol. Rev.* **46**, 281–295.

Dobrzanski, W. T., Bardowski, B., Kozak, W., and Zajdel, J. (1982). Lactostrepcins: bacteriocins of lactic streptococci. *In* "Microbiology-1982" (D. Schlessinger, ed.), pp. 225–229. American Society for Microbiology, Washington, D.C.

Driessen, A. J. M., Zhen, T., In't Veld, G., op den Kamp, J., and Konings, W. N. (1988). The lipid requirement of the branced amino acid carrier of *Streptococcus cremoris*. *Biochem.* **27**, 865–872.

Dufour, A., Thuault, D., Boulliou, A., Bourgeois, C. M., and Le Pennec, J.-P. (1991). Plasmid-encoded determinants for bacteriocin production and immunity in a *Lactococcus lactis* strain and purification of the inhibitory peptide. *J. Gen. Microbiol.* **137**, 2423–2429.

Ebina, Y., Kishi, G., and Nakazawa, A. (1982). Direct participation of LexA protein in repression of colicin E1 synthesis. *J. Bacteriol.* **150**, 1479–1481.

Ebina, Y., Takahara, Y., Kishi, F., Nakazawa, A., and Brent, R. (1983). *lex*A protein is a repressor of the colicin E1 gene. *J. Biol. Chem.* **258**, 13258–13261.

Endicott, J. A., and Ling, V. (1989). The biochemistry of P-glycoprotein-mediated multidrug resistance. *Annu. Rev. Biochem.* **58**, 137–171.

Foote, S. J., Thompson, J. K., Cowman, A. F., and Kemp, D. J. (1989). Amplification of multidrug resistance gene in some chloroquine-resistant isolates of *P. falciparum*. *Cell* **57**, 921–930.

Galvez, A., Maqueda, M., Martinez-Bueno, M., and Valdivia, E. (1991). Permeation of bacterial cells, permeation of cytoplasmic and artificial membrane vesicles, and channel formation in lipid bilayers by peptide antibiotic AS-48. *J. Bacteriol.* **173**, 886–892.

Gao, F. H., Abee, T., and Konings, W. N. (1991). The mechanism of action of the peptide antibiotic nisin in liposomes and cytochrome C oxidase proteoliposomes. *Appl. Environ. Microbiol.* **57**, 2164–2170.

Geis, A., Singh, J., and Tueber, M. (1983). Potential of lactic streptococci to produce bacteriocin. *Appl. Environ. Microbiol.* **45**, 205–211.

Gilson, L., Mahanty, H., and Koltner, R. (1987). Four plasmid genes are required for colicin V synthesis, export and immunity. *J. Bacteriol.* **169**, 2466–2470.

Gilson, L., Mahanty, H., and Koltner, R. (1990). Genetic analysis of an MDR-like export system; the secretion of colicin V. *EMBO J.* **9**, 3875–3884.

Glaser, P., Sakamoto, H., Bellalou, J., Ullmann, A., and Danchin, A. (1988). Secretion of cyclolysin, the calmodulin-sensitive adenylate cyclase-haemolysin bifunctional protein of *Bordetella pertussis*. *EMBO J.* **7**, 3997–4004.

Hastings, J. W., Sailer, K., Johnson, K., Roy, K. L., Vederas, J. C., and Stiles, M. E. (1991). Characterization of Leucocin A-UAL 187 and cloning of the bacteriocin gene from *Leuconostoc gelidium*. *J. Bacteriol.* **173**, 7491–7500.

Henderson, J. T., Chopko, A. L., and van Wassenaar, D. (1992). Purification and primary structure of the bacteriocin PA-1 produced by *Pediococcus acidilactici* PAC 1.0 *Arch. Biochem. and Biophys.* **295**, 5–12.

Higgins, C. F., Hiles, I. D., Salmond, P. C., Gill, D. R., Downie, J. A., Evans, I. J., Holland, I. B., Gray, L., Buckel, S. D., Bell, A. W., and Hermodson, M. A. (1986). A family of related ATP-binding subunits coupled to many distinct biological processes in bacteria. *Nature (London)* **323**, 448–450.

Holo, H., Nilssen, Ø., and Nes, I. F. (1991). Lactococcin A, a new bacteriocin from *Lactococcus lactis* subsp. *cremoris*: isolation and characterization of the protein and its gene. *J. Bacteriol.* **173**, 3879–3887.

Kaletta, C., and Entian, K.-D. (1989). Nisin, a peptide antibiotic: cloning and sequencing of the *nis*A gene and post-translational processing of its peptide product. *J. Bacteriol.* **171,** 1597–1601.
Klaenhammer, T. R. (1988). Bacteriocins in lactic acid bacteria. *Biochimie* **70,** 337–350.
Klein, P., Kanehisa, M., and De Lilsi, C. (1985). The detection and classification membrane-spanning proteins. *Biochem. Biophys. Acta* **815,** 468–476.
Kojic, M., Svircevic, J., Banina, A., and Topisirovic, L. (1991). Bacteriocin-producing strain of *Lactococcus lactis* subsp. *diacitilactis* S50. *Appl. Environm. Microbiol.* June 1991, 1835–1837.
Kok, J. (1991). Special purpose vectors for lactococci. *In* "Genetics and Molecular Biology of Streptococci, Lactococci and Enterococci" (G. M. Dunny, P. P. Cleary, and L. L. McKay, eds.) pp. 97–102. American Society of Microbiology, Washington D.C.
Kok, J., van der Vossen, J. M. B. M., and Venema, G. (1984). Construction of plasmid cloning vectors of lactic streptococci which also replicate in *Bacillus subtilis* and *Escherichia coli*. *Appl. Environ. Microbiol.* **48,** 726–731.
Konings, W. N., Poolman, B., and Driessen, A. J. M. (1989). Bioenergetics and solute transport in lactococci. *CRC Critical Rev. in Microbiol.* **16,** 419–476.
Kordel, M., and Sahl, H.-G. (1986). Susceptibility of bacterial, encraiotic and artificial membranes to disruptive action of the cationic peptides PEP5 and nisin. *FEMS Microbiol. Lett.* **34,** 139–144.
Kordel, M., Benz, R., and Sahl, H.-G. (1988). Mode of action of the staphylococcin-like peptide Pep 5: voltage-dependent depolarization of bacterial and artificial membranes. *J. Bacteriol.* **170,** 84–88.
Kozak, W., J. Bardowski, and Dobrzanski, W. T. (1977). Lactostrepcin—a bacteriocin produced p *Streptococcus lactis Bull. Acad. Pol. Sci. CL.* 6. **25,** 217–221.
Kozak, W., Bardowski, J., and Dobrzanski, W. T. (1978). Lactostrepcins—acid bacteriocins produced by lactic streptococci. *J. Dairy Res.* **45,** 247–257.
Kyte, J., and Doolittle, R. F. (1982). A simple method for displaying the hydropathic character of a protein. *J. Mol. Biol.* **157,** 105–132.
Létoffé, S., Delepelaire, P., and Wandersman, C. (1990). Protease secretion by *Erwinia chrysanthemi:* the specific secretion functions are analogous to those of *Escherichia coli* α-hemolysin. *EMBO J.* **9,** 1375–1382.
McGrath, J. P., and Varshavsky, A. (1989). The yeast STE6 gene encodes a homologue of the mammalian multidrug resistance P-glycoprotein. *Nature (London)* **340,** 400–404.
McKay, L. L. (1983). Functional properties of plasmids in lactic streptococci. *Antonie van Leeuwenhoek* **49,** 259–274.
Marugg, J. D., Gonzalez, C. F., Kunka, B. S., Lederboer, A. M., Pucci, M. J., Toonen, M. Y., Walker, S. A., Zoetmulder, L. C. M., and Vandenbergh, P. A. (1992). Cloning, expression, and nucleotide sequence of genes involve din production of pediocin PA-1, a bacteriocin from *Pediococcus acidilactici* PAC 1.0. *Appl. Environ. Microbiol.* **58,** 2360–2367.
Masure, H. R., Au, D. C., Gross, M. K., Donovan, M. G., and Storm, D. R. (1990). Secretion of the *Bordetella pertussis* adenylate cyclase from *Escherichia coli* containing hemolysin operon. *Biochemistry* **29,** 140–145.
Mørtvedt, C. I., Nissen-Meyer, J., Sletten, K., Nes, I. F. (1991). Purification and amino acid sequence of lactocin S, a bacteriocin produced by *Lactobacillus sake* L45. *Appl. Environ. Microbiol.* **57,** 1829–1834.
Muriana, P. M., and Klaenhammer, T. R. (1991). Cloning, phenotypic expression, and DNA sequence of the gene for lactacin F, an antimicrobial peptide produced by *Lactobacillus* spp. *J. Bacteriol.* **173,** 1779–1788.
Neve, H., Geis, A., and Teuber, M. (1984). Conjugal transfer and characterization of bacteriocin plasmids in Group N (lactic acid) streptococci. *J. Bacteriol.* **157,** 833–838.
Nissen-Meyer, J., Håvarstein, L. S., Holo, H., Sletten, K., and Nes, I. F. (1992). Association of lactococcin A immunity factor with the cell membrane: Purification and characterization of the immunity factor. *J. Gen. Microbioloogy* In press.
Ojcius, D. M., and Young, J. D.-E. (1991). Cytolytic pore-forming proteins and peptides: is there a common structural motif. *J. Trends in Biochem. Sci.* **16,** 225–229.

Oxford, A. E. (1944). Diplococcin, an anti-microbial protein elaborated by certain milk streptococci. *Biochem. J.* **38**, 178–182.
Poolman, B., Smid, E. J., and Konings, W. N. (1987). Kinetics of a phosphate-bond driven glutamate-glutamine transport system in *Streptococcus lactis* and *Streptococcus cremoris*. *J. Bacteriol.* **169**, 1460–1468.
Pugsley, A. P. (1984). The ins and outs of colicins. Part II. Lethal action, immunity and ecological implications. *Microbiol. Sci.* **1**, 203–205.
Pugsley, A. P. (1988). Protein secretion across the outer membrane of Gram-negative bacteria. *In* "Protein Transfer and Organelle Biogenesis" (R. C. Das and P. W. Robins, eds.), pp. 607–652. Academic Press, Inc., Orlando, Florida.
Randall, L. L., Hardy, S. J. S., and Thom, J. R. (1987). Export of protein: A biochemical view. *Annu. Rev. Microbiol.* **41**, 507–541.
Rao, J. K. M., and Argos, P. (1986). A conformational preference parameter to predict helices in integral membrane proteins. *Biochim. Biophys. Acta* **869**, 197–214.
Riordan, J. R., Rommens, J. M., Korem, B., Alon, N., Rozmahel, R., Grzelczak, Z., Zielenski, J., Lok, S., Plasvic, N., Chou, J. L., Drumm, M. L., Iannuzzi, M. C., Collins, F. S., and Tsui, L. C. (1989). Identification of the cystic fibrosis gene: Cloning and characterization of complementary DNA. *Science* **245**, 1066–1073.
Schaller, F., Benz, R., and Sahl, H.-G. (1989). The peptide antibiotic subtilin acts by formation of voltage-dependent multi-state pores in bacterial and artificial membranes. *Eur. J. Biochem.* **182**, 182–186.
Scherwitz, K., and McKay, L. L. (1987). Restriction enzyme analysis of lactose and bacteriocin plasmids from *Streptococcus lactis* subsp. *diacetylactis* WM4 and cloning of *BclI* fragments coding for bacteriocin production. *Appl. Environ. Microbiol.* **53**, 1171–1174.
Scherwitz, K. M., Baldwin, K. A., and McKay, L. L. (1983). Plasmid linkage of a bacteriocin-like substance in *Streptococcus lactis* subsp. *diacetylactis* strain WM4: transferability to *Streptococcus lactis*. *Appl. Environ. Microbiol.* **45**, 1506–1512.
Schiffer, M., and Edmundson, A. B. (1967). Use of helical wheels to represent the structures of proteins and to identify segments with helical potential. *Biophys. J.* **7**, 121–135.
Simon, D., and Chopin, A. (1988). Construction of a vector plasmid and its use for molecular cloning in *Streptococcus lactis*. *Biochimie* **70**, 559–566.
Stoddard, G. W., Petzel, J. P., Van Belkum, M. J., Kok, J., and McKay, L. L. (1992). Molecular analyses of the lactococcin A gene cluster from *Lactococcus lactis* subsp. *lactis* biovar. *diacetylactis* WM4. *Appl. Environ. Microbiol.* **58**, 1952–1961.
Strathdee, C. A., and Lo, R. Y. C. (1989). Cloning, nucleotide sequence, and characterization of genes encoding the secretion function of *Pasteurella haemolytica* leukotoxin determinant. *J. Bacteriol.* **171**, 916–928.
Tagg, J. R., Dajani, A. S., and Wannamaker, L. W. (1976). Bacteriocins of Gram-positive bacteria. *Bacteriol. Rev.* **10**, 799–756.
van Belkum, M. J. (1991). Lactococcal bacteriocins: genetics and mode of action, Ph.D. thesis, University of Groningen, Groningen, The Netherlands.
van Belkum, M. J., Hayema, B. J., Geis, A., Kok, J., and Venema, G. (1989). Cloning of two bacteriocin genes from a lactococcal bacteriocin plasmid. *Appl. Environ. Microbiol.* **55**, 1187–1191.
van Belkum, M. J., Hayema, B. J., Jeeninga, R. E., Kok, J., and Venema, G. (1991a). Organization and nucleotide sequence of two lactococcal bacteriocin operons. *Appl. Environ. Microbiol.* **57**, 492–498.
van Belkum, M. J., Kok, J., Venema, G., Holo, H., Nes, I. F., Konings, W. N., and Abee, T. (1991b). The bacteriocin lactococcin A specifically increases the permeability of lactococcal cytoplasmic membranes in a voltage-independent, protein-mediated manner. *J. Bacteriol.* **173**, 7934–7941.
van Belkum, M. J., Kok, J., and Venema, G. (1992). Cloning, sequencing, and expression in *Escherichia coli* of *lcnB*, a third bacteriocin determinant from the lactococcal bacteriocin plasmic p9B4-6. *Appl. Environ. Microbiol.* **58**, 572–577.
Van den Elzen, P. J. M., Matt, J., Walters, H. H. B., Veltkamp, E., and Nijkamp, H. J. J. (1982).

The nucleotide sequence of bacteriocin promoters of plasmids Clo DF13 and Col E1: role of LexA repressor and CAMP in regulation of promoter activity. *Nucleic Acids Res.* **10,** 1913–1928.

Von Heijne, G. (1983). Pattern of amino acids near signal-sequence cleavage sites. *Eur. J. Biochem.* **133,** 17–21.

Von Heijne, G. (1986). The distribution of positively charged residues in bacterial inner membrane proteins correlates with the trans-membrane topology. *EMBO J.* **5,** 3021–3027.

Wagner, W., Vogel, M., and Goebel, W. (1983). Transport of hemolysin across the outer membrane of *Escherichia coli* requires two functions. *J. Bacteriol.* **154,** 200–210.

Wandersman, C., and Delepelaire, P. (1990). TolC, an *Escherichia coli* outer membrane protein required for haemolysin secretion. *Proc. Natl. Acad. Sci. U.S.A.* **87,** 4776–4780.

Whitehead, H. R. (1933). A substance inhibiting bacterial growth, produced by certain strains of lactic streptococci. *Biochem. J.* **27,** 1793–1800.

Zajdel, J. K., and Dobrzanski, W. T. (1983). Isolation and preliminary characterization of *Streptococcus cremoris* (strain 202) bacteriocin. *Acta Microbiol. Pol.* **32,** 119–129.

Zajdel, J. K., Ceglowski, P., and Dobrzanski, W. T. (1985). Mechanism of action of lactostrepcin 5, a bacteriocin produced by *Streptococcus cremoris* 202. *Appl. Environ. Microbiol.* **49,** 969–974.

CHAPTER 7

Molecular Biology of Bacteriocins Produced by Lactobacillus

T. R. KLAENHAMMER
C. FREMAUX
C. AHN
K. MILTON

I. Introduction

Lactobacilli are ubiquitous in the environment, while simultaneously occupying a multitude of specialized environmental niches, including many fermenting foods (vegetables, meat, cereals, and dairy products) and the intestinal tracts of humans and animals. The diverse metabolic capabilities and physiological characteristics of this genus have proven problematic in the proper classification of many *Lactobacillus* strains, a dilemma only recently alleviated by the development of more sophisticated biochemical and molecular classification systems.

Lactobacilli are fastidious organisms requiring nutritionally complex environments in which they derive energy via homo- or heterofermentative catabolism of carbohydrates. Organic acids (lactic and acetic), the primary endproducts of fermentation, serve directly as antagonists to other competing microflora by lowering the pH of the surrounding environment (Kashet, 1987). Acidification of their environment allows the lactobacilli to effectively compete and ultimately dominate fermenting ecosystems, since they are frequently more acid tolerant than other organisms, including many pathogenic and spoilage species, as well as other lactic acid bacteria. In addition, lactobacilli produce a number of other antimicrobial substances: hydrogen peroxide, various poorly characterized compounds (Vincent et al., 1959; Hamdan and Mikolajcik, 1974; Silva et al., 1987), antimicrobials (Talarico and Dobrogosz, 1989), and most notably, bacteriocins (Klaenhammer, 1988; 1990).

The lactic acid bacteria are well recognized for their production of proteinaceous antimicrobials (Klaenhammer, 1988; Lindgren and Dobrogosz, 1990); however, lactobacilli have proven to be particularly bacteriocinogenic. Barefoot and Klaenhammer (1983) determined that 63% of the *Lactobacillus acidophilus* strains they surveyed produced bacteriocinlike inhibition. Nineteen different bacteriocins have been reported (Table I) since the first descriptions of "bacteriocins" produced by homo- and heterofermentative lactobacilli appeared (de Klerk and Coetzee, 1961; de Klerk, 1967). Included among these are the bacteriocins produced by *Carnobacterium piscicola* (previously designated as *Lactobacillus carnis*), a heterofermentative, nonaciduric strain naturally associated with meats (Ahn and Stiles, 1990).

Bacteriocins of the lactobacilli have only recently been the focus of biochemical and genetic characterization (reviewed by Klaenhammer, 1988; 1990). At present, limited information is available on the bactericidal mechanism(s) of action and immunity or on receptors or routes of access in sensitive bacteria. Substantial attention is now being directed to define the genetic organization, regulation (processing, immunity, and transport), and structure-function relationships of *Lactobacillus* bacteriocins.

II. Evidence and Roles

The bacteriocins of *Lactobacillus* species usually prove inhibitory to bacteria that are closely related to the producer strain or that compete for the same ecological niche (Table I; Klaenhammer, 1988). For example, *Lactobacillus delbrueckii* produces two bacteriocins, designated lacticins A and B, which target other *L. delbrueckii*-related subspecies associated with fermenting dairy products (Toba et al., 1991c), while plantaricin A, produced by *Lactobacillus plantarum*, is bactericidal to various lactic acid bacteria, including other *L. plantarum* strains, *Pediococcus pentosaceus*, and *Leuconostoc paramesenteroides*, normally associated with fermenting vegetables (Daeschel et al., 1990). Other bacteriocins appear to have an extended host range and can include *Enterococcus* species, *Listeria monocytogenes*, *Leuconostoc* species, *Pediococcus* species, and, recently reported, *Clostridium botulinum*, *Staphylococcus aureus*, and *Aeromonas hydrophila* (Lewus et al., 1991; Okereke and Montville, 1991). Because of renewed interest in "natural" and biodegradable preservation systems by the consumer and food manufacturers, the inhibition of food-borne pathogens by *Lactobacillus* bacteriocins could prove highly significant, especially in the area of minimally processed foods.

TABLE I
Bacteriocins of Lactobacillus Species

Producer strain	Bacteriocin	Spectrum of activity	Characteristics	References[a]
L. acidophilus	Lactacin B	Lactobacillus delbrueckii Lactobacillus helveticus Listeria monocytogenes (?)	High molecular weight complex, purified to 6.3 kDa; chromosomal determinants	a, b, c, d
	Lactacin F	Lactobacillus fermentum Enterococcus faecalis Lactobacillus delbrueckii Lactobacillus helveticus Aeromonas hydrophila Staphylococcus aureus	High molecular weight complex, purified, cloned, expressed, and sequenced; 57 amino acids, 6.3-kDa peptide, 18 amino acid N-terminal extension; heat stable at 121°C for 15 min; episomal and conjugative genetic determinants	e, f, g, d
	Acidophilucin A	Lactobacillus delbrueckii Lactobacillus helveticus	Proteinaceous, inactivated by trypsin and actinase; heat labile at 60°C for 10 min	h
L. brevis	Brevicin 37	Pediococcus damnosus Lactobacillus brevis Leuconostoc oenos	Proteinaceous; heat stable, 121°C for 1 hour; active over pH range 2–10; inactivated by chloroform	i
L. casei	Caseicin 80	Lactobacillus casei	40-kDa Protein; pI = 4.5; heat stable, 60°C for 10 min; active over pH range 3–9; inducible by mitomycin C	i
L. carnis[b]	Bacteriocin(s)	Lactobacillus	Proteinaceous; heat stable, 100°C for 30 min; pH stable, pH 2–11; membrane active; plasmid linked	j, k
	Carnocin U149	Carnobacterium Pediococcus Enterococcus Listeria	Lanthionine, 4635 daltons, heat stable	y
L. delbrueckii	Lacticin A Lacticin B	L. delbrueckii subsp. lactis L. delbrueckii subsp. bulgaricus L. delbrueckii subsp. delbrueckii	Proteinaceous; heat labile at 60°C for 10 min	l

(*continued*)

TABLE I (Continued)

Producer strain	Bacteriocin	Spectrum of activity	Characteristics	References[a]
L. fermenti	Bacteriocin	*Lactobacillum fermenti*	Macromolecular complex; lipocarbohydrate moiety	m
L. gasseri	Gassericin A	*Lactobacillus acidophilus* *Lactobacillus delbrueckii* *Lactobacillus helveticus* *Lactobacillus casei* *Lactobacillus brevis*	Proteinaceous, trypsin sensitive; stable at 120°C for 20 min; characteristics similar to lactacin B and F	n
L. helveticus	Lactocin 27	*Lactobacillus acidophilus* *Lactobacillus helveticus*	Protein-lipopolysaccharide complex at >200 kDa, purified to 12.4 kDa	o
	Helveticin J	*Lactobacillus helveticus* *Lactobacillus delbrueckii* subsp. *lactis* and *bulgaricus*	Complex aggregate at >300 kDa, purified to 37 kDa, cloned, expressed, sequenced	p
L. plantarum	Plantaricin A	*Lactobacillus plantarum* *Lactobacillus* spp. *Leuconostoc* spp. *Pediococcus* spp. *Lactococcus lactis* *Enterococcus faecalis*	Proteinaceous; >8 kDa; stable at 100°C for 30 min; active over pH range of 4–6.5	q
	Plantacin B	*Lactobacillus plantarum* *Leuconostoc mesenteroides* *Pediococcus damnosus*	Proteinaceous	r
	Bacteriocin	*Leuconostoc* *Lactobacillus* *Pediococcus* *Lactococcus* *Streptococcus*	Sensitive to proteases, α-amylase, and lipase; heat stable at 100°C for 30 min; phenotype unstable, suggesting plasmid-borne determinants	s

Strains 75 and 592	Bacteriocin	Lactobacillus sake Lactobacillus curvatis Lactobacillus plantarum Lactobacillus divergens Listeria monocytogenes Clostridium botulinum spores Aeromonas hydrophila Staphylococcus aureus	Proteinaceous	t, d, u, v
L. sake	Sakacin A	Carnobacterium piscicola Enterococcus spp. Lactobacillus sake Lactobacillus curvatus Leuconostoc paramesenteroides Listeria monocytogenes Aeromonas hydrophila Staphylococcus aureus	Plasmid borne; 28 kb; proteinaceous	t, d
	Lactocin S	Lactobacillus Leuconostoc Pediococcus	Lanthionine containing 33-amino-acid protein; active over pH 4.5–7.5; 50% nonpolar residues; lactocin S production and immunity linked to 50-kb plasmid pCIM1	w, x, z

[a], Barefoot and Klaenhammer (1983); b, Barefoot and Klaenhammer (1984); c, Harris et al. (1989); d, Lewus et al. (1991); e, Muriana and Klaenhammer (1987); f, Muriana and Klaenhammer (1991a); g, Muriana and Klaenhammer (1991b); h, Toba et al. (1991b); i, Rammelsberg and Radler (1990); j, Ahn and Stiles (1990); k, Schillinger and Holzapfel (1990); l, Toba et al. (1991c); m, de Klerk and Smit (1967); n, Toba et al. (1991a); o, Upreti and Hinsdill (1975); p, Joerger and Klaenhammer (1986); q, Daeschel et al. (1990); r, West and Warner (1988); s, Jimenez-Diaz et al. (1990); t, Schillinger, Lucke (1989); u, Okereke and Mortville (1991); v, Schillinger et al. (1991); w, Mortvedt and Nes (1990); x, Mortvedt et al. (1991a); y, Stoffels et al. (1992); z, McCormick and Savage (1983).
[b]Current taxonomy is Carnobacterium fiscicola.

The lactobacilli, particularly *L. acidophilus* and *L. gasseri*, are purported to suppress undesirable putrefactive and pathogenic microorganisms in the human gastrointestinal ecosystem. Bacteriocins are thought to play an important role in the competition of *L. acidophilus* for colonization of the intestinal tract of humans. Definition of this role and determination of the means through which bacteriocins facilitate competition of lactobacilli in the gut will provide the information base necessary for selection of strains for clinical evaluation of probiotics or dietary adjuncts. Breakthroughs in this area should promote rapid development of new and improved dairy products as vehicles for the oral delivery of lactobacilli to the intestinal tract.

III. Classification and Biochemical Characteristics

Characterization of the bacteriocins produced by lactic acid bacteria has historically focused on a generalized definition of a proteinaceous inhibitor, crude estimation of molecular weight (by ultrafiltration or retention in dialysis membranes), and determination of susceptible host range. Recent efforts to purify and biochemically characterize these compounds are beginning to provide information concerning their structure, processing, and antimicrobial mechanism of action.

Three general classes of antimicrobial proteins from lactic acid bacteria have been characterized to date: (I) lantibiotics; (II) small (<13 kDa) hydrophobic heat-stable peptides; and (III) larger (>30 kDa) heat-labile proteins. Table II shows categorization on the basis of size, hydrophobic propensity, and heat stability of those bacteriocins that have been purified and characterized.

TABLE II
Lactobacillus Bacteriocins Biochemical Types and Classes

Class I	Lantibiotics	■Lactocin S	------------ 3.7 kDa (33aa)
		■Carnocin U149	------------ 4.6 kDa (35-37aa)
Class II	Small hydrophobic peptides		
	Moderately heat stable (>30 min @ 100°C–15 min @ 121°C) molecular weight <13 kDa		
	100°	■Lactocin 27	------------ 12.4 kDa, glycoprotein
		■Carnobacteriocins	---------- 4.9 kDa
		■Lactacin B	------------ 6.3 kDa
	121°C	■Lactacin F	------------ 6.3 kDa (57aa)
		■Brevicin 37	
Class III	Large heat-labile proteins		
	Inactivated within 10–15 min @ 60°C–100°C		
		Helveticin J	------------ 37 kDa (334aa)
		Acidophilucin A	
		Lacticin A & B	

7. Molecular Biology of Bacteriocins Produced by *Lactobacillus* 157

A separate classification for bacteriocins reported as being heterogeneous (de Klerk and Smit, 1967; Upreti and Hinsdill, 1975; Jimenez-Diaz et al., 1990; Toba et al., 1991b) may be forthcoming. However, these compounds are not included in Table II, since it has yet to be established if the carbohydrate and lipid moieties are intrinsic to the primary structure and composition of the active bacteriocin, or whether they are nonproteinaceous contaminants of the bacteriocin purification process. The hydrophobic nature of many bacteriocins could lead to interactions with lipids and other materials (Muriana and Klaenhammer, 1991a).

Nisin, produced by *Lactococcus lactis* subsp. *lactis*, is the most predominant member of the lantibiotics. The 34 amino acids that comprise nisin's structure include 2 unique sulfur-containing amino acids, lanthionine and β-methyllanthionine. Classification of bacteriocins from lactobacilli had previously not included any lantibiotics; however, Mørtvedt et al. (1991) recently reported that lactocin S, a bacteriocin produced by *Lactobacillus sake*, contains lanthionine residues. In addition, another bacteriocin (35 to 37 amino acids) produced by a *Carnobacterium* species isolated from fish has been found to contain lanthionine residues (Stoffels et al., 1992). Typically, lantibiotics (nisin, subtilin, and epidermin) possess a broader host range compared to nonlanthionine-containing bacteriocins; therefore, the first reports of lantibiotics among the *Lactobacillaceae* may imply expanded roles for these bacteriocins in future food preservation systems. Thus far, however, the bacteriocins produced by *Lactobacillus* species have generally been classified as nonlantibiotic compounds.

De Klerk and Smit (1967) purified and characterized the first bacteriocin from lactobacilli. *Lactobacillus fermenti* produced a proteinaceous (sensitive to trypsin and pepsin) macromolecular antagonist containing a lipocarbohydrate component. The absence of net charge and lack of mobility in an electrical field suggested a hydrophobic protein or involvement of a lipidlike material. Separation of the carbohydrate component, composed of 56.7% mannose, completely inactivated the bacteriocin. The bacteriocin was relatively heat stable (96°C, 30 minutes) and contained a high percentage of glycine (11.1%) and alanine (13.4%) residues. Other hydrophobic heat-stable bacteriocins have since been purified, including lactocin 27, produced by *L. helveticus* (Upreti and Hinsdill, 1973; 1975). Similar to the *L. fermenti* bacteriocin, lactocin 27 was isolated initially as a protein-lipopolysaccharide complex with a molecular weight in excess of 200 kDa. Purification of the bacteriocin to homogeneity, however, yielded a glycoprotein of 12.4 kDa. Lactocin 27 was sensitive to trypsin and pronase, but not ficin, and unusually high concentrations of glycine (15.1%) and alanine (18.1%) were present in this heat-stable bacteriocin (100°C, 60 minutes). The action of lactocin 27 against the indicator strain *Lactobacillus helveticus* LS18 was considered to be bacteriostatic, even though treatment with the highest concentration of the bacteriocin resulted in a 2-log reduction in CFUs per milliliter. Although gross physical disruption of the bacterial membrane was not detected, lactocin 27 caused a leakage of potassium ions, allowed an influx of sodium ions, and terminated protein synthesis. The cytoplasmic membrane was thus proposed as the target of lactocin 27 (Upreti and Hinsdill, 1975).

Lactacins B and F produced by *L. acidophilus* N2 and 11088, respectively, are extremely heat stable and retain activity upon autoclaving at 121°C for 15 minutes (Barefoot and Klaenhammer, 1983; Muriana and Klaenhammer, 1987). Both bacteriocins have molecular weights of approximately 6.3 kDa and share similar solubility characteristics. However, the host range of lactacin B is a subset of that of lactacin F (Table I). This difference in activity spectra implies a structural variation between these two closely related antimicrobial peptides that may alter activity or effectiveness.

Other small, heat-stable peptides that have recently been identified and purified from lactobacilli include lactocin S produced by *L. sake* (Mørtvedt et al., 1991a,b) and several carnobacteriocins produced by *Carnobacterium piscicola* (Ahn and Stiles, 1990). Both bacteriocins withstand a heat treatment of 100°C for 30 minutes, although lactocin S appears to be the more heat labile of the two, with a 50% reduction in activity after one hour. Two other recently reported bacteriocins, brevicin 37 (Rammelsberg and Radler, 1990) and gassericin A (Toba et al., 1991a), are highly heat stable (121°C, 20 minutes) and appear to belong within the general class of small, heat-stable antimicrobial peptides.

Heat-labile proteins of large molecular weight comprise the third general class of *Lactobacillus* bacteriocins. There appear to be numerous members representative of this class among the lactobacilli, including helveticin J (Joerger and Klaenhammer, 1986), acidophilucin A (Toba et al., 1991b), lactacin A and B (Toba et al., 1991c), and caseicin 80 (Rammelsberg and Radler, 1990). At the present time, however, only helveticin J has been purified and characterized at the genetic level (Joerger and Klaenhammer, 1990). Helveticin J is a 37-kDa, heat-sensitive protein (loss of activity within 30 minutes at 100°C) that retains activity after treatment with various dissociating agents, suggesting the formation of small active subunits (Joerger and Klaenhammer, 1986). Little information exists on the modes of action and the biochemical properties of the larger, heat-labile bacteriocins. Because of their size and heat sensitivity, changes in conformation and secondary structure may be important in the bactericidal activities of this class of bacteriocins.

The amino acid compositions of those bacteriocins characterized to date are presented in Table III and clearly illustrate that the bacteriocins of the *Lactobacillaceae* are a highly heterogenous group. It appears that the smaller, hydrophobic peptides retain a high proportion of nonpolar residues, consisting largely of alanine, glycine, and valine. Muriana and Klaenhammer (1991a,b) determined that the lactacin F peptide was composed primarily of hydrophobic and polar neutral residues (87.3%), including glycine (21.6%), alanine (15.8%), and valine (8.8%). The amino acid composition and N-terminal sequencing of lactocin S (Mørtvedt et al., 1991b) revealed that 50% of the approximate 33 amino acids were hydrophobic and nonpolar resi-

TABLE III
Amino Acid Composition of *Lactobacillus* Bacteriocins

Amino acids	Lactacin F	Lactocin S	Lactocin 27	Helveticin J	*L. fermenti* Bacteriocin
Hydrophobic					
Alanine	15.8	24.0	15.1	5.7	13.4
Isoleucine	7.0	0	5.2	8.1	3.8
Phenylalanine	0	3.0	3.3	3.0	2.8
Leucine	3.5	12.0	5.1	6.3	6.4
Methionine	1.7	3.0	0	1.2	0.5
Proline	5.3	6.0	3.3	2.7	4.7
Valine	8.8	15.0	8.3	4.8	5.6
Tryptophan	3.5	0	0.8	1.5	nd
Charged					
Arginine	3.5	0	3.3	3.6	2.9
Aspartate	0	3.0	8.9[a]	6.3	10.0[a]
Glutamate	0	3.0	6.8[b]	6.6	9.2[b]
Histidine	1.7	6.0	2.3	3.0	1.9
Lysine	3.5	6.0	5.8	6.9	5.7
Polar neutral					
Asparagine	5.3	0	nd[c]	8.7	nd
Cysteine	3.5	0	0	0.6	nd
Glycine	21.0	3.0	18.1	8.7	11.1
Glutamine	1.7	0	nd	4.2	nd
Serine	1.7	0	5.3	7.2	10.1
Threonine	10.6	3.0	5.8	5.1	8.7
Tyrosine	1.7	6.0	2.9	5.7	3.1
Estimated size in kDa	7.5	3.3	12.4	37.5	nd

[a] Including asparagine.
[b] Including glutamine.
[c] nd, Not determined.

dues of alanine, valine, and glycine. The lower glycine content and hydrophobicity of lactocin S compared to lactacin F correlates with their relative heat stabilities (lactacin F > lactocin S).

Amino acid comparisons between lactacin F and the available portions of lactacin B, lactocin S, and the carnobacteriocins have not yet revealed conserved or identical regions (Henkel et al., 1991; Mørtvedt et al., 1991b; Muriana and Klaenhammer, 1991b; Nettles et al., 1991). These particular bacteriocins are among the group of small, heat-stable hydrophobic peptides that are thought to target the cytoplasmic membrane (Ahn and Stiles, 1990; Mørtvedt et al., 1991b; Muriana and Klaenhammer, 1991a). A comparison of the N-terminal regions of lactacin B (Nettles et al., 1991) and lactacin F (Muriana and Klaenhammer, 1991a) illustrates the heterogeneity of their sequences:

lactacin F [N-Arg–Asn–Asn–Trp–Gln–Thr–Asn–Val–Gly–Gly–Ala . . .]
lactacin B [N-Arg–Gln–Pro–Gly–Phe–Ile–Leu–Phe–Pro–Thr–Val . . .]

Lactacin B and F have similar molecular weights, heat stabilities, and solubility characteristics. They also inhibit the same basic indicator group. Barefoot and Klaenhammer (1983) reported that lactacin B was produced by a majority of *L. acidophilus* strains and inhibits *L. delbrueckii* subsp. *lactis* and *bulgaricus* strains. *L. acidophilus* 11088 produces lactacin F, whose inhibitory spectrum includes the same host range of *L. delbrueckii* strains as lactacin B and, additionally, other *L. acidophilus* strains, *Lactobacillus fermentum*, and *Enterococcus faecalis*. Although these two bacteriocins share many properties in common, no DNA homology or amino acid similarities have been detected between them (Nettles et al., 1991; K. Milton and T. Klaenhammer, unpublished data). Therefore, the possibility exists that their suspected membrane-active functions could be related to their hydrophobic and nonpolar amino acid content, and not to the conservation of specific DNA or amino acid sequences.

IV. Genetic Organization of Bacteriocin Operons

A. Helveticin J

Helveticin J represents the least predominant group of bacteriocins produced by lactic acid bacteria. These bacteriocins, including helveticin J, acidophilucin A (Toba et al., 1991b), lactacin A and B (Toba et al., 1991c), and caseicin 80 (Rammelsberg and Radler, 1990), are large, inactivated by heat treatment (Table I), and their mechanisms of action have not yet been determined.

1. Cloning of Helveticin J Structural Gene (hlvJ)

At neutral pH, *L. helveticus* 481 produces a proteinaceous aggregate with a molecular weight in excess of 300,000 daltons, that is resolved by SDS-PAGE to a single 37,000 Da band with bacteriocin activity (Joerger and Klaenhammer, 1986). This purified protein was used to obtain polyclonal antibodies specific for helveticin J. Two recombinant phages producing helveticin J-lacZ fusion proteins were obtained (Joerger and Klaenhammer, 1990) by immunoscreening of a λ-gt11 expression library (Huynh et al., 1984). They contained DNA inserts of 350 and 600 bp, homologous to each other and specific to the strain producing helveticin J. These inserts, too small to encode the entire helveticin J protein, were found subsequently to be internal to the helveticin J structural gene. Therefore, the 600-bp cloned fragment was used to screen a second *L. helveticus* 481 genomic library prepared in λEMBL3. Southern hybridization analysis revealed that the helveticin J region was located on a 5.5-kb *Hin*dIII restriction fragment. Attempts to subclone this fragment into high-copy-number vectors in *E. coli* proved unsuccessful because of frequent plasmid deletions and rearrangements, suggesting that the helveticin J region was either unstable or its gene products were lethal for *E. coli*.

2. Expression of Helveticin J in Heterologous Hosts

A 3.8-kb *Bgl*II restriction fragment contained within the 5.5-kb *Hind*III restriction fragment was ligated into pGK12, a lactic acid bacteria-*E. coli* shuttle vector present in low copy number in *E. coli* (Kok et al., 1984). Five *Lactobacillus* hosts, closely related to *L. helveticus*, have been electroporated with the resulting recombinant plasmid (pTRK135): *L. gasseri* NCK101; *L. helveticus* 481-C; *L. fermentum* 1750; and *L. acidophilus* NCK64 and NCK89. Transformants were obtained only with *L. acidophilus* strains. *L. acidophilus* NCK64 (Laf$^-$, Laf^r), a derivative of the lactacin F producer *L. acidophilus* NCK88, does not produce the bacteriocin but retains immunity (Muriana and Klaenhammer, 1987). The isogenic derivative of *L. acidophilus* NCK88, *L. acidophilus* NCK89 (Laf$^-$, Laf^s), does not produce lactacin F or exhibit immunity, apparently because of a deletion of at least 2.2 kb in the lactacin F region (Muriana and Klaenhammer, 1991b). Alternatively, *L. acidophilus* NCK64 retains this 2.2-kb region, suggesting that loss of lactacin F production resulted from a point mutation in the structural gene that does not disturb lactacin F-related immunity functions.

The failure to express helveticin J in the NCK89 strain suggests that all of the essential processing and immunity information for the production of helveticin J are either not present in pTRK135 or are not expressed properly in *L. acidophilus*. In contrast, since the *L. acidophilus* NCK64 (pTRK135) transformants express helveticin J, some essential functions must be provided *in trans*, resulting in heterologous expression of helveticin J. However, only few of the transformants, obtained with the NCK64 strain, harbored the expected plasmid profile and expressed the helveticin J phenotype (Joerger and Klaenhammer, 1990). The other transformants incurred large deletions in pTRK135 (from 3 to 5 kb). Considering this high level of spontaneous deletions observed in pTRK135 cloned into *L. acidophilus*, possibly short deletions, rearrangements, or point mutations may have conferred genetic stability to pTRK135 in a heterologous host without affecting helveticin J expression.

3. Molecular Organization of the Helveticin J Region

Using restriction fragments cloned in pBluescript II KS + (Stratagene, La Jolla, California), the complete sequence of the helveticin J region present in pTRK135 (3.8-kb *Bgl*II fragment) has been determined in addition to a 0.7-kb flanking region. In the sequence presented here (Figure 1), three corrections are noted in the previously published helveticin J region sequence (Joerger and Klaenhammer, 1990; GenBank M30121):

> @ nt-1141 GGATCAT instead of GGATC*c*AT
> @ nt-1196 CGG*c*TT instead of CGGTT
> @ nt-3029 GCGAT instead of GCG*g*AT

```
                EcoRI
      GAATTCCAAAGATGCAGATCCTATTTATGTAGGAAAAAACAACTACAAATATGCTTTAAC  60
       N  S  K  D  A  D  P  I  Y  V  G  K  N  N  Y  K  Y  A  L  T

      GCATTATGAAACCTTCAAGGGCAAGACAATTAGTCCAGCTAAGGTTCAAAACGTTAAATT 120
       H  Y  E  T  F  K  G  K  T  I  S  P  A  K  V  Q  N  V  K  F

      TAGAGTTGAAAAGATAGTCAGATTCCACGGCAAAATTAGTGGTGCTCCATTGTACTTAGT 180
       R  V  E  K  I  V  R  F  H  G  K  I  S  G  A  P  L  Y  L  V

      GGCTTCTAAGGATAAGAAGTACAGCTGCTGGACTACGCAAGCAATGCTTCAATATTATTA 240
       A  S  K  D  K  K  Y  S  C  W  T  T  Q  A  M  L  Q  Y  Y  Y

      CTTCAATAGCAAGGGGATGCGCGGAGTAGTAAATCCATTGAAGAGAATTGCTAATAGAAG 300
       F  N  S  K  G  M  R  G  V  V  N  P  L  K  R  I  A  N  R  S

      TGCTGATAAGAATATTATTAGTCTAAAGAATAAACAAAATAAACGTGACTTTAATGCAGC 360
       A  D  K  N  I  I  S  L  K  N  Q  N  K  R  D  F  N  N  A

      AATGAAGGCTGCTAATAAGCTTAAGGGCAGTCAGAAGAAGTTTGTTGTAAATTCTTTGAA 420
       M  K  A  A  N  K  L  K  G  S  Q  K  K  F  V  V  N  S  L  K

      GCAACTTAAGAAAGATAACAATATTGGCGTTGAAGGCGACAACTTGCTTTTGTTTGGTTT 480
       Q  L  K  K  D  N  N  I  G  V  E  G  D  N  L  L  L  F  G  F
      End ORF1◄
      TTAAAGTGAAAAAAATACGTTAAAAGAACATAGAAAAACTGATGCATCTAAGCTATGAAG 540

                          ►End ORF6
      CTAGGAGCATCAGTTTTTTGTTACATTAGATCGTCATTGTCTTTAAAAAGTTGCTTGGTT 600
                             M  L  D  D  N  D  K  F  L  Q  K  T

      GCAGCGGTCATTCTTTCGGGGATGCCATAGTTCTGCATAAAGGTTCCTGGTTCTTGAATG 660
       A  A  T  M  R  E  P  I  G  Y  N  Q  M  F  T  G  P  E  Q  I
                                                       BglII
      TAGTCACCATTTTTGCAGGAATAGTCGTAAATAATCTTAAGAGAAAATAGATCTGCTTCC 720
       Y  D  G  N  K  C  S  Y  D  Y  I  I  K  L  S  F  L  D  A  E

      TCTTCTTCCGAGTTTTGTCTTGAAAAGCTAGGCCAATAGGCGATACCTGAATCACCTAAC 780
       E  E  E  S  N  Q  R  S  F  S  P  W  Y  A  I  G  S  D  G  L

      ATTAAGTGACCAATTTCGTGACCAATGATAAAAGGCAATTCATTAGGATTATGCCAATTA 840
       M  L  H  G  I  E  H  G  I  I  F  P  L  E  N  P  N  H  W  N

      GTATTGATTACCATTTTATGAGCATTCTTAAATGAAAGTGCTGGGTCGTATGGCTCTCCT 900
       T  N  I  V  M  K  H  A  N  K  F  S  L  A  P  D  Y  P  E  G

      TTAACCAAGATGTATGAAAGACCGTGCTTAAAAGCATAATTGAGTAAGTACTCAATTAAC 960
       K  V  L  I  Y  S  L  G  H  K  F  A  Y  N  L  L  Y  E  I  L
                 Start ORF6◄►End ORF5
      TCTTCTTTACCTAAGTTTCTCATTTCTTGTCTGCACGACCTTCCTTAATGTCATCTTCCA 1020
       E  E  K  G  L  N  R  M  K  K  D  A  R  G  E  K  I  D  D  E  M

      TTAGTCCTCGAATCATATTGAGATATCTCTCAGGAACGTTGTAACCATGATATGAATAAG 1080
       L  G  R  I  M  N  L  Y  R  E  P  V  N  Y  G  H  Y  S  Y  P

      GCTTTTCTTCGTCTAGTGGAATAGAAGTTGGTTGTCCTTTTGCAGGTATAGGGGTAGGAT 1140
       K  E  E  D  L  P  I  S  T  P  Q  G  K  A  P  I  P  T  P  D

      CATCAGTTTTACCTTTTAGGTAATCGATTGAAGTATTTAGTACTTCGGCAACGGCTTCTA 1200
       D  T  K  G  K  L  Y  D  I  S  T  N  L  V  E  A  V  A  E  L
```

Figure 1 Nucleotide sequence of the helveticin J operon region of *L. helveticus* 481. The derived amino acid sequence of ORFs is shown below the nucleotide sequence. *Eco*RI and *Bgl*II restriction sites are indicated. The -10 and -35 promoter regions of the helveticin J operon and ribosomal binding sites are underlined. Stem-loop structures in the promoter region and the terminator of the helveticin J operon are indicated by facing horizontal arrows.

7. Molecular Biology of Bacteriocins Produced by *Lactobacillus*   163

```
AGGCATCGCCACCAGGTCTTTTATTCTTCCAATTATAGATAGAATTAGTACCTAACCCAG 1260
  A  D  G  G  P  R  K  N  K  W  N  Y  I  S  N  T  G  L  G  A

CCTTATCATTCACTTCACGTAAGTTCATTTTATATTTTTTCGCTAATTCTTTAGTACGCT 1320
  K  D  N  V  E  R  L  N  M  K  Y  K  K  A  L  E  K  T  R  E
                                                Start ORF5◄
CTAATTCAATCATGGCAAGCATTCTCCGAATATTTTAAATTATTAAATAAGTATTTTTAC 1380
  L  E  I  M
        "-35"                    "-10"
TAAAAATGTATTGACAATATTAGTCTTTGGACGTAATATAAGAACTGTCAACGAGATATG 1440


AAAGAAATAAGAAAAAGCCTAGAAACACGTGTGTAAGTAAGAAAGGAATCATGTGCTTAA 1500
                  ----------►  ◄-----------
GGTTATTTCTTATACCATTATATTAGTATAAATACTAATATAAGTCAACGAATATTAGTA 1560
                         ---------►               ◄--------

TCTAGTTGATAAATTTTATTCAAAGAGGTGGTGCCAATGACTGTGCAATAAGGAAAAAGA 1620
-
               ►Start ORF2
GAGGTGAAGAAGAAATCATGGATATTCATGATTACGTTGAATTGATAGCTTTAGCGTTTT 1680
                   M  D  I  H  D  Y  V  E  L  I  A  L  A  F

GGGTTATTAGTGTTGTAAGTGTTGGTATCTTGAGTCATGTTCATTTTAAGAATAAGAGGC 1740
  W  V  I  S  V  V  S  V  G  I  L  S  H  V  H  F  K  N  K  R

TGGAACAGTTTCGTATTACTGCTGATGATTTGATGAAAAACTACGTTGGTTTGTACAACA 1800
  L  E  Q  F  R  I  T  A  D  D  L  M  K  N  Y  V  G  L  Y  N

AAGAAAGTTTAGCCAGCGATCAAAAAATCAATCGGATTGTCAATGCAGTAGTAGACGGAC 1860
  K  E  S  L  A  S  D  Q  K  I  N  R  I  V  N  A  V  V  D  G

TAGAAGCTAAAGGTTTTAAAGTGGAAGACCAAGATGTAAAGGATATTTTTGCAAAGGTCG 1920
  L  E  A  K  G  F  K  V  E  D  Q  D  V  K  D  I  F  A  K  V
                                    End ORF2◄
CAAAAATTATTAATGAAAATTCTTCTAAGTAGGAAGATAGAGATTTTTTCGGAGGTTTTA 1980
  A  K  I  I  N  E  N  S  S  K
  ►Start HlvJ
TTATGAAGCATTTAAATGAAACAACTAATGTTAGAATTTTAAGTCAATTTGATATGGATA 2040
   M  K  L  H  N  E  T  T  N  V  R  I  L  S  Q  F  D  M  T

CTGGCTATCAAGCAGTAGTTCAAAAAGGCAATGTAGGTTCAAAATATGTATATGGATTAC 2100
  T  G  Y  Q  A  V  V  Q  K  G  N  V  G  S  K  Y  V  Y  G  L

AACTTCGCAAAGGTGCTACTACTATCTTCCGTGGTTACCGTGGAAGTAAAATTAATAACC 2160
  Q  L  R  K  G  A  T  T  I  L  R  G  Y  R  G  S  K  I  N  N

CTATTCTTGAATTATCTGGTCAAGCAGGTGGTCACACACAGACATGGGAATTTGCTGGTG 2220
  P  I  L  E  L  S  G  Q  A  G  G  H  T  Q  T  W  E  F  A  G

ATCGTAAAGACATTAATGGTGAAGAAAGAGCAGGTCAATGGTTTATAGGTGTTAAACCAT 2280
  D  R  K  D  I  N  G  E  E  R  A  G  Q  W  F  I  G  V  K  P

CGAAAATTGAAGGAAGCAAAATTATTTGGGCAAAGCAAATTGCAAGAGTTGATCTTAGAA 2340
  S  K  I  E  G  S  K  I  I  W  A  K  Q  I  A  R  V  D  L  R

ATCAAATGGGACCTCATTATTCAAATACTGACTTTCCTCGATTATCCTACTTGAATCGCG 2400
  N  Q  M  G  P  H  Y  S  N  T  D  F  P  R  L  S  Y  L  N  R

CCGGTTCTAATCCATTTGCTGGTAATAAGATGACGCATGCCGAAGCCGCAGTATCACCTG 2460
  A  G  S  N  P  F  A  G  N  K  M  T  H  A  E  A  A  V  S  P
```

Figure 1 (*Continued*)

```
ATTATACTAAGTTTTTAATTGCTACTGTTGAAAATAACTGTATTGGTCATTTTACTATAT 2520
 D   Y   T   K   F   L   I   A   T   V   E   N   N   C   I   G   H   F   T   I

ACAATTTAGATACAATTAATGAAAAACTTGATGAAAAGGGAAATAGTGAAGATGTTAATC 2580
 Y   N   L   D   T   I   N   E   K   L   D   E   K   G   N   S   E   D   V   N

TCGAAACTGTTAAATACGAAGATAGTTTTATCATTGATAATTTATATGGTGATGATAATA 2640
 L   E   T   V   K   Y   E   D   S   F   I   I   D   N   L   Y   G   D   D   N

ATTCTATTGTAAATTCAATTCAAGGGTATGATTTGGATAATGATGGAAATATTTATATTT 2700
 N   S   I   V   N   S   I   Q   G   Y   D   L   D   N   D   G   N   I   Y   I

CCAGTCAAAAAGCGCCAGATTTTGATGGCTCTTATTATGCACATCATAAGCAGATTGTTA 2760
 S   S   Q   K   A   P   D   F   D   G   S   Y   Y   A   H   H   K   Q   I   V

AGATTCCATATTATGCTCGGTCTAAAGAAAGCGAAGACCAATGGAGAGCTGTAAATTTAA 2820
 K   I   P   Y   Y   A   R   S   K   E   S   E   D   Q   W   R   A   V   N   L

GCGAATTCGGTGGCTTGGATATTCCAGGTAAACATAGTGAAGTTGAAAGCATCCAAATTA 2880
 S   E   F   G   G   L   D   I   P   G   K   H   S   E   V   E   S   I   Q   I

TTGGTGAGAATCATTGTTACTTAACTGTTGCATATCATTCTAAAAATAAAGCGGGTGAAA 2940
 I   G   E   N   H   C   Y   L   T   V   A   Y   H   S   K   N   K   A   G   E
                                                           End HlvJ◄
ATAAAACTACTTTGAATGAGATTTATGAATTATCTTGGAATTAGATTCTTGTTAGTGGTC 3000
 N   K   T   T   L   N   E   I   Y   E   L   S   W   N

TCGATTTAGATATAAACTAACAAAAGCGATGAAATATTCATTATTGAAATTCATCGCTTT 3060
         ------------   --->   ◄---   ----------

TTATTTTTAATTAAATTATTGGATATACTTATAATATATATTGCTGGATATATTGCTGGG 3120
-
                                                    ►Start ORF4
ATAAGAGTAAAATAATTATAGGCATTATTTCTAAATTAAAAGGACAATTATTATGATAAA 3180
                                                         M   I   K

AAACAAGATTATATCAGCTTCAATTGCAGCATTAATGGCTGTAAGTCCAGTGCTTCCACT 3240
 N   K   I   I   S   A   S   I   A   A   L   M   A   V   S   P   V   L   P   L

TAGCTCACAGGCTCATACGGTTCAAGCTGCAGATAATTCTGTCAGAAAGACAGTTATGCA 3300
 S   S   Q   A   H   T   V   Q   A   A   D   N   S   V   R   K   T   V   M   H

TAATTCAATTGCTTATGATAAAGATGGCAATTCAACAGGTCAAAAGTATTACGCTTACGG 3360
 N   S   I   A   Y   D   K   D   G   N   S   T   G   Q   K   Y   Y   A   Y   G

ATCAATCAGTGTTGATCCAACCCCTGTAACTATTAACGGTAACCAATATTACAAGATTTC 3420
 S   I   S   V   D   P   T   P   V   T   I   N   G   N   Q   Y   Y   K   I   S

AGGTAAAAACCAATATGTTAAAGTAACTAATATTGATGGTGTAAGACGTAGAGTAACCCA 3480
 G   K   N   Q   Y   V   K   V   T   N   I   D   G   V   R   R   R   V   T   H

CAATGCCTATATTTATCGTACTTCTACTCAAAAAACGCCTTACGGTATGACTGCAAGCAG 3540
 N   A   Y   I   Y   R   T   S   T   Q   K   T   P   Y   G   M   T   A   S   S

TAAGAAATGGAAGTTATACAAAGGCGAAATAGTAACTACTTATGGTGGCTATTACACCTT 3600
 K   K   W   K   L   Y   K   G   E   I   V   T   T   Y   G   G   Y   Y   T   F

TAAAAATGGTAAGCACTACTTCAAGGTAGGCGGACCAAGAAAGCAATATGTTAGAACTGC 3660
 K   N   G   K   H   Y   F   K   V   G   G   P   R   K   Q   Y   V   R   T   A

TAACTTAGGTCCAGTTATCGGAACTAATACTTCTGGTAAATATGAAACCACTGTAACAGT 3720
 N   L   G   P   V   I   G   T   N   T   S   G   K   Y   E   T   T   V   T   V
```

Figure 1 (*Continued*)

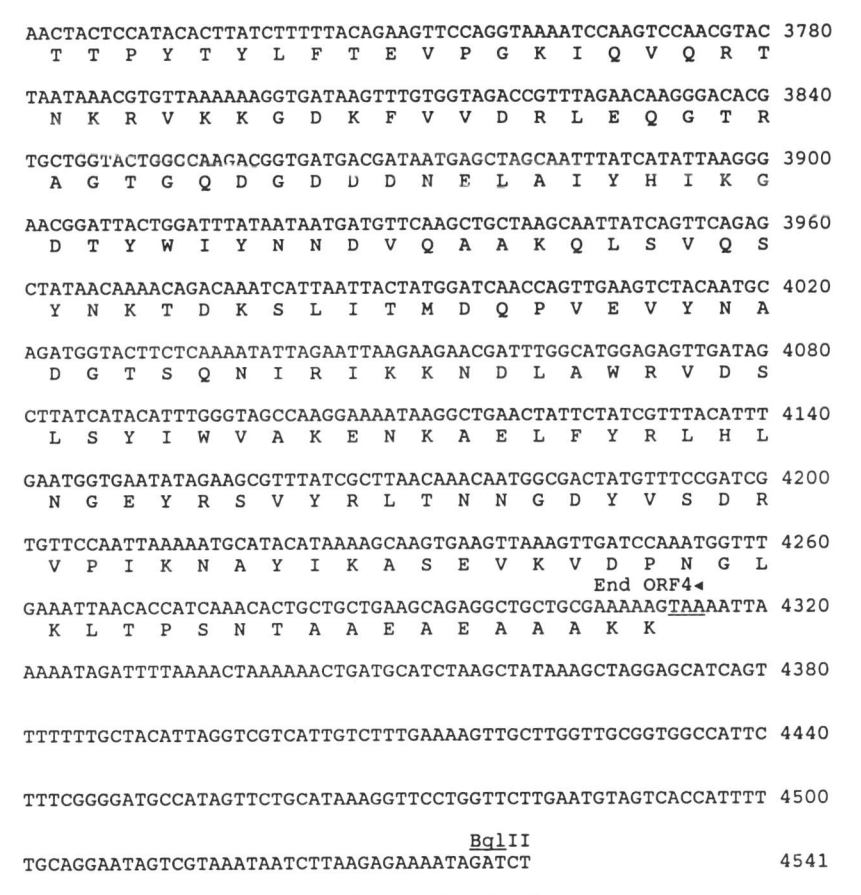

7. Molecular Biology of Bacteriocins Produced by *Lactobacillus*

```
AACTACTCCATACACTTATCTTTTTACAGAAGTTCCAGGTAAAATCCAAGTCCAACGTAC 3780
  T  T  P  Y  T  Y  L  F  T  E  V  P  G  K  I  Q  V  Q  R  T

TAATAAACGTGTTAAAAAAGGTGATAAGTTTGTGGTAGACCGTTTAGAACAAGGGACACG 3840
  N  K  R  V  K  K  G  D  K  F  V  V  D  R  L  E  Q  G  T  R

TGCTGGTACTGGCCAAGACGGTGATGACGATAATGAGCTAGCAATTTATCATATTAAGGG 3900
  A  G  T  G  Q  D  G  D  D  D  N  E  L  A  I  Y  H  I  K  G

AACGGATTACTGGATTTATAATAATGATGTTCAAGCTGCTAAGCAATTATCAGTTCAGAG 3960
  D  T  Y  W  I  Y  N  N  D  V  Q  A  A  K  Q  L  S  V  Q  S

CTATAACAAAACAGACAAATCATTAATTACTATGGATCAACCAGTTGAAGTCTACAATGC 4020
  Y  N  K  T  D  K  S  L  I  T  M  D  Q  P  V  E  V  Y  N  A

AGATGGTACTTCTCAAAATATTAGAATTAAGAAGAACGATTTGGCATGGAGAGTTGATAG 4080
  D  G  T  S  Q  N  I  R  I  K  K  N  D  L  A  W  R  V  D  S

CTTATCATACATTTGGGTAGCCAAGGAAAATAAGGCTGAACTATTCTATCGTTTACATTT 4140
  L  S  Y  I  W  V  A  K  E  N  K  A  E  L  F  Y  R  L  H  L

GAATGGTGAATATAGAAGCGTTTATCGCTTAACAAACAATGGCGACTATGTTTCCGATCG 4200
  N  G  E  Y  R  S  V  Y  R  L  T  N  N  G  D  Y  V  S  D  R

TGTTCCAATTAAAAATGCATACATAAAAGCAAGTGAAGTTAAAGTTGATCCAAATGGTTT 4260
  V  P  I  K  N  A  Y  I  K  A  S  E  V  K  V  D  P  N  G  L
                                                End ORF4◄
GAAATTAACACCATCAAACACTGCTGCTGAAGCAGAGGCTGCTGCGAAAAAGTAAAATTA 4320
  K  L  T  P  S  N  T  A  A  E  A  E  A  A  A  K  K

AAAATAGATTTTAAAACTAAAAAACTGATGCATCTAAGCTATAAAGCTAGGAGCATCAGT 4380

TTTTTTGCTACATTAGGTCGTCATTGTCTTTGAAAAGTTGCTTGGTTGCGGTGGCCATTC 4440

TTTCGGGGATGCCATAGTTCTGCATAAAGGTTCCTGGTTCTTGAATGTAGTCACCATTTT 4500
                           BglII
TGCAGGAATAGTCGTAAATAATCTTAAGAGAAAATAGATCT                    4541
```

Figure 1 (*Continued*)

The general organization of the helveticin J region is presented in Figure 2A. The *Bgl*II fragment spans four complete open reading frames (ORF) named ORF2, 3, 4, and 5 and partially ORF6. Among them ORF2, 3, 4, and 5 have a high calculated coding probability (Fickett, 1982) of 92%. ORF3 (999 bp) defines the helveticin J structural gene (*hlv*J). First, it encodes a protein of the expected size of purified helveticin J, 37,511 daltons (Joerger and Klaenhammer, 1986). Second, the reactive λ-gt11 clones, which contained inserts of 350 and 600 bp from the helveticin J producer, *L. helveticus* 481, were internal to ORF3. ORF2 (315 bp), which ends 30 bp upstream from the *hlv*J gene, could encode an 11,809-Da protein. The function of the putative ORF2 protein remains unknown at this time, but its position relative to *hlv*J suggests that it is an important component of an apparent bacteriocin operon. In addition to ORF2, three additional complete ORFs surrounding the *hlv*J gene have been defined. Downstream, ORF4 (1140 bp) starts 172 bp behind the *hlv*J gene and ends at 229 bp from the *Bgl*II site. Upstream, at 296 bp on the opposite DNA strand starts ORF5 (360 bp),

A

Helveticin J Operon

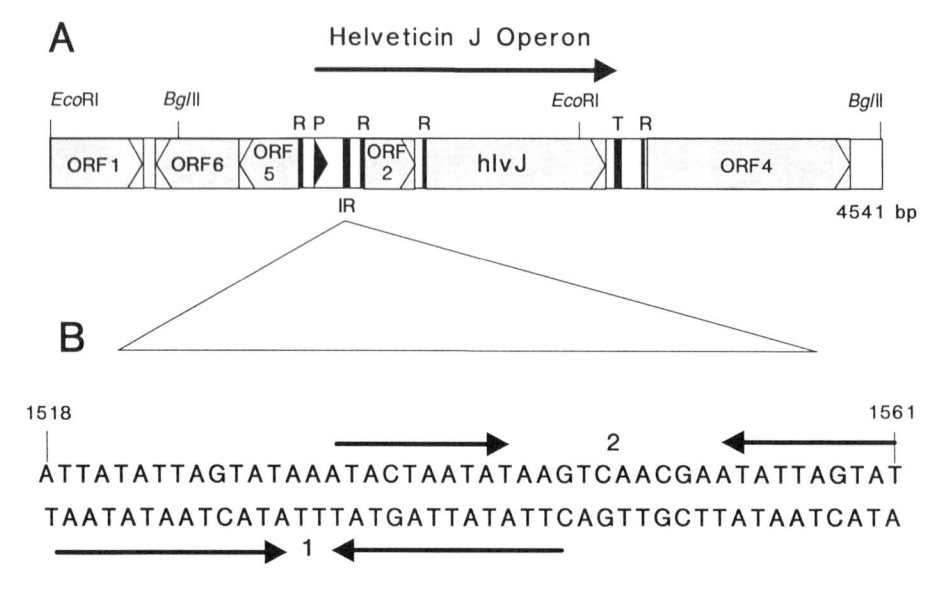

```
EcoRI      BglII                     R P  R    R           EcoRI         T R                          BglII
  |          |                                               |                                          |
  [ ORF 1  ][  ORF6  ][ORF 5][  ][ORF 2][        hlvJ        ][  ][         ORF4          ]
                             IR
                                                                                                  4541 bp
```

B

```
        1518                                                                        1561
         |                      ──────────▶              2      ◀──────────          |
         ATTATATTAGTATAAATACTAATATAAGTCAACGAATATTAGTAT
         TAATATAATCATATTTATGATTATATTCAGTTGCTTATAATCATA
         ──────────▶         1        ◀──────────
```

Figure 2 (A) Organization of the helveticin J operon within a 4541-bp *Eco*RI-*Bgl*II restriction fragment from *L. helveticus* 481. P, promoter; T, ρ-independent terminator; R, ribosomal binding site; IR shows the position of two inverted repeats that could form two stem-loop structures. (B) Sequence of the region defining two palindromic structures within the helveticin J operon promoter region. The free energy of folding for hairpin 1 is −11.4 kcal and for hairpin 2, −8.4 kcal.

which overlaps by one base pair ORF6 (420 bp). It remains to be defined what role, if any, these genes play in the immunity, processing, or expression of helveticin J.

For all these open reading frames, primary and secondary structures of the deduced protein sequences have been analyzed (C. Fremaux and T. Klaenhammer, unpublished data). When compared with the Swiss-prot data bank, none of them has significant homology with previously described proteins. Secondary structure analysis demonstrates that all the deduced proteins, except the predicted product of ORF2, are peripheral globular proteins. The protein encoded by ORF2 could be classified as an integral membrane protein. It contains, at its amino-terminus, a highly hydrophobic sequence predicted as a membrane-spanning segment using the method of Klein et al. (1985). Each ORF, except ORF6, is preceded by a putative ribosomal binding site. Surrounding ORF2-*hlv*J we have found putative expression signals: a typical promoterlike sequence upstream from ORF2 and a ρ-independent terminator (−18.7 kcal/mole) 37 bp below the *hlv*J stop codon. These data strongly suggest that these two genes are organized in an operonlike structure. Northern blot analysis of *L. helveticus* RNA has confirmed this hypothesis (C. Fremaux and T. Klaenhammer, unpublished data).

No promoters or terminators were found for the other ORFs. An additional feature of the ORF2-*hlv*J promoter region is a set of two overlapped, inverted repeats (−11.4 and −8.4 kcal/mole) positioned between the putative "−35 and −10 regions" and the ORF2 ribosomal binding site (Figure 2B). Holo et al. (1991) had also pointed out an inverted repeat in the regulatory region of lactococcin A (LCN-A), a bacteriocin produced by *Lactococcus lactis* subsp. *cremoris*. Palindromic structures in regulatory regions of a number of inducible *E. coli* bacteriocins have been shown to be SOS-boxes (van den Elzen et al., 1982; Akutsu et al., 1989). These regions serve as binding sites for *lex*A repressor, which blocks transcription of the bacteriocin operons. Under inducible conditions that activate SOS functions, the bacteriocin operon is transcribed. In the case of helveticin J, it is not clear what role, if any, the inverted repeats play in the regulation or expression of helveticin J. Neither helveticin J nor lactococcin A have been reported to be inducible.

B. Lactacin F

Lactacin F was the first nonlanthionine bacteriocin characterized from lactic acid bacteria for which both DNA and protein sequence information were available. Originally the lactacin F producer strain, *L. acidophilus* 11088 (NCK88), was isolated in a survey for bacteriocins produced by *L. acidophilus*, which had an extended spectrum of activity beyond the *L. delbrueckii* species to include two other residents of the gastrointestinal tract, *L. fermentum* and *Enterococcus faecalis* (Barefoot and Klaenhammer, 1983). One of the most interesting features of lactacin F is its heat stability. Since it retains full activity after autoclaving for 15 minutes at 121°C, the bacteriocin can be easily used as a selective component in bacteriological media. This characteristic of lactacin F was employed to select transconjugants in matings, where the genetic determinants for production (Laf$^+$) and immunity (Laf^r) were conjugally transferred as an episomal element (Muriana and Klaenhammer, 1987).

1. Cloning of Genes for Lactacin F Production and Expression

Lactacin F has been purified and biochemically characterized as a 6.3-kDa hydrophobic peptide (Muriana and Klaenhammer, 1991a). Compositional analysis originally estimated the peptide to be 54–57 amino acids in length and protein sequencing defined 25 amino acids at the N-terminus of the active bacteriocin. A 63-mer oligonucleotide probe was deduced and used to select a clone bearing a 2.3-kb *Eco*RI DNA fragment from a genomic library of the lactacin F producer, *L. acidophilus* 11088 (Muriana and Klaenhammer, 1991b). This fragment was subcloned onto an *E. coli-Lac-*

tobacillus shuttle vector and the recombinant plasmid (pTRK162) was introduced into two expression hosts that were deficient in lactacin F production, NCK64 (Laf⁻ Laf^r) and NCK89 (Laf⁻ Laf^S). With the introduction of pTRK162, transformants of both strains produced lactacin F. However, the lactacin F-producing colonies of NCK89 were small, variable in size, and produced significantly less bacteriocin than pTRK162 transformants of NCK64. Recent work (C. Ahn and T. Klaenhammer, unpublished data) has shown that repeated propagation of the Laf⁺ NCK89 (pTRK162) clones under bacteriocin-producing conditions (pH 7.6) eventually results in loss of culture viability. In contrast, NCK89 (pTRK162) shows healthy colony formation and can be propagated continuously under conditions where lactacin F is not produced (pH 5.5 or less). The Laf^r expression host (NCK 64) supports excellent lactacin F production and can be propagated without adverse effects under optimal conditions for bacteriocin production. Collectively, these observations indicate that the 2.3-kb fragment that encodes genetic determinants for lactacin F production does not express, or fully express, immunity to the bacteriocin, suggesting that functional complementation occurs *in trans* in the Laf^r expression host NCK64. However, it remains possible that the immunity gene is expressed and the gene product is present in NCK89, but it fails to function properly in this host due to host-specific factors.

2. Molecular Organization of the Lactacin F Operon

The complete DNA sequence of the lactacin F structural gene determined previously (Muriana and Klaenhammer, 1991b) has been extended to confirm and define flanking upstream and downstream regions, spanning the whole 2312-bp *Eco*RI fragment (Figure 3). During extended sequencing, four corrections in the previously sequenced 873-bp region (Muriana and Klaenhammer, 1991b) were made:

@ nt-426 (30) AACA instead of AAA*g*CA
@ nt-1011 (616) TGGCG instead of TGG*g*CG
@ nt-1235 (840) TC*t*AA instead of TC*a*AA
@ nt-1244 (848) TAT*t*G instead of TATG

The numbers in parentheses are the nucleotide positions from the 873-bp sequence published previously (Muriana and Klaenhammer, 1991b; GenBank M57961). The correction at nt-1011 affects the reading frames of ORFX and Y, predicted by the previous sequence data. This alters the predicted products of ORFX from 32 to 62 amino acids and ORFY from 59 to 44 amino acids.

DNA sequence analysis on the 2312-bp *Eco*RI fragment has currently identified four complete open reading frames (LAF-ORF, ORFX, ORFY, and ORFZ) and two incomplete ORFs, 1 and 2. Further subcloning work and mRNA analysis (C. Ahn and T. Klaenhammer, unpublished data) have shown that the four open reading frames span 1046 bp in length and are

EcoRI
GAATTCTATTCCTATAAGTATAGAAATTTTTCGACTATGATTATTATTCCAGCAGCTATT 60
 E F Y S Y K Y R N F S T M I I I P A A I

TTTGTCTTACTTTTATTTGTTGGATCTTTTTTTGCAGTTAGGCAGAATACAGTCTCTTCG 120
 F V L L L F V G S F F A V R Q N T V S S

GTTGGCGTTGTTGAACCAACGGTTGTAATTAAGCGGAAAAATGTTAACTATGATGAGGGA 180
 V G V V E P T V V I K R K N V N Y D E G

CAAGTCGTTACGAAATATGGTCAAAAGTGGGTAGCTCATGTTGATCAAGGTGGTGGGATT 240
 Q V V T K Y G Q K W V A H V D Q G G G I

AGTTTAATGCCTGTTATGAAACCTAAAAGCAAAGTAAAAATTGTAACTTATGTGACAAGT 300
 S L M P V M K P K S K V K I V T Y V T S

GATAAAATCTCGACTATTAAAAAGGGACAGTCGTTAACCTTTTCTGTACCTACTGGAGAT 360
 D K I S T I K K G Q S L T F S V P T G D

GGCTTAACTAGACACTTGACTGGTAAGGTAGAGAAAATTGGAGTCTATCCTGTGAACATG 420
 G L T R H L T G K V E K I G V Y P V N M

AATAAACAAAATATATACGAAATTATTTCGACTGCAAAAGTTAATGATGAGAATATTAAA 480
 N K Q N I Y E I I S T A K V N D E N I K

TATGGGATGCAGGGGAATGTGACGATAATGACCGGTAGGAGTACGTATTTAAAATATCTT 540
 Y G M Q G N V T I M T G R S T Y L K Y L
              End ORF1◄   "-45"
TTGGATAAGGTAAGGAATAATAAGTAAAAAATACTAATTTAGTTAATAAAAGTAATTTTA 600
 L D K V R N N K -
"-35"   "-10"   "-5"
GACACAAATAGAACAATATTGGTCAATTTTATATCTTAAGATGATAACTTTAGTAAGCTA 660

                          ►Start laf
TGCATATAAATAAAATTTTAGGAGGTTTCTATCATGAAACAATTTAATTATTTATCACAT 720
                                  M K Q F N Y L S H

AAAGATTTAGCAGTCGTTGTTGGTGGAAGAAATAATTGGCAAACAAATGTGGGAGGAGCA 780
 K D L A V V V G G R N N W Q T N V G G A

GTGGGATCAGCTATGATTGGGGCTACAGTTGGTGGTACAATTTGTGGACCTGCATGTGCT 840
 V G S A M I G A T V G G T I C G P A C A
                    ▲
GTAGCTGTGGCCATTATCTTCCTATTTTATGGACAGGGGTTACAGCTGCAACAGGTGGT 900
 V A C A H Y L P I L W T G V T A A T G G
                    End laf◄   ►St. ORFX ►Start ORFY
TTTGGCAAGATAAGAAAGTAGGATTTTGACAATGAAATTAAATGACAAACAATTATCAAA 960
 F G K I R K   M K L N D K E L S K
                                   M T K N Y Q

GATTGTTGGTGGAAATCGATGGGGAGATACTGTTTTATCAGCTGCTAGTGGCGCAGGAAC 1020
  I V G G N R W G D T V L S A A S G A G T
 R L L V E I D G E I L F Y Q L L V A Q E
                                                 End ORFY◄
TGGTATTAAAGCATGTAAAAGTTTTCGCCCATGGGGAATGGCAATTTGTGGTGTAGGAGG 1080
   G I K A C K S F G P W G M A I C G V G G
 L V L K H V K V L A H G E W Q F V V -
                                       End ORFX◄
TGCAGCAATAGGAGGTTATTTTGGCTATACTCATAATTAAACTATAGTCAATTAAAGTAA 1140
  A A I G G Y F G Y T H N -

Figure 3 Nucleotide sequence of the lactacin F operon region and flanking region. The derived amino acid sequence of open reading frames is shown below the corresponding nucleotide sequence. The promoter regions of the lactacin F operon and ribosomal binding sites are underlined. Stem-loop structures in the transcription terminator of the lactacin F operon are indicated by facing horizontal arrows. (*Figure continues*)

```
AACAGTGATGATTTGATATTTAGCACTGCATTACTTTATTCATAGATTCATTAGTAGGTA   1200
         ►Start ORFZ
GTTTAAATGACTAAACATTATAAAATTATTGGTCTAAGGATATTGTCATGGGTAATTACG   1260
      M  T  K  H  Y  K  I  I  G  L  R  I  L  S  W  V  I  T
ATTACTGGATTGATAATATTTATCGGAAATGTTCATGAGTATGGCTTACATTTTACTTAT   1320
 I  T  G  L  I  I  F  I  G  N  V  H  E  Y  G  L  H  F  T  Y
AATCAAGTATTGGCAATAATTATTGTGATTTTACTTTTAGTTACTACTATGTATTTTTCA   1380
 N  Q  V  L  A  I  I  I  V  I  L  L  L  V  T  T  M  Y  F  S
GTTACACCCAAATTATTGAAATGGAATGAAAAATATGAGCTTGTATGGTGGTTATGCTAT   1440
 V  T  P  K  L  L  K  W  N  E  K  Y  E  L  V  W  W  L  C  Y
TTATGTGCTCCTATATATCTCTTTTTAACTAATTTATATAATTCAACAGATCTTGGTTAT   1500
 L  C  A  P  I  Y  L  F  L  T  N  L  Y  N  S  T  D  L  G  Y
ACTATCAAGTTTTGGCTTTTCTTTGGTGGAGGAGCAGCTTTAATAATCATAAGTAAGTAT   1560
 T  I  K  F  W  L  F  F  G  G  G  A  A  L  I  I  I  S  K  Y
         End ORFZ◄
ATTTTGAAAAACAAAAAATAAAGTCTAAATAATTTATCTATTTCAATCGTATATAAATAA   1620
 I  L  K  N  K  K  -

AAATTAGAATACCAAATTGGTCTGTAAAACGACTAATTTGGTATTTTTATCTTATGAAA    1680
     -----------  --►       ◄--  -----------

TTTTGAACCTGATAAAATTTTTAATAAGGTAAAAAAGAGCACGTGATAAAAATGGAAATA   1740

ATTAAAAATAATGAAAAGATAGCATCAAATAAAGAAAAATGGATAACACTAATTCTAGCT   1800

GCTGCTTTTGGATATTCAATGAATAAATTGTTTTCGCTGATTGTAAAATTAAATTGAACA   1860

CTTGATAAAAAATAAAAAGTCCATTCGGACCTGTTTGATAGAATGTTAATAACCACAAAA   1920
                 ►Start ORF2
ACTTTCTATTGGAGGTCAAAATGGACTCTTTACATTCTACCATGAACCAGCACGTTAAAG   1980
                      M  D  S  L  H  S  T  M  N  Q  H  V  K
GCAAGCATTTATCATTTGAAGAGCGAGTTATTATTCAATTGCGTTTGAAAGATGGCTATT   2040
 G  K  H  L  S  F  E  E  R  V  I  I  Q  L  R  L  K  D  G  Y
CTTTGCGTGCAATTGCCCGTGAACTTAACTGTTCTCCTTCTACTATCAGCTATGAGGTTA   2100
 S  L  R  A  I  A  R  E  L  N  C  S  P  S  T  I  S  Y  E  V
AGCGTGGCACTGTAAAACTGTATCATGGTAAAGTCAAAAAATATAAGGCTACTCAAGGGC   2160
 K  R  G  T  V  K  L  Y  H  G  K  V  K  K  Y  K  A  T  Q  G
ATGATGCATATAAAGCTCATCGTAAAAATTGTGGGCGCAAATCAGACTTTCTCAGGAAAG   2220
 H  D  A  Y  K  A  H  R  K  N  C  G  R  K  S  D  F  L  R  K
CTCAATTCATGCGCTATGTCCACAAGCATTTTTTTAAAGATGGCTGGTCGCTTGATGTGT   2280
 A  Q  F  M  R  Y  V  H  K  H  F  F  K  D  G  W  S  L  D  V
                       EcoRI
GCAGTAATCGTGCTACTGCTGTTGGCGAATTC                              2310
 C  S  N  R  A  T  A  V  G  E  F
```

Figure 3 (*Continued*)

7. Molecular Biology of Bacteriocins Produced by *Lactobacillus*

organized in an apparent operon structure with a putative promoter (Plaf) and a ρ-independent transcription terminator (Figure 4). The Plaf promoter drives the promoterless chloramphenicol acetyl transferase (CAT) gene in pGKV210 (van der Vossen et al., 1985) resulting in expression of chloramphenicol resistance when it is cloned in *E. coli* and various lactobacilli. The terminator (ΔG = −21.0 kcal/mole) blocks expression of the CAT gene when positioned between the lactacin F promoter and the promoterless CAT gene (C. Ahn and T. Klaenhammer, unpublished data).

The DNA sequence of LAF-ORF predicts that the gene product is a 75-amino-acid peptide with the following sequence:

> 1 Met Lys Gln Phe Asn Tyr Leu Ser His Lys Asp Leu Ala Val Val Val Gly Gly* **Arg Asn Asn Trp Gln Thr Asn Val Gly Gly Ala Val Gly Ser Ala Met Ile Gly Ala Thr Val Gly Gly Thr Ile** Cys Gly Pro Ala Cys Ala Val Ala Gly Ala His Tyr Leu Pro Ile Leu Trp Thr Gly Val Thr Ala Ala Thr Gly Gly Phe Gly Lys Ile Arg Lys 75

The sequence of the first 25 amino acids of mature lactacin F, as determined from purified bacteriocin (Muriana and Klaenhammer, 1991a), was identified following the Arg (19) residue (underlined above). The initial data supported the following conclusions:

1. lactacin F is translated as a prepeptide with an 18-amino-acid N-terminal extension;
2. the bacteriocin is posttranslationally processed by cleavage at a specific site [Val-Val-Gly-Gly * Arg (+1)];
3. the mature hydrophobic peptide is 57 amino acids in length and predicts the presence of a transmembrane helix.

The 18 amino acids comprising the N-terminal extension of the lactacin F prepeptide have several features that resemble signal sequences (von Heijne, 1986): a positively charged residue at the N-terminus (Arg); a central region of hydrophobic amino acids (Leu–Ala–Val–Val–Val); and a C-terminal cleavage site that shows small uncharged amino acids at positions −1 (Gly) and −3 (Val). It does not appear, however, that the N-terminal extension functions as a signal sequence. Secondary structure predictions

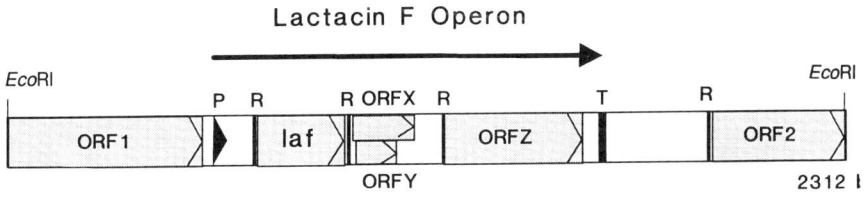

Figure 4 Organization of the lactacin F operon within a 2312-bp *Eco*RI restriction fragment from *L. acidophilus* NCK88. P, promoter; T, ρ-independent terminator; R, ribosomal binding site.

via Garnier et al. (1978) identified an α-helical structure at the N-terminus and a β turn at the processing site, which could place the splicing region into juxtaposition with a peptidase. The "processed" lactacin F peptide retains two hydrophobic β-sheets. Hydropathy plots based on the methods of Kyte and Doolittle (1982) and Klein et al. (1985) are shown in Figure 5 and illustrate the hydrophobic character of the N-terminal sequence and the lactacin F peptide itself. The processing site, defined by the arginine residue at position 19, is denoted by the negative hydropathy index in this region. The small size of the lactacin F peptide and the extent of β structure implicate the cell membrane as the probable target of this antimicrobial peptide. Purified lactacin F was active against *Lactobacillus* and *Enterococcus* protoplasts in preliminary experiments (Muriana and Klaenhammer, 1991a) and we are currently investigating this further.

Following the gene encoding pre-lactacin F are two superimposed open reading frames, ORFX (62 amino acids) and Y (44 amino acids). The actual translation products produced from one or both ORFs have not yet been evaluated. However, the computer-analyzed hydropathogram of ORFX and Y indicates both ORFs would encode hydrophobic proteins, suggesting their interaction sites would be located in the cytoplasmic membrane. Recent subcloning and expression studies have shown that disruption of ORFX/Y eliminates the Laf+ phenotype (C. Ahn and T. Klaenhammer, unpublished data). Therefore, although their specific roles remain to be

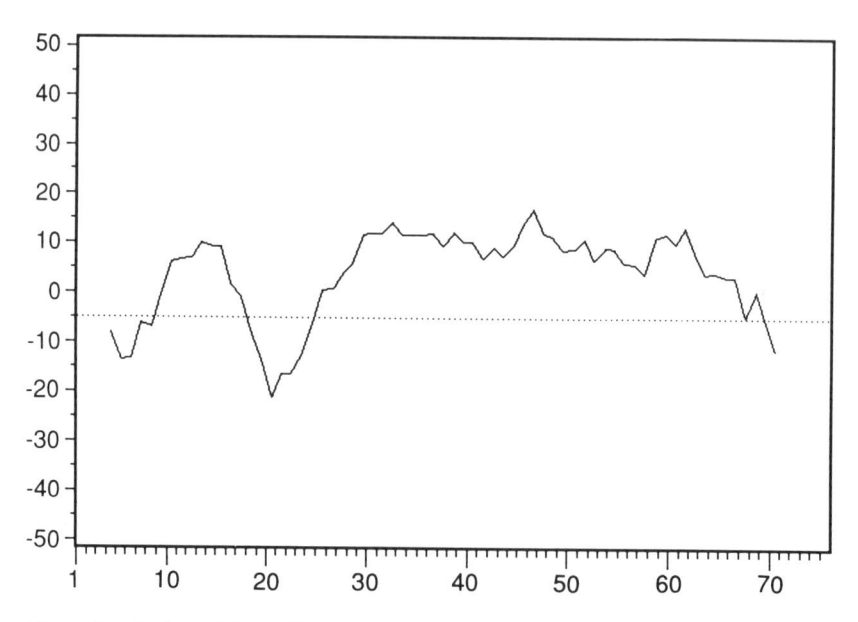

Figure 5 Hydropathic profile of the pre-lactacin F peptide computed using the PC GENE-based programs of Kyte and Doolittle (1982) and Klein et al. (1985). Total number of amino acids is 75. Computed using an interval of 9 amino acids. (GRAVY = 2.7.)

further identified, one or both of these gene products are essential for expression of the Laf⁺ phenotype.

Expanded sequencing efforts downstream from ORFX/Y have recently identified a fourth open reading frame (ORFZ) that could encode a protein of 124 amino acids. The role of ORFZ remains unknown at this time. However, hydropathicity of the computer-simulated gene product of ORFZ predicts a membrane-integrated protein that has four transmembrane segments in its structure. This suggests that ORFZ could encode a membrane protein that may function in immunity, processing, or export of lactacin F. ORFZ is followed by a ρ-independent terminator that forms a stem-loop structure of 38 nucleotides and is followed by a poly-T tail.

The general organization of the genes, expression signals, and terminator over 1046 bp establishes an operonlike structure. This structure is most similar to the lactococcin determinants (now designated *lcn*Ma, *lcn*Mb, *lcn*M; van Belkum et al., 1992) organized in an operonlike structure of ORF-A1, ORF-A2, and ORF-A3 (van Belkum et al., 1991). As with lactacin F, both ORF-A1 and ORF-A2 were required for lactococcin production. In contrast to the ORFZ of the lactacin F operon, ORF-A3 appeared to be responsible for bacteriocin immunity. Additional experiments are being conducted to determine the roles of ORFX/Y and ORFZ in lactacin F production, immunity, and processing. The evidence to date strongly suggests that lactacin F, like helveticin J and the lactococcins, is organized in an operon structure, and a number of gene products from the operon are likely involved in the expression, processing, export, and immunity functions of this *Lactobacillus* bacteriocin.

V. Common Processing Sites in Peptide Bacteriocins

There are six peptide bacteriocins from lactic acid bacteria where the DNA or protein sequences are now described. These include the published sequences for lactacin F (Muriana and Klaenhammer, 1991b) and the lactococcins (van Belkum et al., 1991; 1992), pediocin PA-1 (Gonzalez and Kunka, 1987; Marugg et al., 1991; van Belkum, 1991), and leucocin A-UAL 187 produced by *Leuconostoc gelidum* UAL 187 (Hastings et al., 1991). A comparison of the N-terminal extensions of lactacin F, leucocin A, pediocin PA-1, and lactococcins A, B, and Ma is shown in Figure 6. The arrow denotes the processing site for the bacteriocins for which both the DNA sequence and amino acid sequence have been independently determined. The residues boxed in the lactacin F sequence are found in the identical position in one or more of the other five peptide bacteriocins. In all six cases there are N-terminal extensions of either 18 or 21 amino acids in length with a methionine and lysine at the amino terminus of the leader peptide.

Common Processing Sites in Peptide Bacteriocins

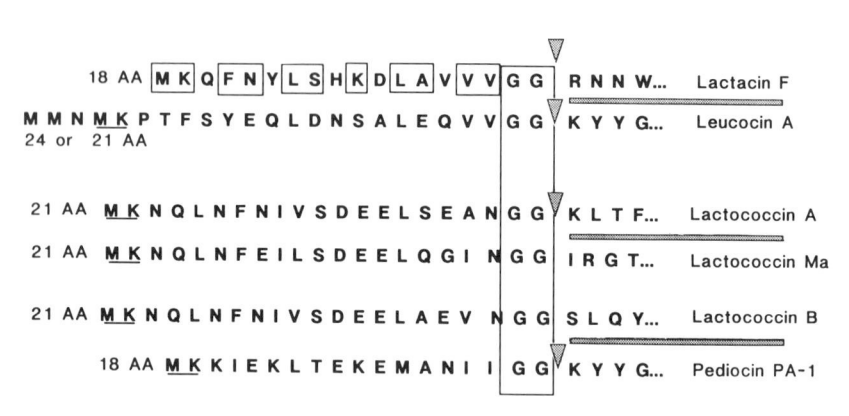

```
           18 AA  M K Q F N Y L S H K D L A V V V G G | R N N W...    Lactacin F
      M M N M K P T F S Y E Q L D N S A L E Q V V G G | K Y Y G...    Leucocin A
      24 or  21 AA

           21 AA  M K N Q L N F N I V S D E E L S E A N G G | K L T F...   Lactococcin A

           21 AA  M K N Q L N F E I L S D E E L Q G I N G G | I R G T...   Lactococcin Ma

           21 AA  M K N Q L N F N I V S D E E L A E V N G G | S L Q Y...   Lactococcin B

           18 AA  M K K I E K L T E K E M A N I I G G | K Y Y G...         Pediocin PA-1
```

Figure 6 N-terminal amino acid sequences of six prepeptide bacteriocins produced by different lactic acid bacteria representing the genera *Lactobacillus* (lactacin F), *Leuconostoc* (leucocin A), *Lactococcus* (lactococcin A, B, and Ma), and *Pediococcus* (pediocin). The processing sites, indicated by the inverted triangles, were determined by comparison of the DNA sequences with the known amino acid sequences of the mature peptides. Boxed residues illustrated in lactacin F are found in one or more of the other prepeptides at the identical position within the N-terminal extension.

The DNA sequence of leucocin A harbors two methionine residues that could start two possible ORFs generating either a 24- or 21-amino-acid extension. Within all six prepeptides are two glycine residues in the -1 and -2 positions of the processing site. This strongly indicates that the Gly–Gly residues are a common feature of the processing site for prepeptide bacteriocins in lactic acid bacteria.

In the four cases in which both the DNA sequence and N-terminal amino acid sequence of the processed peptide are known, the $+1$ position is occupied by a positively charged amino acid (see Figure 6). The N-terminal extension and processing sites of lactacin F are consistent with those of signal sequences (von Heijne, 1986). The leader peptide of lactacin F also exhibits good hydropathicity within its core region (Figure 5), supporting a possible membrane interaction. However, the 18-amino-acid N-terminal extension is short relative to most signal peptides. The other prepeptides bearing the Gly–Gly processing site do not show features characteristic of signal peptides. For example, lactococcins A, B, Ma, and leucocin A exhibit low hydropathic indices over their leader regions (data not shown). Isoleucine and serine occupy the $+1$ position in lactococcins Ma and B, respectively, suggesting that a charged amino acid is not essential for processing at the Gly-Gly site; however, the N-terminus of these bacteriocins has not yet been determined by amino acid sequencing of the mature peptide. The -3 position of lactococcins A and Ma is also occupied by a large and polar amino acid, asparagine, which is not found in sites processed by signal peptidases (von Heijne, 1986). Although the N-terminal extension of lactacin F shares some important features that are characteristic of signal peptides, the data accumulated thus far suggests that these small prepeptide bacteriocins are subject to a processing mechanism that is distinct of signal peptidases.

The common features of prepeptide bacteriocins from lactic acid bacteria characterized to date are the following:

1. an N-terminal extension;
2. the −1 and −2 positions from the processing site are occupied by glycine–glycine;
3. the two residues at the N-terminus of the leader peptide are methionine and lysine;
4. the +1 position is occupied by a positively charged amino acid (in those cases in which the N-terminus of the mature peptide has been determined).

The presence of these common features in prepeptide bacteriocins suggests that there may be a general posttranslational processing mechanism for the maturation of small hydrophobic bacteriocins in lactic acid bacteria, as also noted by van Belkum (1991). This hypothesis could be experimentally borne out through expression of structural bacteriocin determinants in heterologous strains that harbor the general prepeptide processing mechanism. Based upon these speculations, we have conducted experiments on the expression of lactacin F in leucocin A-producing *Leuconostoc gelidum* 187-22 which has the intact genetic system for leucocin A production (Hastings et al., 1991). Lactacin F production was observed in the leucocin-producing host (C. Ahn, M. Stiles, and T. Klaenhammer, unpublished data). These findings indicate that prepeptide bacteriocins with common features may share a common processing mechanism.

VI. Perspectives and Conclusions

Antimicrobial proteins and peptides produced by a variety of *Lactobacillus* strains are antagonistic toward other competitive microorganisms. Such natural inhibitors could also contribute to the margin of safety in processed foods where the bacteriocins are effective against targeted food-borne pathogens.

The information accumulated thus far on the genetics of bacteriocin production in lactobacilli indicates that the genes involved are organized in operons. However, there is little information available on the function of many of the gene products encoded within bacteriocin operons. Therefore, the targets of genetic and protein engineering of *Lactobacillus* bacteriocins lie in: (1) the operon promoter region; (2) the structural gene; (3) the bacteriocin leader sequence; (4) the immunity gene; and (5) the effects of any genetic changes on the accompanying export and processing mechanisms. Studies in these areas will further our understanding of the expression and regulation of bacteriocins in lactobacilli and improve the potential for future applications in the food and agricultural industries.

Three different approaches could be used to study and potentially alter promoter specificity to improve bacteriocin expression. First, modification of essential nucleotides in the promoter region could alter regulation constraints and increase affinity for RNA polymerase. Second, interchange of promoters could enhance expression of a bacteriocin in heterologous hosts, or promote expression of heterologous proteins in *Lactobacillus* hosts. Other exciting developments will occur in elucidating bioenvironmental signals that regulate and control bacteriocin production in lactic acid bacteria. Although limited data are available, production of both helveticin J and lactacin F is dependent on pH. Experiments are in progress to determine whether pH affects gene expression or bacteriocin processing and export. The helveticin J promoter shares common features with SOS-inducible promoters, suggesting that it may be derepressed under various environmental stresses. Promoters that are reactive to bioenvironmental signals could be useful to design expression systems that can be turned "on" and "off" at specific stages of industrial bioprocessing and fermentation.

Understanding pre/pro protein structures and posttranslational processing enzymes produced from these operons will elucidate important structure–function relationships of these bacteriocins. Small hydrophobic proteins with no apparent tertiary structure are more amenable to structural analysis based on alteration of the nucleotide sequence. Structural modifications are being engineered pursuing three main objectives: (1) definition of the minimal length required for an active peptide bacteriocin by creating new termination codons in the carboxy-terminus; (2) modification of the bacteriocin hydrophobicity by replacing residues in hydrophobic regions with charged amino acids; and (3) position and occurrence of intramolecular disulfide bonding by introducing or removing cysteine residues. Site-directed mutagenesis can also be applied to analyze the posttranslational modifications affecting the prepeptide by changing hydrophobicity of the peptide leader core region or by exchanging amino acids at the processing site. In addition, leader peptide sequences could be used to direct the processing or export of heterologous bacteriocins or proteins cloned in *Lactobacillus*.

Target sites, mechanisms of action, and immunity systems have yet to be defined. Information about bacteriocin resistance mechanisms and their development within sensitive populations will be critical in order to optimize the effectiveness and longevity of food preservation and fermentation systems.

The usefulness of these compounds could further be expanded once their structure-function relationships have been elucidated. Genes for bacteriocin production and immunity could be directed into desired industrial starter strains. Increased bacteriocin production, expanded host range, and combinations of heterologous bacteriocins to produce novel antimicrobials could target spoilage and pathogenic organisms that were not previously susceptible.

Bacteriocin production and immunity determinants will be important phenotypic markers for construction of food-grade cloning and expression vectors. Utilization of these gene systems will greatly facilitate the application of genetic technologies to lactic acid bacteria used in industrial food and dairy fermentations.

Acknowledgments

The research efforts on the lactacin F and helveticin J systems have been funded in part by the National Dairy Promotion and Research Board, Arlington, Virginia, and Nestle, Ltd., Switzerland. We acknowledge and thank P. Muriana and M. Joerger for their initial contributions, which have led to the further genetic characterization of lactacin F and helveticin J. We gratefully thank S. Barefoot of Clemson University, and M. Stiles and J. Vederas of the University of Alberta, Canada, for access to their unpublished results on lactacin B, leucocin A-UAL 187, and the carnobacteriocins.

References

Ahn, C., and Stiles, M. E. (1990). Plasmid-associated bacteriocin production by a strain of *Carnobacterium piscicola* from meat. *Appl. Environ. Microbiol.* **56,** 2503–2510.
Akutsu, A., Masaki, H., and Ohta, T. (1989). Molecular structure and immunity specificity of colicin E6, an evolutionary intermediate between E-group colicins and cloacin DF13. *J. Bacteriol.* **171,** 6430–6436.
Barefoot, S. F., and Klaenhammer, T. R. (1983). Detection and activity of lactacin B, a bacteriocin produced by *Lactobacillus acidophilus*. *Appl. Environ. Microbiol.* **45,** 1808–1815.
Barefoot, S. F., and Klaenhammer, T. R. (1984). Purification and characterization of the *Lactobacillus acidophilus* bacteriocin lactacin B, *Antimicrobial Agents Chemother.* **26,** 328–334.
Daeschel, M. A., McKenney, M. C., and McDonald, L. C. (1990). Bactericidal activity of *Lactobacillus plantarum* C11. *Food Microbiol.* **7,** 91–98.
de Klerk, H. C. (1967). Bacteriocinogeny in *Lactobacillus fermenti*. *Nature (London)* **214,** 609.
de Klerk, H. C., and Coetzee, J. N. (1961). Antibiosis among lactobacilli. *Nature (London)* **192,** 340–341.
de Klerk, H. C., and Smit, J. A. (1967). Properties of a *Lactobacillus fermenti* bacteriocin. *J. Gen. Microbiol.* **48,** 309–316.
Fickett, J. W. (1982). Recognition of protein coding in DNA sequence. *Nucleic Acids Res.* **10,** 5303–5318.
Garnier, J., Osguthorpe, D. R., and Robson, B. (1978). Analysis of the accuracy and implications of simple methods for predicting the secondary structure of globular proteins. *J. Mol. Biol.* **120,** 97–120.
Gonzalez, C. F., and Kunka, B. S. (1987). Plasmid-associated bacteriocin production and sucrose fermentation in *Pediococcus acidilactici*. *Appl. Environ. Microbiol.* **53,** 2534–2538.
Hamdan, I. Y., and Mikolajcik, E. M. (1974). Acidolin: an antibiotic produced by *Lactobacillus acidophilus*. *J. Antibiotics* **27,** 631.

Harris, L. J., Daeschel, M. A., Stiles, M. E., and Klaenhammer, T. R. (1989). Antimicrobial activity of lactic acid bacteria against *Listeria monocytogenes*. *J. Food Prot.* **52**, 384–387.

Hastings, J. W., Sailer, M., Johnson, K., Roy, K. L., Vederas, J. C., and Stiles, M. E. (1991). Characterization of leucocin A-UAL 187 and cloning of the bacteriocin gene from *Leuconostoc gelidum*. *J. Bacteriol.* **173**, 7491–7500.

Henkel, T., Sailer, M., Vederas, J. C., Worobo, R. W., Quandri, L., and Stiles, M. E. (1991). Purification and characterization of bacteriocins produced by *Carnobacterium piscicola* LV17. EMBO-FEMS-NATO Symposium Poster, September, Island of Bendor France.

Holo, H., Nilssen, O., and Nes, I. F. (1991). Lactococcin A, a new bacteriocin from *Lactococcus lactis* subsp. *cremoris:* isolation and characterization of the protein and its gene. *J. Bacteriol.* **173**, 3879–3887.

Huynh, T. V., Young, R. A., and Davis, R. W. (1984). Constructing and screening cDNA libraries in lambda-gt10 and lambda-gt11. *In* "DNA cloning techniques: a practical approach" (D. Glover, ed.), pp. 46–78. IRL Press, Oxford.

Jimenez-Diaz, R., Piard, J. C., Ruiz-Barba, J. L., and Desmazeaud, M. J. (1990). Isolation of a bacteriocin-producing *Lactobacillus plantarum* from a green olive fermentation. *FEMS Microbiol. Rev.* 87, p91.

Joerger, M. C., and Klaenhammer, T. R. (1986). Characterization and purification of helveticin J and evidence for a chromosomally determined bacteriocin produced by *Lactobacillus helveticus* 481. *J. Bacteriol.* **167**, 439–446.

Joerger, M. C., and Klaenhammer, T. R. (1990). Cloning, expression, and nucleotide sequence of the *Lactobacillus helveticus* 481 gene encoding the bacteriocin helveticin J. *J. Bacteriol.* **171**, 6339–6347.

Kashet, E. R. (1987). Bioenergetics of lactic acid bacteria: cytoplasmic pH and osmotolerance. *FEMS Microbiol. Rev.* **46**, 233–244.

Klaenhammer, T. R. (1988). Bacteriocins of lactic acid bacteria. *Biochimie* **70**, 337–349.

Klaenhammer, T. R. (1990). Antimicrobial and bacteriocin interactions of the lactic acid bacteria. *In* "Proceedings of the 6th International Symposium on Genetics of Industrial Microorganisms" (H. Heslot, J. Davies, J. Florent, L. Bobichon, G. Durand, and L. Penasse, eds.), Vol. 1, pp. 433–445. Societe Francaise de Microbiologie, Strasbourg France.

Klein, P., Kanehisa, M., and DeLisi, C. (1985). The detection and classification of membrane-spanning proteins. *Biochim. Biophys. Acta* **815**, 468–476.

Kok, J. J., van der Vossen, J. M. B. M., and Venema, G. (1984). Construction of plasmid cloning vectors for lactic streptococci which also replicate in *Bacillus subtilis* and *Escherichia coli. Appl. Environ. Microbiol.* **48**, 726–731.

Kyte, J., and Doolittle, R. F. (1982). A simple method for displaying the hydropathic character of a protein. *J. Mol. Biol.* **157**, 105–132.

Lewus, C. B., Kaiser, A., and Montville, T. J. (1991). Inhibition of food-borne bacterial pathogens by bacteriocins from lactic acid bacteria isolated from meat. *Appl. Environ. Microbiol.* **57**, 1683–1688.

Lindgren, S. E., and Dobrogosz, W. J. (1990). Antagonistic activities of lactic acid bacteria in food and feed fermentations. *FEMS Microbiol. Rev.* **87**, 149–164.

Marugg, J., Gonzalez, C. S., Kunka, B. S., Ledeboer, A. M., Pucci, M. J., Toonen, M., Walker, S. A., Zoctmulder, L., and Vandenbergh, P. (1992). Cloning, expression, and nucleotide sequence of genes involved in production of pediocin PA-1, a bacteriocin from *Pediococcus acidilactici* PAC1.0. *Appl. Environ. Microbiol.* **58**, 2360–2367.

McCormick, E. L., and Savage, D. C. (1983). Characterization of *Lactobacillus* sp. strain 100-37 from the murine gastrointestinal tract: ecology, plasmic content, and antagonistic activity toward *Clostridium ramosum* H1. *Appl. Environ. Microbiol.* **46**, 1103–1112.

Mørtvedt, C. I., and Nes, I. (1990). Plasmid-associated bacteriocin production by a *Lactobacillus sake* strain. *J. Gen. Microbiol.* **136**, 1601–1607.

Mørtvedt, C. I., Nissen-Meyer, J., Sletten, K., and Nes, I. F. (1991b). Purification and amino acid sequence of lactocin S, a bacteriocin produced by *Lactobacillus sake* L45. *Appl. Environ. Microbiol.* **57**, 1829–1834.

Muriana, P. M., and Klaenhammer, T. R. (1987). Conjugal transfer of plasmid-encoded determinants for bacteriocin production and immunity in *Lactobacillus acidophilus* 88. *Appl. Environ. Microbiol.* **53**, 553–560.

Muriana, P. M., and Klaenhammer, T. R. (1991a). Purification and partial characterization of lactacin F, a bacteriocin produced by *Lactobacillus acidophilus* 11088. *Appl. Environ. Microbiol.* **57**, 114–121.

Muriana, P. M., and Klaenhammer, T. R. (1991b). Cloning, phenotypic expression, and DNA sequence of the gene for lactacin F, an antimicrobial peptide produced by *Lactobacillus* spp. *J. Bacteriol.* **173**, 1779–1788.

Nettles, C. G., Barefoot, S. F., and Bodine, A. B. (1991). Purification and partial sequence of the *Lactobacillus acidophilus* bacteriocin lactacin B. Proceedings of the Annual Meeting of the Society for Industrial Microbiology, Philadelphia PA, (August 3–9).

Okereke, A., and Montville, T. J. (1991). Bacteriocin inhibition of *Clostridium botulinum* spores by lactic acid bacteria. *J. Food Protection* **54**, 349–353.

Rammelsberg, M., and Radler, F. (1990). Antibacterial polypeptides of *Lactobacillus* species. *J. Appl. Bacteriol.* **69**, 177–184.

Schillinger, U., and Holzapfel, W. H. (1990). Antibacterial activity of carnobacteria. *Food Microbiol.* **7**, 305–310.

Schillinger, U., and Lucke, F. K. (1989). Antibacterial activity of *Lactobacillus sake* isolated from meat. *Appl. Environ. Microbiol.* **55**, 1901–1906.

Schillinger, U., Kaya, M., and Lucke, F. K. (1991). Behavior of *Listeria monocytogenes* in meat and its control by a bacteriocin-producing strain of *Lactobacillus sake*. *J. Appl. Bacteriol.* **70**, 473–478.

Stoffels, G., Nissen-Meyer, J., Gudmundsdottir, A., Sletten, K., Holo, H., and Nes, I. F. (1992). Purification and characterization of a new bacteriocin isolated from a *Carnobacterium sp*. *Appl. Environ. Microbiol.* **58**, 1417–1422.

Silva, M., Jacobus, N. V., Deneke, C., and Gorbach, S. L. (1987). Antimicrobial substance from a human *Lactobacillus* strain. *Antimicrobial Agents Chemother.* **31**, 1231–1233.

Talarico, T. L., and Dobrogosz, W. J. (1989). Chemical characterization of an antimicrobial substance produced by *Lactobacillus reuteri*. *Antimicrobial Agents Chemother.* **33**, 674–679.

Toba, T., Yoshioka, E., and Itoh, T. (1991a). Potential of *Lactobacillus gasseri* isolated from infant faeces to produce bacteriocin. *Letters in Appl. Microbiol.* **12**, 228–231.

Toba, T., Yoshioka, E., and Itoh, T. (1991b). Acidophilucin A, a new heat-labile bacteriocin produced by *Lactobacillus acidophilus* LAPT 1060. *Letters in Appl. Microbiol.* **12**, 106–108.

Toba, T., Yoshioka, E., and Itoh, T. (1991c). Lacticin, a bacteriocin produced by *Lactobacillus delbrueckii* subsp. *lactis*. *Letters in Appl. Microbiol.* **12**, 43–45.

Upreti, G. C., and Hinsdill, R. D. (1973). Isolation and characterization of a bacteriocin from a homofermentative *Lactobacillus*. *Antimicrobial Agents Chemother.* **4**, 487–494.

Upreti, G. C., and Hinsdill, R. D. (1975). Production and mode of action of lactocin 27: bacteriocin from a homofermentative *Lactobacillus*. *Antimicrobial Agents Chemother.* **7**, 139–145.

van Belkum, M. (1991). Lactococcal bacteriocins: genetics and mode of action. Ph.D. Thesis, University of Groningen, Gronigen, The Netherlands.

van Belkum, M., Hayema, B. J., Jeeninga, R. E., Kok, J., and Venema, G. (1991). Organization and nucleotide sequences of two lactococcal bacteriocin operons. *Appl. Environ. Microbiol.* **57**, 492–498.

van Belkum, M., Kok, J., and Venema, G. (1992). Cloning, sequencing, and expression in *Escherichia coli* of *LCNB*, a third bacteriocin determinant from the lactococcal bacteriocin plasmid p9B4-6. *Appl. Environ. Microbiol.* (in press).

van den Elzen, P. J. M., Maat, J., Walters, H. H. B., Velkamp, E., and Nijkamp, H. J. J. (1982). The nucleotide sequence of the bacteriocin promoters of plasmids Col DF13 and Col EI: role of *lex*A repressor and cAMP in the regulation of promoter activity. *Nucleic Acids Res.* **10**, 1913–1928.

van der Vossen, J. M. B. M., Kok, J., and Venema, G. (1985). Construction of cloning, promoter-screening, and terminator-screening shuttle vectors for *Bacillus subtilis* and *Streptococcus lactis*. *Appl. Environ. Microbiol.* **50,** 540–542.

Vincent, J. G., Veomett, R. C., and Riley, R. F. (1959). Antibacterial activity associated with *Lactobacillus acidophilus*. *J. Bacteriol.* **78,** 477–484.

von Heijne, G. (1986). A new method for predicting signal sequence cleavage sites. *Nucleic Acids Res.* **14,** 4683–4690.

West, C., and Warner, P. J. (1988). Plantacin B, a bacteriocin produced by *Lactobacillus plantarum* NCDO 1193. *FEMS Microbiol. Lett.* **49,** 163–165.

CHAPTER 8

Pediocins

BIBEK RAY
DALLAS G. HOOVER

I. Introduction

The activity of pediococci in cultured foods such as pickles and fermented sausages is well established. Although nutritionally fastidious in a manner similar to the lactococci, pediococci are isolatable from sources other than fermented meats and vegetables. Members of the genus *Pediococcus* reside on plants (Mundt et al., 1969), in plant products (e.g., silage), and can also be isolated from humans (Facklam et al., 1989). Activities and uses outside the area of bacteriocins include (Raccach, 1987): the bioprocessing and biopreservation of muscle-derived foods (Bacus and Brown, 1981) and plant-derived foods (Fleming and McFeeters, 1981; Steinkraus, 1983); flavor development in ripened cheeses (Thomas et al., 1985; Thomas, 1987) and sourdough breads (Sugihara, 1985); and bioanalysis of vitamins (Solberg et al., 1975). Pediococci are also a concern in the spoilage of alcoholic beverages (Rainbow, 1975; Lafon-Lafourcade, 1983). Their occurrence in raw meats is not considered a problem, as pediococci are unable to grow in the absence of a fermentable carbohydrate, which is not present in sufficient levels in meat. Pediococci have been isolated from raw sausages (Reuter, 1970), fresh fish (Blood, 1975), and raw poultry (Teuber and Geis, 1981), as well as from the rumen of cows (Baumann and Foster, 1956) and turkey feces (Harrison and Hansen, 1950).

II. Description of the Genus *Pediococcus*

The latest edition of *Bergey's Manual* (Garvie, 1986) described *Pediococcus* as a group of nonmotile, nonencapsulated, homofermentative, facultative anaerobes whose spherical cells form pairs or tetrads. The catalase-negative pediococci are chemoorganotrophs requiring a rich medium containing nicotinic acid, pantothenic acid, and biotin. *P. cerevisiae,* the species name used for many years to describe varieties of fermentative and beer spoilage pediococci, is no longer valid. The eight defined species of *Pediococcus* can be grouped according to sensitivities to temperature, pH, and NaCl (Table I). *P. damnosus* and *P. parvulus* are described as acid tolerant and require growth temperatures of 35°C or more under anaerobic conditions. *P. pentosaceus* and *P. acidilactici* grow equally well aerobically or anaerobically. The phenotypic difference between these two species used as commercial meat starter cultures is that *P. acidilactici* can grow at 50°C while *P. pentosaceus* cannot. In the laboratory this distinction is not always clear-cut.

There may also be confusion in distinguishing pediococci from lactococci and leuconostocs. If the capability of a laboratory to determine mol% G + C of DNA or DNA–DNA hybridization is lacking (which is usually the case), genus identification of these Gram-positive cocci normally depends on a Gram stain for morphological examination of tetrad formation (*Pediococcus*) or chain formation (*Lactococcus* and *Leuconostoc*). All three genera form pairs, which is often the predominant morphology. Tetrads are not always present in pediococci. The pediococci and lactococci are homofermentative, while the leuconostocs are heterofermentative, that is, leuconostocs will produce carbon dioxide during active growth, but this is not always evident, especially with oxygen-sensitive leuconostocs. Fermentation of carbohy-

TABLE I
Differential Conditions of Growth of Species of the Genus *Pediococcus*

Characteristics	*P. damnosus*	*P. parvulus*	*P. inopinatus*	*P. dextrinicus*	*P. pentosaceus*	*P. acidilactici*	*P. halophilus*	*P. urinaeequi*
Growth at								
35°C	−	+	+	+	+	+	+	+
40°C	−	−	−	+	+	+	−	+
50°C	−	−	−	−	−	+	−	−
Growth at								
pH 4.2	+	+	−	−	+	+	−	−
pH 7.5	−	+	d	+	+	+	+	+
pH 8.5	−	−	−	−	d	d	+	+
Growth in								
4% NaCl	−	+	+	+	+	+	d	+
6.5% NaCl	−	+	d	−	+	+	+	+
18% NaCl	−	−	−	−	−	−	+	−

Source: Garvie, 1986.
Note: +, 90% or more of strains positive; −, 90% or more of strains negative; d, 11 to 89% of strains positive.

drates with production of acid is sometimes used for identification. It is estimated that 11–89% of the strains of *P. acidilactici* and *P. pentosaceus* are capable of metabolizing lactose, and are usually positive for utilization of arabinose and xylose (Garvie, 1986), while lactococci are lactose positive but normally arabinose and xylose negative (Mundt, 1986). However, freshly isolated atypical strains for these phenotypes are not uncommon, and a plasmid-mediated trait has a high chance of intrageneric transfer in the environment.

Pediococci are sometimes confused with micrococci because of similarities in morphology and NaCl tolerance, and a weak catalaselike reaction from some pediococci; however, since the micrococci are often pigmented and not acid tolerant, rapid separation is possible.

III. Bacteriocin Activity in Pediococci

Early indication of an inhibitive effect by pediococci other than by lactic acid production was documented by Etchells et al. (1964), where a mixed-species inoculation of brined cucumbers was devised to generate an early, rapid initial growth and moderate acid production using *P. cerevisiae* FBB-61 and a high final acidity using *Lactobacillus plantarum* FBB-67. However, it was found that some strains of *P. cerevisiae* delayed the growth of the *L. plantarum* up to 10 days in this mixed inoculation. It was realized that lactic acid could not be a factor since *L. plantarum* is a more acid-resistant lactic acid bacterium than *P. cerevisiae*. Fleming et al. (1975) compared the inhibitory properties of 16 isolates of pediococci against lactic acid bacteria associated with brined cucumber fermentations as well as other bacteria and yeasts from the genera *Candida*, *Debaryomyces*, *Pichia*, and *Saccharomyces*. A Trypticase soy agar (TSA) spot assay with soft TSA-seeded overlay was used. *P. cerevisiae* FBB-61 and L-7230 consistently gave zones of inhibition against other pediococci, four strains of *L. plantarum*, and single cultures of *Leuconostoc mesenteroides*, *Micrococcus luteus*, *Streptococcus faecalis*, *Staphylococcus aureus*, and *Bacillus cereus*. None of the Gram-negative bacteria or yeasts tested was inhibited, nor was a strain of *Lactobacillus brevis*. Hydrogen peroxide inhibition was ruled out by the addition of filter-sterilized catalase to the TSA for some of the experiments. It was noted by Fleming et al. (1975) that although Haines and Harmon (1973) reported the antagonism of *S. aureus* by *P. cerevisiae*, the inhibition of other lactic acid bacteria by specific bacteriocin action or pediococci was not determined. Fleming et al. (1975) did recognize a warning that such antagonism can defeat the purpose of controlled mixed-species fermentation of brined cucumbers. The compatibility of multiple starter cultures is an issue in fermented produce, just as it is in dairy fermentations (Klaenhammer, 1988).

The antagonistic activity of FBB-61 was characterized by Rueckert (1979). The inhibitory agent was bacteriocidal, stable to 100°C for 60 minutes and freezing temperatures, sensitive to pronase, and nondialyzable across a semipermeable membrane. These properties strongly indicate that the agent was a bacteriocin.

Since bacteriocin biosynthesis in *Pediococcus* has been demonstrated to be often associated with plasmids, and the antagonistic effect is easily detected using agar plates, bacteriocins can make excellent plasmid markers. Daeschel and Klaenhammer (1985) examined the plasmid biology of bacteriocin activity in FBB-61 and L-7230. The bacteriocin, termed pediocin A, was produced by both strains displaying activity against other lactic acid bacteria as well as *Clostridium botulinum, Clostridium perfringens, Clostridium sporogenes, Staphylococcus aureus,* micrococci, and bacilli using an agar spot test. No effect was found against Gram-negative bacteria or yeasts. Immunity and production of pediocin A was associated with the presence of a 13.6-MDa plasmid as determined by plasmid-curing experiments and agarose gel electrophoresis. The authors noted the value of having a starter culture produce a bacteriocin effective against food-poisoning organisms, especially against *S. aureus* during the manufacture of fermented sausages. In addition, production of pediocin A would aid pure culture vegetable fermentations in which competing, naturally occurring lactic acid bacteria could be inhibited. Finally, since the bacteriocins in *Pediococcus* are plasmid mediated, improved genetic transfer techniques (Chassy et al., 1988) make modification of lactic acid starter organisms more straightforward.

Graham and McKay (1985) also associated the presence of plasmids with bacteriocin activity. *P. cerevisiae* FBB63 contains a 10.5-MDa plasmid, linked to antagonistic activity against other pediococci.

Plasmids in *Pediococcus* were first studied by Gonzalez and Kunka (1983). They demonstrated conjugation in *Pediococcus* using pIP501, a broad host-range antibiotic resistance plasmid from *Streptococcus faecalis.* This plasmid was conjugally transferred in and out of strains of *P. acidilactici* and *P. pentosaceus.* Raffinose utilization and α-galactosidase and sucrose hydrolase activity were linked to plasmids in strains of *P. pentosaceus* (Gonzalez and Kunka, 1986). Characterization of *P. acidilactici* PAC 1.0 (NRRL B-5627) revealed the presence of a bacteriocin designated pediocin PA-1, associated with a 6.2-MDa plasmid (Gonzalez and Kunka, 1987). Activity was shown against other lactic acid bacteria using an agar spot assay with inactivation of PAC 1.0 by chloroform vapor; however, there was no inhibition of lactococci or *Streptococcus thermophilus*. Partially purified PA-1 was not active against *Micrococcus varians, Micrococcus sodonensis, Staphylococcus xylosus, Staphylococcus epidermidis, Staphylococcus carnosus, Lactobacillus acidophilus, Lactobacillus lactis,* and *Lactobacillus bulgaricus*. PA-1 was approximately 16,500 Da and displayed sensitivities to protease, papain, pepsin, and α-chymotrypsin. The protein was stable to heat (100°C for 10 minutes) and to lipase, lysozyme, DNase, and RNase activities. The greatest pH stability was between 4 and 7.

Additional work was done by Pucci et al. (1988) to develop PA-1 for use as a food preservative. PA-1 was prepared from a dried powder of PAC 1.0

culture supernatant fortified with 10% milk powder. Inhibition of *Listeria monocytogenes* LMO1 was demonstrated in cottage cheese, half-and-half cream, and cheese sauce over a pH range of 5.5 to 7.0 and at both 4 and 32°C. In the foods at 40°C and 50 arbitrary units (AU)/ml, PA-1 was effective against an inoculum of approximately 5×10^2 CFU/ml of *L. monocytogenes* for two weeks, indicative of PA-1's effectiveness as a preservative in these foods against this pathogen.

PA-1 has been thoroughly characterized and purified and its gene cloned and plasmid mapped (Vandenbergh, 1991; Henderson et al., 1992). PA-1 is a simple protein of 44 amino acids (4629 Da) with two disulfide bonds. It has a net positive charge and an isoelectric point of 10.0. This coiled protein is stable at temperatures from $-20°C$ to $100°C$. Its minimum inhibitory concentration is 54.7 AU/ml versus *Listeria monocytogenes* and approximately 27 AU/ml versus *Pediococcus pentosaceus*. The protein was difficult to purify. Table II shows the increase in biological activity with purification.

Another strain of *Pediococcus* used in the production of fermented meats and examined for bacteriocin activity is *P. acidilactici* PO2. This strain was used in evaluating the Wisconsin Process for bacon (Tanaka et al., 1985a). In this process bacon is manufactured with 40 or 80 ppm sodium nitrite, *P. acidilactici*, and 0.7% sucrose. It has shelf-life and sensory characteristics similar to conventional bacon made with 120 ppm sodium nitrite. It is equally effective for antibotulinal protection, while having markedly reduced levels of nitrosamines (Tanaka et al., 1985b). If Wisconsin Process bacon is temperature abused, the pediococci grow and utilize the sucrose to produce acid (and other inhibitory compounds), thus preventing growth and toxin production from *Clostridium botulinum*.

Bacteriocin activity in *P. acidilactici* PO2 was associated with a plasmid of approximately 5.5 MDa (Hoover et al., 1988). Activity was demonstrated against lactic acid bacteria as well as three out of five strains of *Listeria monocytogenes*. Similar antagonistic activity was found in *Pediococcus acidilactici* strains B5627 and PC as well as in *P. pentosaceus* MC 03. The bacteriocin effect of PO2 is probably pediocin PA-1, since *P. acidilactici* PAC 1.0 has been

TABLE II
Purification and Activity of Pediocin PA-1

Treatment	Specific activity (AU/ml)	% Yield	Fold purification
Crude extract	107	100	1.0
Gel filtration	166	50	1.5
Ion exchange	1330	9.4	12.5
Dialysis	6700	7.0	63.0
HPLC	50,000	0.6	470.0

Source: Henderson et al., 1992.

shown by rDNA-typing to be genetically the same as *P. acidilactici* PO2 (S. Harlander, personal communication).

In a short survey of bacteriocin activity by 37 cultures of pediococci against eight strains of *Listeria monocytogenes*, 28% of the match-ups were positive as measured by a soft agar overlay/plate diffusion method of Kekessy and Piguet (1970; Hoover et al., 1989). As evidenced by many surveys of bacteriocin activity, the phenomenon is quite widespread in nature and found in many types of bacteria (Ray, 1992a; Tagg et al., 1976). Pediococci appear to be no exception.

Using a liquid assay system described by Anderson (1986), dialyzed growth extracts of *Pediococcus acidilactici* PO2 were evaluated against the growth of *Listeria monocytogenes* 19113 (Hoover et al., 1989). The lag phase of a 10^5-CFU/ml inoculum of 19113 was extended by nearly 30 hours beyond the control when APT broth was supplemented with a 2% growth extract of PO2. APT (control) broth adjusted to pH 5.0 had the same inhibitive effect on 19113 as pH 5.0-adjusted PO2 growth extract, but at pH 6.0 and 7.0 *L. monocytogenes* growth was evident long before growth occurred in PO2 growth extracts at the same pH (Figure 1). However, growth of *L.*

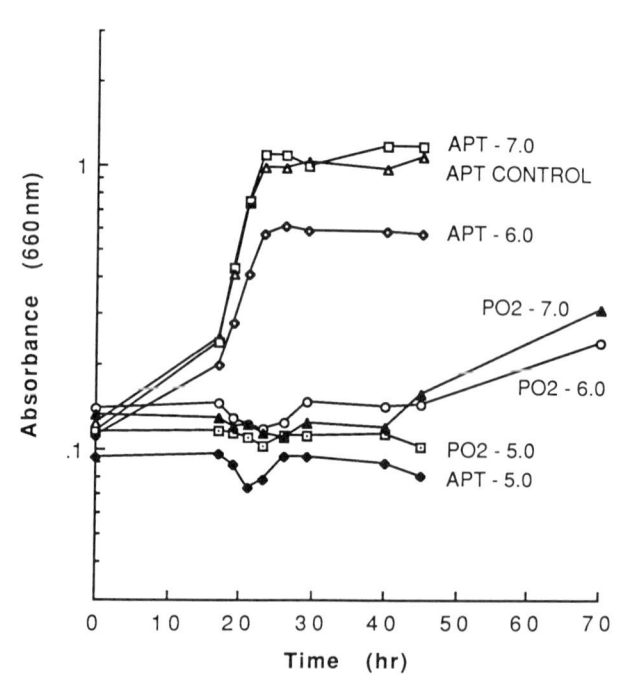

Figure 1 Comparison of inhibitory effect of pH with 100% growth extract of *Pediococcus acidilactici* PO2 on *Listeria monocytogenes* 19113. △, APT broth control; □, APT broth adjusted to pH 7.0; ▲, APT growth extract of *P. acidilactici* PO2 adjusted to pH 7.0; ◊, APT broth adjusted to pH 6.0; ○, APT growth extract of PO2 adjusted to pH 6.0; ♦, APT broth adjusted to pH 5.0; ▣, APT growth extract of PO2 adjusted to pH 5.0. (From Hoover et al., 1989; printed with permission of Marcel Dekker, Inc.)

monocytogenes did occur after 36 hours of exposure to the PO2 growth extract, indicative of either a bacteriocin-resistant subpopulation of 19113 or a loss of active bacteriocin from the growth medium with incubation.

Raccach et al. (1989) examined several strains of lactic acid bacteria for antagonism against *Listeria monocytogenes* using the "spot on the lawn" agar plate method. Among the lactic cultures were three strains each of *P. acidilactici* and *P. pentosaceus;* five of six of these strains demonstrated antibiosis against *Listeria*.

Spelhaug and Harlander (1989) addressed the effect that media composition and assay method has on the detection of specific bacteriocin-mediated inhibition. Along with several *Lactococcus lactis* strains, two *Pediococcus pentosaceus* cultures were evaluated (FBB61 and FBB63-DG2). While various Gram-positive and Gram-negative food-poisoning bacteria were used as indicator organisms, *Listeria* species constituted the greatest number. The agar spot method was found to be more reproducible and reliable than a flip streak method and a flip spot method. FBB61 and FBB63-DG2 had no effect on the Gram-negative strains but did inhibit growth of all 22 strains of *Listeria* (including *L. innocua, L. ivanovii, L. seeligeri,* and *L. welshimeri*), as well as *Bacillus cereus, Clostridium perfringens,* and *Staphylococcus aureus*. It was noted that significantly different results were produced by the agar spot assay, depending on whether brain heart infusion, M17-glucose, or MRS agars were used for the bottom and overlay media.

Harris et al. (1989) also evaluated bacteriocin effect against the viability of *Listeria*. They tested eight strains of *L. monocytogenes* and one strain of *L. innocua* for antagonism from fifteen cultures of lactic acid bacteria, four of which were pediococci. Two types of inhibition assays were used; one by deferred antagonism (Tagg et al., 1976), whereby a BHI agar plate inoculated with the lactic culture was overlaid with indicator-seeded soft BHI agar, and a well diffusion assay (Tagg and McGiven, 1971), in which cell-free, neutralized growth supernatants of lactic cultures were placed into wells cut in *Listeria*-seeded BHI agar. Only *Pediococcus acidilactici* PAC 1.0 was capable of inhibiting *Listeria* by both deferred antagonism and well diffusion assays. *P. pentosaceus* FBB61-1 and L-7230 demonstrated inhibition only by deferred antagonism, while the bacteriocin-negative control strain, *P. pentosaceus* FBB61-2 (Daeschel and Klaenhammer, 1985), was negative in both assays as expected. The other lactic acid bacteria that inhibited the *Listeria* in both assays were *Lactobacillus* sp. UAL11 and *Leuconostoc* sp. UAL14 (original source: Institute of Food Research, Bristol, England). All eight strains of *L. monocytogenes* were identical in their reaction to bacteriocins used in the study.

A study of bacteriocin-positive lactic acid bacteria isolated from retail cuts of meat was conducted by Lewus et al. (1991) for antagonism against varieties of *Listeria monocytogenes, Staphylococcus aureus,* and *Aeromonas hydrophila*. Included in the group of lactic acid bacteria were *Pediococcus acidilactici* H and *P. pentosaceus* 43200 and 43201, all of which demonstrated activity against cultures of *L. monocytogenes* and *A. hydrophila*. The *P. pentosaceus* strains both inhibited *S. aureus,* but *P. acidilactici* H did not affect one of the two *S. aureus* examined. The bacteriocin assay was performed anaerobically in GasPak jars to rule out hydrogen peroxide formation, and yeast extract-supplemented Trypticase soy agar was used to prevent acid production from glucose, which is an ingredient in MRS, the medium most commonly used in plate assays for lactic acid bacteria. The method was a spot deferred antagonism test as noted earlier (Harris et al., 1989). *A. hydrophila* 7965 and K144 were sensitive to nearly all of the bacteriocin-producing strains of lactic acid bacteria. Normally, bacteriocins from Gram-positive bacteria have nominal or no effect against Gram-negative bacteria, and vice versa (Tagg et al., 1976). Earlier, Bhunia et al. (1988) noted the inhibition of *A. hydrophila* by pediocin AcH. A stable, broad-spectrum bacteriocin effective against both Gram-positive and Gram-negative pathogenic bacteria would be a highly desirable preservative compound for use in foods.

IV. Pediocin AcH

Besides nisin, pediocin AcH is the most thoroughly studied bacteriocin of lactic acid bacteria. Recent information revels that both pediocin AcH and pediocin PA-1 are same. The different characteristics of pediocin AcH are presented in this discussion.

A. Properties

Bhunia et al. (1987a) first reported partial purification of a bacteriocin (later designated as pediocin AcH) from *Pediococcus acidilactici* strains that had antibacterial action against different pathogenic and food spoilage bacteria. The antibacterial compound was found to be protein in nature and produced extracellularly. Later studies by this group (Bhunia et al., 1987b) indicated that pediocin AcH is a fairly small protein of approximately 2700 Da. The antibacterial activity of the protein is retained after treatment with many physical and chemical agents, but is destroyed following treatment with proteolytic enzymes (Table III). Retention of antibacterial activity following treatment with lipase, ribonuclease, lysozyme, and organic solvents indicates it is a pure protein, rather than a conjugated protein (Bhunia et al., 1988). The ability to retain antibacterial property after treatment with formaldehyde or high heat also indicates that the molecule may be fairly small. The loss of activity subsequent to exposure to alkaline pH of 10 and above suggests that the molecule probably has some type of secondary structure that is important for its antibacterial efficiency; alkaline treatment destroys that structure resulting in loss of antibacterial activity.

Bhunia et al. (1987b) reported the molecular weight of pediocin AcH

TABLE III
Effect of Several Physical and Chemical Treatments on the Antibacterial Activity of Pediocin AcH

Treatment	Activity
Lipase	+
Lysozyme	+
Ribonuclease A	+
Chloroform (10%)	+
Isopropanol (10%)	+
Ethanol (25%)	+
Hexane (25%)	+
Acetonitrile (70%)	+
Trypsin	−
Chymotrypsin	−
Ficin	−
Papain	−
121°C/15 minutes	+
pH 2.5–9/24 hours at 25°C	+
pH 10/24 hours at 25°C	−
pH 12/2 hours at 25°C	−

Source: Adapted from Bhunia et al., 1991.

produced by *P. acidilactici* H to be 2700. A dialyzed growth medium was used and after cell removal pediocin AcH was precipitated from the broth supernatant with 70% ammonium sulfate. The precipitate, after dialysis to remove ammonium sulfate, was examined for antibacterial activity and electrophoresed in a 10–25% SDS-PAGE gel by applying the samples in duplicate. After electrophoresis the gel was cut in two halves, one was stained with Coomassie blue and the other half was washed in deionized water for at least 4 hours to remove the SDS and then was assayed for antibacterial activity. This was done by placing the gel on the top of an agar medium surface in a petri dish and overlaying the gel surface with melted soft agar seeded with cells of indicator bacteria. After overnight incubation the plate was examined for the location of the zone of growth inhibition and compared with the bands on the stained half of the gel to locate the protein associated with the antibacterial activity (Figure 2). Initial data indicated that the smallest band was between 2510 and 6210 Da and this band of about 2700 Da was associated with the antibacterial property, while two other bands with higher molecular weight were not antibacterial. Similar testing in SDS-PAGE gels revealed that the pediocins produced by other three strains, E, F, and M, were also of the same molecular weight (Ray et al., 1989a; Ray 1992b). Additional purification steps prior to SDS-PAGE included separation in a Superose-12 HR 10/30 column and passage through a Mono Q HR 5/5 column by FPLC (Bhunia et al., 1988). In each step the active fractions were used for subsequent purification. Although this method produced

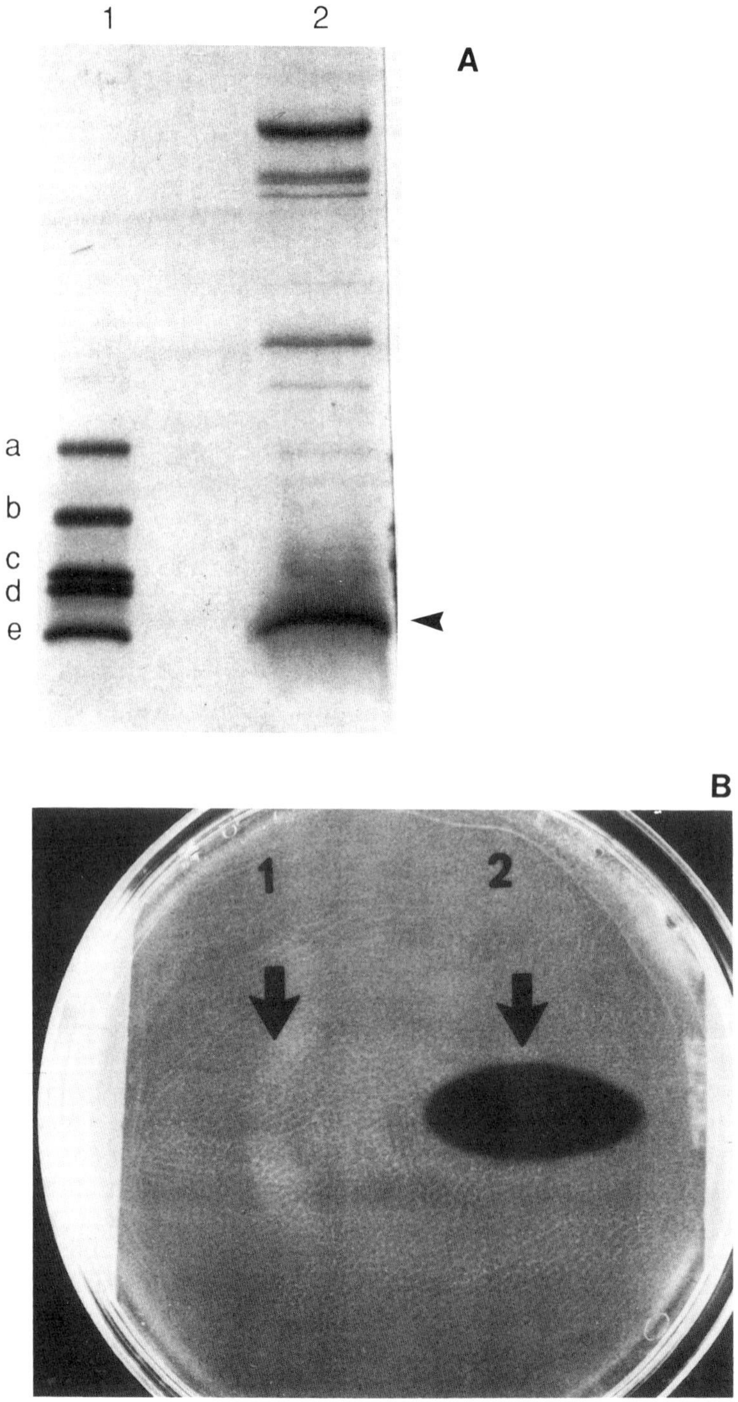
A
1
2
a
b
c
d
e
B
1
2

only a single band (Figure 3), examination of the material in a C-18 column produced more than one peak with activity. Analysis of the materials from the peaks for amino acid composition and amino acid sequencing did not produce consistent results.

Yang et al (1992) developed a novel purification method of bacteriocins of lactic acid bacteria, including pediocin AcH. This is based on pH dependency of bacteriocins to adsorb to and release from the cell surface of Gram-positive bacteria. Purified pediocin AcH, prepared by this method, was used to determine the amino acid composition and sequence of pediocin AcH (Motlagh et al. 1992b). The molecule has 44 amino acids with a mass of 4.6 kDa and a pI of 9.6. It is a cationic amphiphile molecule with a net 6+ charge at pH 6 to 7. The amino acid sequence has been presented in a separate section.

Studies by Henderson et al. (1992) and Lozano et al. (1992) have also shown that pediocin PA-1 of *P. acidilactici* (PAC 1.0) has the same amino acid sequence. Two disulfide bonds between cysteine molecules at positions +9 and +14 and at +23 and +44 gave the molecule a secondary structure.

B. Toxicity

Bhunia et al. (1990) investigated the antibody production ability of the crude pediocin AcH prepared by ammonium sulfate precipitation in mice and rabbit. They injected mice with an oil adjuvant-based pediocin AcH as well as pediocin AcH conjugated with bovine serum albumin, and also injected rabbits with pediocin AcH mixed with incomplete Freund's adjuvant by the recommended protocols used for immunization. Although ELISA indicated the presence of antibody, results of immunoblotting showed that the antibodies were against the larger proteins and not against pediocin AcH.

Current investigations by A. Bhunia (personal communication) with a much purer preparation did not produce antibody in mice. This is further being investigated. Murine hybridoma cells did not also show viability loss following treatment with pediocin AcH.

Injecting mice and rabbit subcutaneously, intravenously, and per-

---

Figure 2 Detection of pediocin AcH band on SDS-PAGE gel. (A) The stained half of the SDS-PAGE gel shows bands formed by several proteins present in the (NH$_4$)$_2$SO$_4$ preparation (Lane 2). The MW of the uppermost band is about 50,000 and the bottom one about 2700 (see arrow). Lane 1 contains the low-MW standards (MW-SDS-17, Sigma): a. 16,950; b. 14,400; c. 8160; d. 6210; e. 2510, which were used to calculate the MW of lower bands (16,000 and below). High-MW standards were used to calculate the MW of the upper bands (above 16,000 not presented). (B) The other half of the gel was overlaid with *Lactobacillus plantarum* to determine which band(s) corresponds to the antimicrobial activity. Lane 2, showing the zone of growth inhibition (see arrow) corresponded with the lowermost band (MW 2700); other bands did not show antimicrobial activity. Lane 1 contained trypsin-treated sample and showed no growth inhibition. (From Bhunia et al., 1987b, with permission of *J. Ind. Microbiol.*)

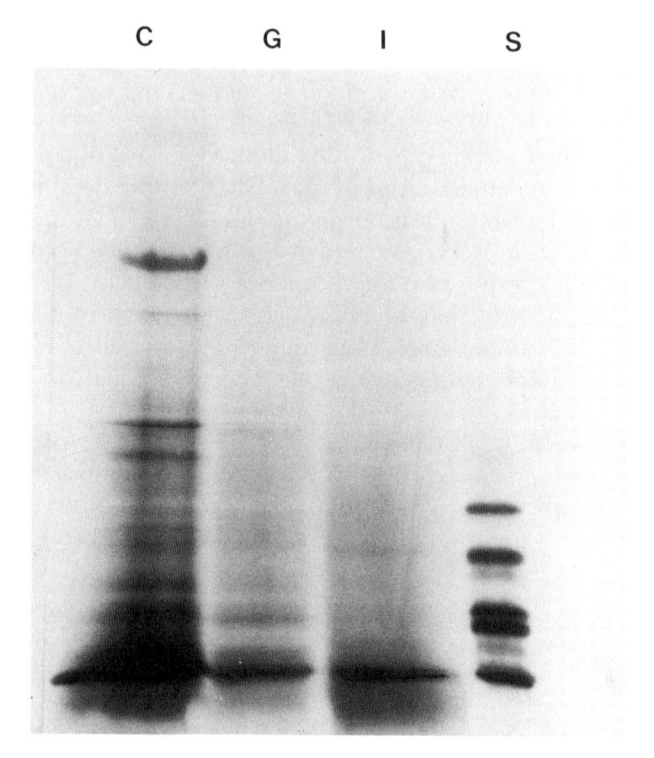

Figure 3 Sodium dodecyl sulfate polyacrylamide gel electrophoresis of pediocin AcH preparations after various stages of purification, silver stained. Lane C, crude pediocin AcH; Lane G, pediocin AcH after elution from a gel filtration column; Lane I, pediocin AcH eluted from ion exchange chromatography; Lane S; molecular weight standards (MW-SDS-17, Sigma), 16,950, 14,400, 8160, 6210, and 2510 Da. (From Bhunia et al., 1988, with permission of *J. Appl. Microbiol.*)

itoneally during immunization studies did not produce any local or general reactions or death to the animals (Bhunia et al., 1990). In addition, the ingestion of pediocin AcH-containing sausages fermented by *P. acidilactici* H by 12–14 taste panel personnel during a six-month period did not cause any physical discomfort during the study (Wu et al., 1991).

C. Mode of Action

Initial studies on pediocins of *Pediococcus pentosaceus* and *P. acidilactici* strains indicated that like other bacteriocins of lactic acid bacteria, they are bacteriocidal to sensitive Gram-positive bacterial cells (Daeschel and Klaenhammer, 1985; Gonzalez and Kunka, 1987; Bhunia et al., 1988). Gonzalez and Kunka (1987) showed that pediocin PA-1 molecules of *P. acidilactici* PAC 1.0 were adsorbed on the sensitive cells prior to killing of the cells; however, they observed that the resistant Gram-positive bacteria, including the producers of pediocin PA-1, adsorbed pediocin PA-1 molecules. Thus, adsorption of pediocin molecules on a cell could not be the only cause of bacteriocidal action. They suggested, referring to Tagg et al. (1976), that the nonlethal binding might be due to high surface activity of the pediocins that resulted in their nonspecific adsorption to various bacteria. This suggestion failed to explain why a bacteriocin from a Gram-positive bacterial strain binds to Gram-positive and not to Gram-negative bacteria. Bhunia et al. (1988) showed that the viability loss of sensitive *Lactobacillus plantarum* WSO-39 cells following exposure to pediocin AcH from *P. acidilactici* H is associated with membrane damage; however, the damage was not severe enough to cause cell lysis or even efflux of large molecules from the cytoplasm, such as β-galactosidase. They also observed that the suspension of a sensitive bacterial strain contained some resistent cells that could multiply in the presence of pediocin AcH (Bhunia et al., 1988). Pucci et al. (1988) reported that treatment of pediocin PA-1 to growing sensitive strains of *Listeria monocytogenes* LMO1 and *P. pentosaceus* FBB63 resulted in cell lysis, determined from the measurement of turbidity; however, they observed a difference in the lysis rate between the two strains and the rates directly corresponded with the minimum inhibitory concentration of the two strains, that is, the more resistant cells lysed at a rate slower than that of the less resistant cells.

Several questions need to be answered to understand the mode of bacteriocidal action of pediocin to sensitive strains. These include: (1) why pediocin molecules are not effective against Gram-negative cells; (2) why Gram-positive cells adsorb pediocin at the cell surface; (3) why some strains are sensitive while others are resistant even though both types adsorb pediocin, and why some cells, even in a sensitive strain, develop resistance to pediocin; (4) why some sensitive strains show lysis but others do not when treated with pediocin; (5) what is the nature of the cell surface receptor for pediocin; and (6) after binding to the receptor, how does the pediocin produce its bacteriocidal effect?

Bhunia et al. (1991) has answered some of these questions. They showed that Gram-negative bacteria generally do not adsorb pediocin AcH, while Gram-positive cells, including sensitive and resistant strains, do adsorb pediocin AcH, with some strains adsorbing more molecules than other strains. The adsorption of pediocin AcH to a sensitive strain can be prevented by anions such as chloride, phosphate, or bicarbonates in a competitive manner. The neutralizing effect of certain concentrations of an anion can be overcome by adding more pediocin AcH; however, it is not known if the neutralizing effect of the anions is due to their binding to the pediocin AcH molecule or due to their blocking of the pediocin AcH receptors on the cell surface. The results indicate that an ionic bond is most probably involved in the binding of pediocin AcH to the cell surface receptor(s). Treatments of sensitive cells with heat, protein-extracting chemicals (SDS, Triton X-100), a reducing agent (mercaptoethanol), lipid solvents (ethanol, hexane), and hydrolytic enzymes (lipases, mucopeptidase, phosphatase) did not result in subsequent loss in adsorption of pediocin AcH to the cells. Treatment of purified cell walls with some of these chemicals did not cause loss in adsorption of pediocin AcH. However, when the purified cell wall preparation from a sensitive Gram-positive strain was treated with a mixture of methanol/chloroform/trichloroacetic acid (2:2:1) at 93°C for 5 minutes, the remaining wall material no longer was able to adsorb or bind pediocin AcH. As the methanol/chloroform/trichloroacetic acid treatment at high temperature is used to extract lipoteichoic acid (LTA) from Gram-positive bacteria (Wallinder and Neujahr, 1971), the possible binding site of pediocin AcH on the cell surface of Gram-positive bacteria was considered to be LTA. To confirm this possibility, they measured the binding of pediocin AcH to *L. plantarum* NCDO 955 in the presence of LTA from two different Gram-positive bacteria. In the presence of LTA, pediocin failed to adsorb on the cells. Thus, it is most likely that LTA molecules are the receptor(s) for binding of pediocin AcH. This explains why the Gram-negative bacteria, which do not have LTA, do not bind pediocin. In the next several experiments these researchers showed that after binding to the receptors, pediocin AcH caused destabilization of the cytoplasmic membrane in sensitive cells. This was evident by efflux of K^+ from the cells, and influx of ortho-nitrophospho-β-galactoside (ONPG) into the cells. The consequence of the membrane damage brought about by the hydrophobic nature of pediocin AcH molecules probably disrupted the proton motive force and resulted in cell death. Following pediocin AcH treatment, some sensitive strains showed cell lysis,, as evidenced by a drop in optical density (OD); other strains did not demonstrate lysis, but did show growth inhibition (Figure 4). The OD reduction data were in agreement with the electron microscopy data (Figure 5). The strains that were lysed showed the presence of ghost cells, while the strains that did not lyse showed darkening of the

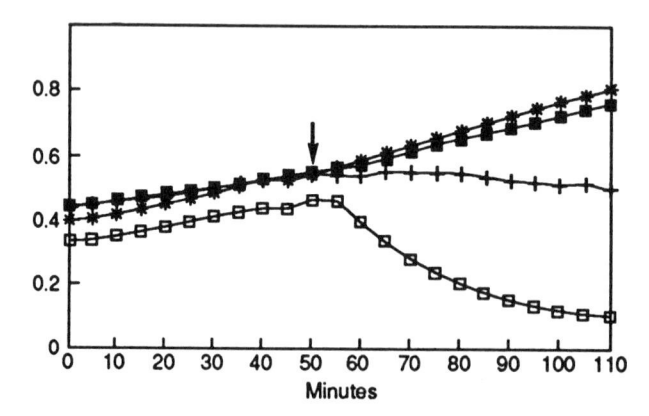

Figure 4 Effect of treatment with pediocin AcH on OD_{600} of growing cultures of *Lactobacillus plantarum* NCDO 955 (+, treated; ■, untreated) and *Leuconostoc mesenteroides* Ly (□, treated; *, untreated). Pediocin AcH was added at 50 minutes (arrow). Untreated controls continued to grow, while treated samples either stopped growth (+) or lysed (□). (From Bhunia et al., 1991, with permission of *J. Appl. Microbiol.*)

8 Pediocins

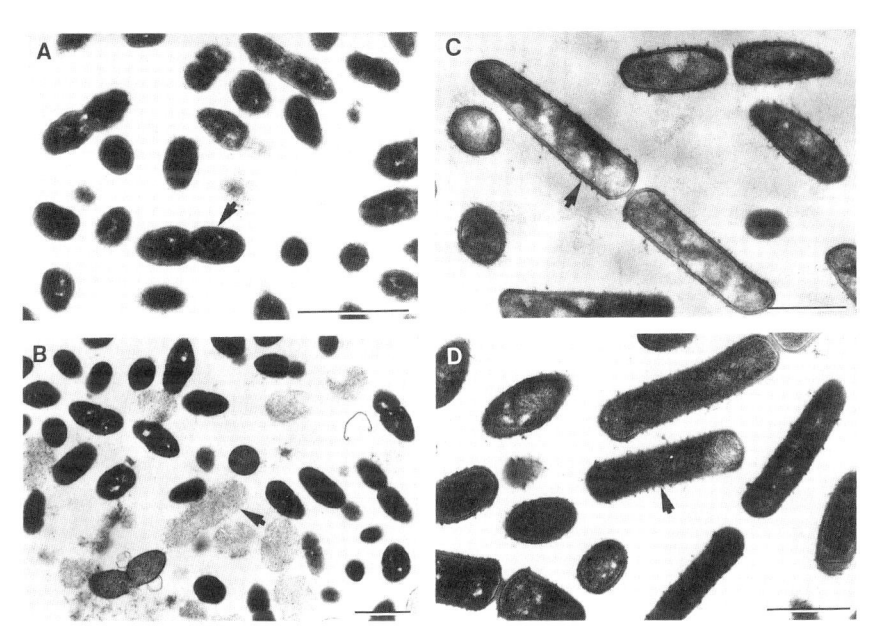

Figure 5 Transmission electron micrographs of cells of *Leuconostoc mesenteroides* Ly (A and B) and *Lactobacillus plantarum* NCDO 955 (C and D). Pediocin-treated samples showed lysed ghost cells (B) and darker cytoplasm (D). (From Bhunia et al., 1991, with permission of *J. Appl. Microbiol.*)

cytoplasm. Following the killing by pediocin AcH, probably the strains that are lysed have autolytic systems that once activated result in cell lysis. In a separate study, Motlagh et al. (1991) showed that of the seven pediocin AcH-sensitive *Listeria monocytogenes* strains, three showed lysis and four did not lyse following exposure to pediocin AcH. Apparently, relative sensitivity of the strains to pediocin AcH was not related to the cell lysis. Thus, a highly sensitive strain, Ohio₂, did not lyse by pediocin AcH while a less sensitive strain, Camp⁺/Beta⁺ showed lysis.

To explain why pediocin AcH can cause damage to cytoplasmic membrane of the sensitive strains only after binding to the cell wall, Bhunia et al. (1991) proposed, following the hypothesis of Tagg et al. (1976), that resistant cells have only nonspecific receptors, while the sensitive cells have both nonspecific and specific receptors. The nonspecific receptors are probably LTA and binding of pediocin AcH to LTA does not allow the molecules to come in contact with the cytoplasmic membrane. In contrast, binding in the specific sites in the sensitive cells allows pediocin AcH to come in contact with the cytoplasmic membrane. Ray (1992b, 1993) has proposed that binding of pediocin AcH (or other bacteriocins) to the specific cell surface receptors of the sensitive cells, but not to resistant cells, causes a loss in cell wall permeability barrier. This in turn allows other bacteriocin molecules to pass through the wall, come in contact with the membrane, and destabilize its functions.

D. Influence of Growth Conditions on Production of Pediocin AcH

Pediocin-producing *Pediococcus acidilactici* strains have been cultured in MRS broth (Gonzalez and Kunka, 1987), APT broth (Hoover et al., 1988), and casein-glucose broth (Bhunia et al., 1987b; 1988) to determine the ability of the strains to produce pediocins. Recently the influence of several growth parameters on the production of pediocin AcH by *P. acidilactici* H has been reported (Biswas et al., 1991). The active pediocin AcH is produced as a secondary metabolite and a lower terminal pH range favors an increased level of production. These results (Figure 6) show that in TGE (Trypticase or Tryptose, 1%; glucose, 1%; yeast extract, 1%; Tween 80, 0.2%; Mn^{2+} and Mg^{2+}, each 0.005%; pH 6.5) broth, the activity units (AU)/ml of pediocin AcH is about 3.6×10^4 at terminal pH of 3.6, compared to about 1.0×10^3 at terminal pH 5.0. Above pH 5.0 no pediocin AcH activity is detected (Figure 6); however, at higher terminal pHs the cell mass, as determined by OD at 600 nm, was relatively higher than the cell mass at pH 4.0 or below. The need for an acidic environment for production of active pediocin AcH could be associated with posttranslation processing of the inactive prepediocin AcH molecules to the active form. This aspect is currently being investigated. Production of pediocin AcH is relatively less in media that are highly buffered, such as MRS broth, but higher when the cells are grown in a nonbuffered broth, such as TGE broth. Increasing concentrations of Trypticase/Tryptone, glucose, and yeast extract from 1 to 2% results in

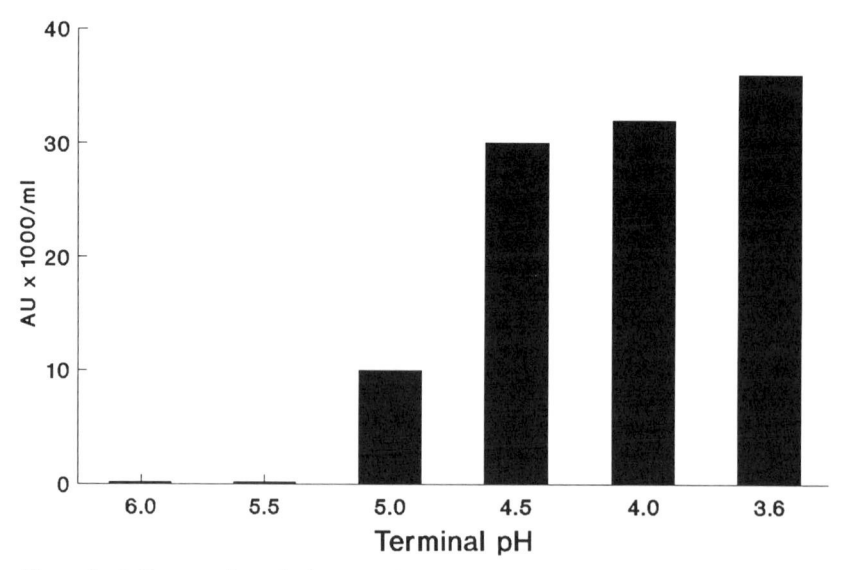

Figure 6 Influence of terminal pH on the concentration of active pediocin AcH produced by *Pediococcus acidilactici* H in TGE broth during incubation at 37°C for 22 hours. Initial pH for each study was 6.5. The terminal pH was controlled by addition of NaOH solution automatically in a fermentor. (Adapted from Biswas et al., 1991.)

higher production of pediocin AcH by about 15%. Among fermentable carbohydrates, maximum production is observed with glucose followed by sucrose, xylose, and galactose. With arabinose, raffinose, and trehalose little cell growth is found with no detectable pediocin AcH. The level of pediocin AcH production is higher at incubation temperature of 37°C and lower at both 30°C and 40°C. Also, with a 1% initial inoculum and 37°C incubation, the level of pediocin AcH in broth reaches maximum between 16 and 22 hours; beyond 22 hours the level of pediocin AcH decreases, probably because of the action of proteolytic enzymes from the dead cells. The supplementation of growth media with several B vitamins did not increase pediocin AcH production (B. Ray, unpublished data).

E. Genetics

Pediocin AcH production in *Pediococcus acidilactici* strains H, E, F, and M has been shown to be encoded by plasmid pSMB74 of about 8.9 kb. This has been determined by plasmid-curing experiments as well as by plasmid transfer techniques (Ray et al., 1989a,b; Kim et al., 1992). The plasmid has been mapped and found to have unique sites for *Eco*RI, *Sal*I, *Sac*I, and *Bgl*I. It also has three *Hin*dIII sites producing fragments of approximately 3.5 kb, 2.7 kb, and 2.7 kb (Ray et al. 1991, 1992). The three fragments have been cloned in vector pUC 119. Using primer extension techniques, plasmid pSMB74 has been completely sequenced and found to contain 8877 bp (A. Motlagh and B. Ray, unpublished data). In pSMB74 a segment of DNA, between *Hin*dII and Sau3A, of 3570 bp has a cluster of four open reading frames (ORF), ORF 1, 2, 3 and 4. The cluster has a common promoter upstream of ORF1 and an inverted repeat of 8 bp, downstream of ORF4. Between each pair of ORF there are noncoding spacer sequences of 23 to 98 bp, and each ORF has individual ribosome binding sites (rbs) and terminating codon. They encode for proteins of 62, 112, 174, and 724 amino acid, respectively. While the 62 amino acid protein is found to be prepediocin and 724 amino acid protein has excretory (secretory) functions, the functions of the other two proteins are yet not known. They might have functions related to immunity against pediocin AcH and proteolytic cleavage of prepediocin (Ray et al., 1989c).

Prepediocin at the translation level is a 62-amino acid peptide with an N-terminus leader segment of 18 amino acids (−1 to −18) and a C-terminus propediocin (+1 to +44) (Motlagh et al., 1992b):

```
                                          ↓
M  K  K  I  E  K  L  T  E  K  E  M  A  N  I  I  G  G   K  Y  Y  G  N  G
-18                     -10                         -1  +1

V  T  C  G  K  H  S  C  S  V  D  W  G  K  A  T  T  C  I  I  N  N  G
            +10                     +20

A  M  A  W  A  T  G  G  H  Q  G  N  H  K  C . . .
+30                        +40         +44
```

Following translation the molecules undergo maturation (probably disulfide bonds between cysteines, one at position +9 and +14 and another at positions +23 and +44, to give the molecule a secondary structure) and processing (removal of 18 amino acid leader segment by a specific protease). The processing occurs at low pH by an acid protease that converts to inactive prepediocin to active pediocin AcH. However, as pediocin AcH molecules have a destabilizing function on the cytoplasmic membrane, the molecules are immediately excreted, by secretory proteins, outside in the environment. Current studies have confirmed that phenotypes for immunity against pediocin AcH of the producer cells and processing as well as excretory functions are all encoded in pSMB74 (B. Ray, unpublished data).

Marugg et al. (1992) has also reported that pediocin PA-1 is encoded in plasmid pSRQ11. There is a cluster of four ORF that are associated with the production of pediocin PA-1 and ORF1 encodes for prepediocin of 62 amino acid. Other findings are also similar to the observations made with pediocin AcH.

F. Antibacterial Effectiveness of Pediocin AcH

The antibacterial spectrum of pediocin AcH has been studied using pure cultures by assaying zone of growth inhibition, by measuring viability loss from the enumeration of CFUs (Bhunia et al., 1988; 1991), or by examining turbidity in a broth culture (Kalchayanand, 1990). These results have indicated that pediocin AcH is effective against vegetable cells of many strains of *Lactobacillus* spp., *Leuconostoc* spp., *Enterococcus* spp., *Pediococcus* spp., *Lactococcus* spp., *Propionibacterium* spp., *Bacillus* spp., *Clostridium* spp., *Listeria* spp., *Staphylococcus* spp., *Brochothrix thermosphacta*, and spores of some *Bacillus* and *Clostridium* spp. The antibacterial effectiveness of pediocin AcH against a sensitive strain increases as the pediocin AcH concentration increases; in addition, the antibacterial effect of pediocin AcH increases as the population of a sensitive strain decreases. In each sensitive strain there are some cells that are resistant to pediocin AcH, a percentage of which is strain variable. Also, sensitivity to pediocin AcH varies greatly among the species and strains in the same species. As examples, a 1-hour exposure of about 10^8 CFU/ml of *Listeria monocytogenes* Scott A and *L. ivanovii* ATCC 19119 to 1500 AU/ml pediocin AcH reduced the CFU/ml by 2.2 and 5.6 logs, respectively; also, under similar conditions the $\log_{10}$ reduction of 9 strains of *L. monocytogenes* ranged from 1.5 to 4.1 (Motlagh et al., 1991; Motlagh, 1991; Holla, 1990). While pediocin AcH produces a bacteriocidal effect on sensitive vegetative cells by destabilizing the membrane function, its inhibitory effect on sensitive bacterial spores is produced by inhibiting outgrowth following germination (Kalchayanand, 1990).

Bhunia et al. (1988) showed that a growing *Lactobacillus plantarum* WSO-39 culture on exposure to pediocin AcH in a broth lost over 92% CFU within 1 hour but the surviving cells continued to multiply at the same rate as the pediocin AcH-untreated control. By increasing the pediocin AcH concentrations, the CFU could be reduced by 4 logs (Bhunia et al., 1988; 1991). Isolates of lactobacilli, leuconostocs, and *Brochothrix thermosphacta* when treated with pediocin AcH (2000 AU/ml) showed reduction in CFU ranging from 1 to over 4 logs within 30 minutes. 10^5 CFU/ml of a psychrotrophic clostridial species, *Clostridium laramie*, associated with spoilage of vacuum-packaged refrigerated beef, failed to grow in a broth containing 10,000 AU/ml of pediocin AcH. Treatment of the *C. laramie* spores in broth with pediocin AcH prevented outgrowth of the spores (Ray et al., 1989b; Kalchayanand, 1990).

Similar studies also showed that pediocin AcH (1500 AU/ml) treatment of *L. monocytogenes* and other *Listeria* species reduces CFU from 1.5 to over 7 logs within 1 hour in tryptic soy broth (Motlagh et al., 1991, 1992a; Motlagh, 1991). During storage at 4°C the surviving cells multiply at a rate characteristic of the species and strain. For two *L. monocytogenes* strains the survivors of the more sensitive strain, Ohio$_2$, multiply at a faster rate than the survivors of less sensitive strain, Scott A (Motlagh et al., 1991; Holla 1990).

G. Pediocin AcH in Foods

The antibacterial effectiveness of pediocin AcH against spoilage and pathogenic Gram-positive bacteria in sterile and nonsterile food systems has been studied. Sterile foods from meat and dairy groups consisting of ground beef, sausage mix, cottage cheese and ice cream in liquid suspension (1:1), and milk were inoculated with cells of *Leuconostoc mesenteroides* Ly, *Listeria monocytogenes*, or *L. ivanovii* and pediocin AcH (Holla, 1990). During storage at 4°C or 10°C, the five foods did not interfere with antibacterial effectiveness of pediocin AcH and, in general, not much variation in viability loss was observed between five food systems under a given set of conditions, that is, with same concentrations of cell of a species or strain for a given amount of pediocin AcH. During storage the surviving cells multiplied at a rate that was characteristic of the species and strains. The multiplication occurred in the presence of residual pediocin AcH. Representative results in cottage cheese for *L. mesenteroides*, *L. monocytogenes* Scott A, and *L. ivanovii* ATCC 19119 are presented in Figure 7. In the pediocin AcH-treated (1300 AU/ml) samples, CFU of *L. mesenteroides* were reduced by 3 logs within 1 hour and the surviving cells stated growth at 4°C after 1 week (Figure 7A). In untreated samples, cell multiplication started after 1 day. The CFU of *L. monocytogenes* Scott A was reduced by about 1 log in 1 hour by pediocin AcH (1250 AU/ml) and the surviving cells did not start growth at 4°C until after the third week, although in the untreated sample growth started within 1 week (Figure 7B). Pediocin AcH (750 AU/ml) treatment reduced the CFU of *L. ivanovii* by 5 logs and the surviving cells did not multiply for up to 4 weeks at 4°C; in the untreated sample growth started within 1 week (Figure 7C). However, when the storage temperature of the samples was increased to 10°C, the surviving cells in the pediocin AcH-treated samples started

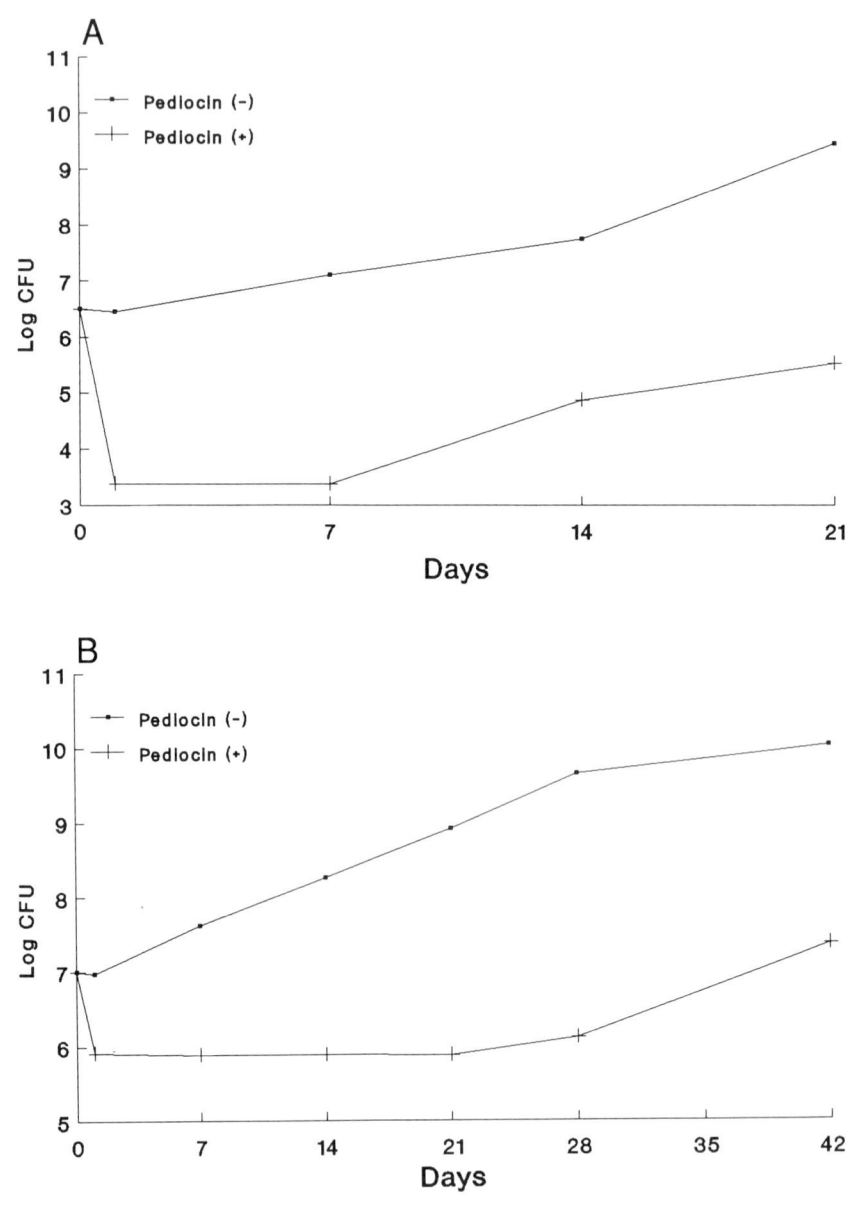

Figure 7 Viability loss and growth of survivors of *Leuconostoc mesenteroides* Ly (A), *Listeria monocytogenes* strains Scott A (B), and *Listeria ivanovii* ATCC 19116 (C) in cottage cheese at 4°C with or without (control) pediocin AcH during storage. In the untreated samples growth was initiated earlier. At 10°C growth of *L. ivanovii* ATCC 19116 (D) was initiated earlier and proceeded at a faster rate than at 4°C. Similar data were observed for other foods, namely, ground beef, sausage mix, milk, and ice cream. (From Holla, 1990.) (*Figure continues*)

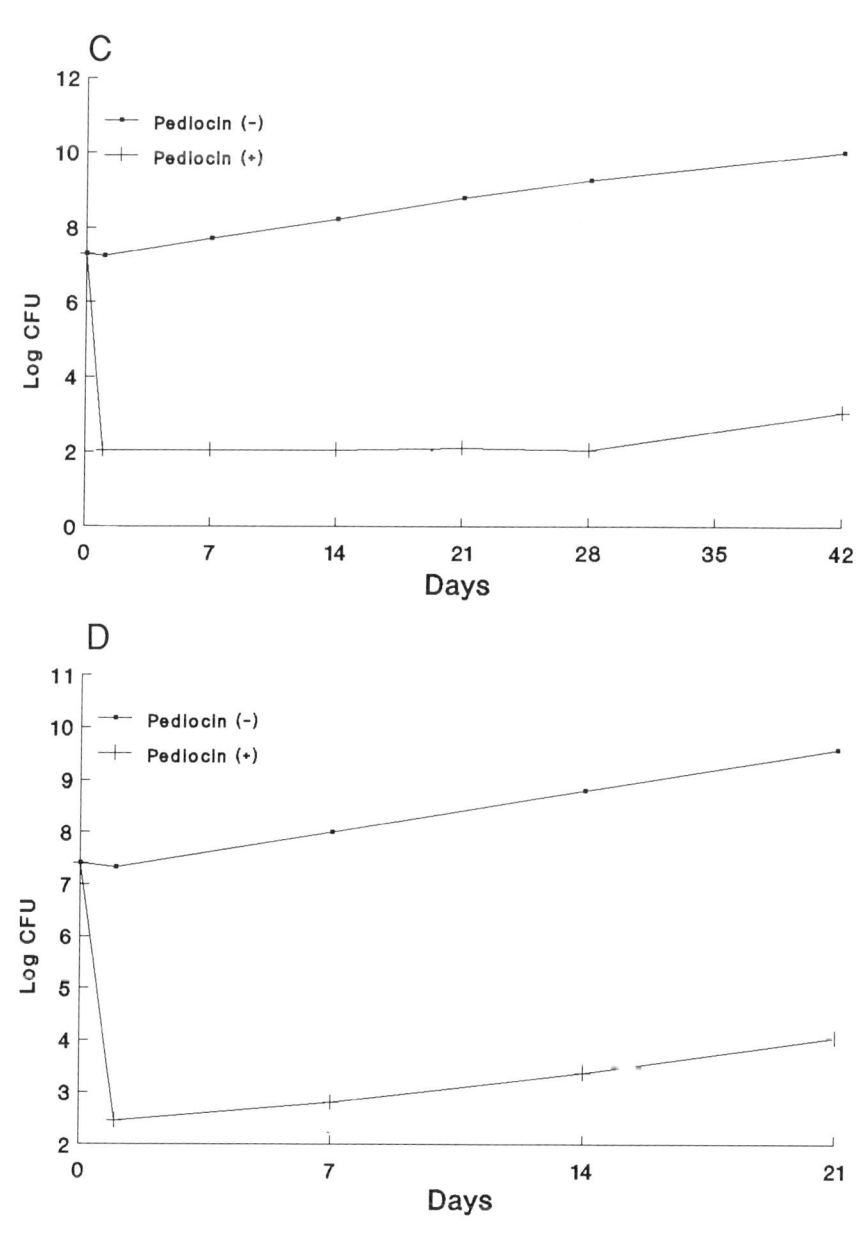

Figure 7 (*Continued*)

growth before 1 week (Figure 7D). Similar early growth initiation was observed for other bacterial strains in all foods tested (Holla, 1990). These results point out two important aspects about effective use of pediocin AcH as a secondary barrier to control spoilage and pathogenic bacteria: the initial load of the sensitive strains in the product should be low, so that a given pediocin AcH concentration produces maximum viability loss; also, the storage temperature (refrigeration) should be properly controlled to prevent early growth initiation and rapid growth of the resistant, surviving population.

To determine the effectiveness of pediocin AcH in controlling growth of spoilage bacteria, Kalchayanand (1990) inoculated psychrotrophic *Leuconostoc mesenteroides* Ly cells (about 5×10^3 ml) in fresh beef, treated with pediocin AcH (1240 AU/g), vacuum packaged the meat, stored it at 3°C, and enumerated for psychrotrophs during storage. The results (Figure 8) showed that initially in the pediocin AcH-treated samples there was more than a 2.5-log reduction in CFU. Up to 8 weeks the treated samples had 2 logs lower CFU than the untreated samples and at 12 weeks the treated

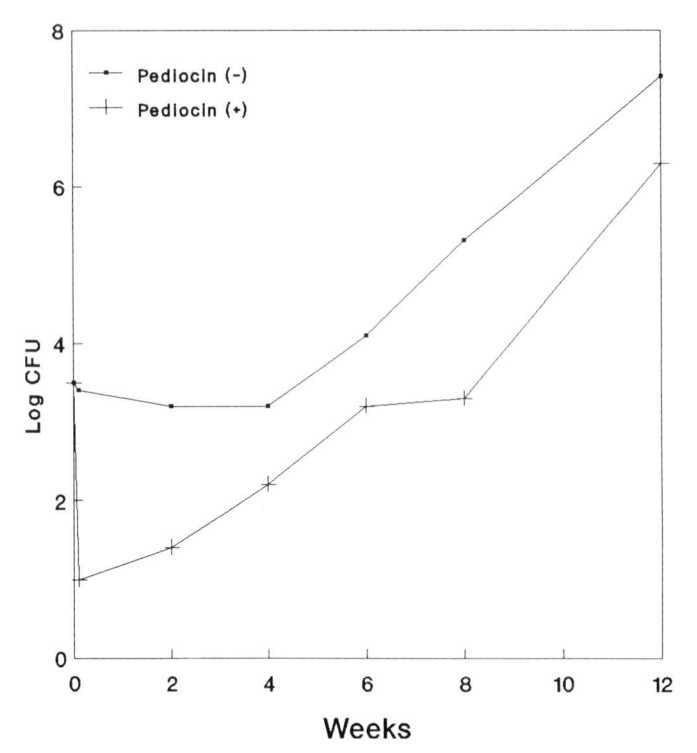

Figure 8 Viability loss and growth of *Leuconostoc mesenteroides* Ly inoculated in vacuum-packaged fresh beef containing pediocin AcH and stored at refrigeration temperature during storage for 12 weeks. Initial reduction due to cell death by pediocin AcH. Cells in untreated control samples and survivors in the treated samples multiplied during storage. (From Kalchayanand, 1990.)

samples had 1 log lower counts. But most importantly, the counts in the treated samples remained within the "spoilage level" ($\simeq 5 \times 10^6/g$). This indicates that to evaluate the actual benefit of adding pediocin AcH to vacuum-packaged fresh meat, several factors have to be carefully controlled. These include oxygen permeability of the bag, level of vacuum drawn, uniform mixing of pediocin AcH and bacterial cells, storage temperature, and absence of leaks. In a later study (Rozbeh et al., 1991), the effect of pediocin AcH treatment (2000 AU/g) on the psychrotrophic counts of commercially produced vacuum-packaged fresh beef during refrigeration storage was studied. The uninoculated beef was vacuum packaged in low-oxygen permeable bags after careful treatment with a pediocin AcH preparation and enumerated for psychrotrophs during a 10-week storage at 3°C. The results presented in Figure 9 show that initial reduction in CFU due to pediocin AcH treatment was low (only about 0.5 log). This was because only a small fraction of the normal microflora were Gram-positive pediocin AcH-sensitive bacteria. Many were Gram-negative and Gram-positive pediocin AcH-resistant bacteria. At the end of 10 weeks the population in the treated samples was about 1 log less than in the untreated

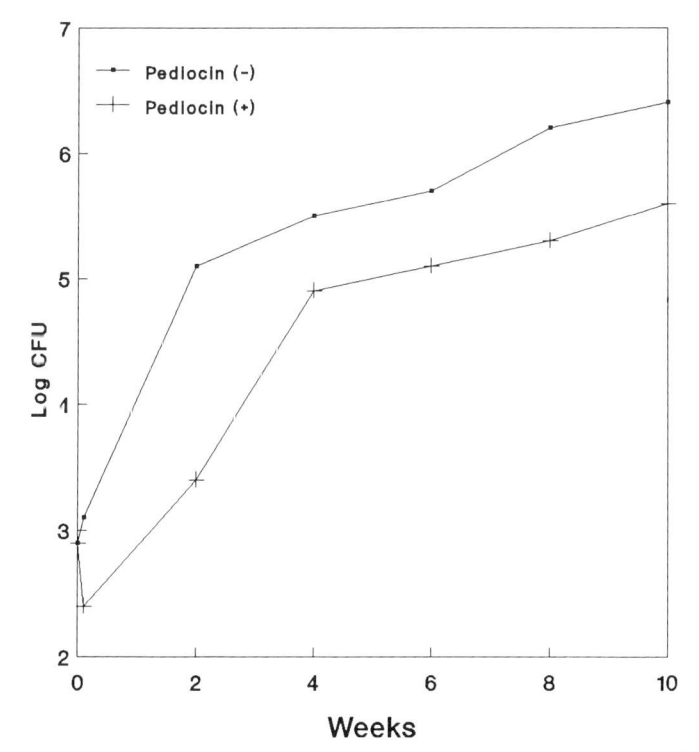

Figure 9 Viability loss and growth of psychrotrophic bacterial population in vacuum-packaged commercial beef treated with pediocin AcH during refrigeration storage at 4°C for 10 weeks. Viability loss was less, as a greater proportion of the normal bacteria are Gram-negative, with some resistant Gram-positive. (From Kalchayanand, 1990.)

controls and remained below $\log_{10} 6$. The quality of the treated meat samples remained acceptable.

Similar studies in which Gram-positive spoilage and pathogenic bacteria have been inoculated into vacuum-packaged refrigerated fresh beef byproducts and partially processed, cooked meat products (beef, pork, and poultry) treated with pediocin AcH preparations have been found to be highly effective.

V. Additional Studies of Pediocins in Meat Systems

Listeria monocytogenes Scott A is a strain that is popular in the evaluation of food process systems to be tested against *Listeria*. Scott A is a milk-borne clinical isolate of serotype 4B. It was used by Berry et al. (1990) in a study of the antilisterial effectiveness of a bacteriocin-positive meat starter, *Pediococcus* JD1-23, in summer sausage, a fermented semidry meat product. Another meat starter culture of *Pediococcus,* designated MP1-08, was used as a bacteriocin-negative control. Bacteriocin activity against Scott A was determined by both direct and deferred methods. In 13.6-kg batches of commercial sausage formulation, JD1-23 caused an approximate 2 $\log_{10}$-CFU/g reduction in an original inoculum of 10^6 cells/g of *L. monocytogenes* Scott A, while MP1-08 reduced the *Listeria* by less than 1 $\log_{10}$ CFU/g. Recovery of Scott A after sausage storage was variable and indicated that the bacteriocin was not totally effective in the inactivation of *L. monocytogenes* that survived the heating process. It was also noted that bacteriocin production was not linked to carbohydrate fermentation because in one study in which adequate acid production did not occur (as indicated by a pH of 5.5), Scott A was inactivated nonetheless.

Fresh meat was the test vehicle in a study conducted by Nielsen et al. (1990). Scott A was again the *Listeria monocytogenes* strain of choice and a commercial culture, *Pediococcus acidilactici* LACTACEL 110, was supplied by Microlife Technics (Sarasota, Florida). Fresh, lean beef muscle was cut into 0.5-cm-thick pieces with a final surface area of 4 cm^2, γ irradiation sterilized, vacuum packaged, and stored at $-15°C$ until use. The thawed pieces were soaked in buffer containing *L. monocytogenes* so that the extent of attachment of the pathogen to the meat surface could be quantified. The bacteriocin preparation consisted of supernatant from growth of LACTACEL 110 in MRS broth that was neutralized and filter sterilized and exposed to the inoculated meat pieces in solutions placed in sterile beakers at 5°C. Within 2 minutes a 0.5- to 2.2-log cycle decrease was witnessed in the number of attached *L. monocytogenes*. If the meat was pretreated with the bacteriocin solution, a drop of 1 to 2.5 log cycles was found over control for the number of attached bacteria. These effects were related to the inhibitory and bactericidal effects of the pediocin.

A wiener exudate derived from all-beef frankfurters was the model system used by Yousef et al. (1991) and Degnan et al. (1992) to evaluate the effect of *Pediococcus acidilactici* H or pediocin AcH on a three-strain mixture of *Listeria monocytogenes* composed of Scott A, V7, and 101M. While the ability of the wiener exudates to support growth of *L. monocytogenes* varied, it was concluded that the presence of either the bacteriocin-producing pediococci or a pediocin preparation (a supernatant that was heated, filtered, lyophilized, and rehydrated) inhibited *L. monocytogenes* at 25°C, even though there was growth of the listeriae in controls. The presence of the pediocin preparation resulted in an immediate drop in levels of *L. monocytogenes*, while antilisterial effect was not found with *P. acidilactici* H until after 64 hours of incubation, since the pediococcal cells required time to synthesize and release the bacteriocin for expression of antibiosis.

Foegeding et al. (1992) developed antibiotic-resistant cultures of *Pediococcus acidilactici* PAC 1.0 (which was originally designated as strain PO2) and five strains of *Listeria monocytogenes* to improve enumeration of these cultures when inoculated into defatted pork used in the production of dry fermented sausages. Plating on media containing the appropriate antibiotics allowed for selection of the added strains over the natural meat microflora. An isogenic pediocin-negative derivative was used as a control. Their study indicated that populations of *L. monocytogenes* at starting concentrations of approximately 10^5 CFU/g were reduced by a combination of both the pediocin produced by the starter culture and the dry sausage fermentation process when the final pH was less than 4.9. Most of the 5-log unit reduction occurred as a result of the fermentation, heat, and drying of the sausage-making. Nonetheless, the results did show that the presence of the pediocin produced by the meat starter culture contributed to an increase in product safety evident during both the fermentation and drying portions of sausage manufacturing.

Overall these contributions add further evidence that the presence of a bacteriocin can be a significant deterrent to growth of a bacteriocin-sensitive pathogen in a meat system, especially a fermented product in which a lactic acid starter culture is a necessary additive. This supports the justification that to improve the safety of these products it would be advisable to include as many effective bacteriocins as possible, either by the addition of purified bacteriocin as a food preservative, or by developing and using starter cultures capable of synthesizing a range of bacteriocins. Of course, an important issue is the regulatory or legal aspects for approval of such use. These aspects often overshadow the scientific or technological issues.

VI. Other Potential Applications for Pediocins

Preservatives are used in a wide variety of products other than foods and beverages. Such items as cosmetics, personal hygiene products, disinfectants and cleansers, and animal feeds all have storage requirements and need preservatives that are effective, economical, and environmentally responsible. Those preservatives that lack the harshness of some of the chemical agents will be in great demand.

Another application in which bacteriocin production is a potentially important factor is the use of pediococci as an inoculant in silage production, with the subsequent factor of probiotic effect in animals fed such silage. Pediococci are commonly part of the natural flora of ensiled plant products, but their activity in the rumen or gastrointestinal tract of mammals is not usually considered probiotic, since pediococci are not normally found there. In a growth study conducted with heifers fed corn silage inoculated with *Pediococcus acidilactici* and *Lactobacillus xylosus*, no significant advantage was found in growth and efficiency of feed utilization for the inoculated silage versus control, although the aerobic stability of the silage appeared to be improved by inoculation (Cleale et al., 1990). However, it may still be desirable to move genes for pediocin biosynthesis and immunity into a recipient gut bacterium that is indigenous to the intestinal tract of a specific host. In poultry, such an intergeneric transfer should not be necessary, since pediococci are found in the caeca of newly hatched chicks (Barnes et al., 1980). Here, pediococci are potential candidates for probiotic therapy because in addition to producing bacteriocins, they can colonize the GI tract, producing hydrogen peroxide and organic acids, while possibly promoting deconjugation of bile acids and salts to yield more inhibitory free bile acids (Juven et al., 1991).

A species of *Pediococcus* that receives scant attention in the West, which has been studied for many years in Japan because of its economic importance in the production of miso and soy sauce is *P. halophilus*. Although the synthesis and expression of pediocins in *P. halophilus* have not been studied, the plasmid biology has (Kayahara et al., 1989; Toda et al., 1989). Given the ubiquitous nature of plasmids and bacteriocins, one can assume plasmid-encoded bacteriocin genes will some day be characterized, and with further development of host–vector systems in *P. halophilus*, bacteriocins of *P. halophilus* may have a potential for commercial applications similar to that of *P. acidilactici* and *P. pentosaceus*.

Acknowledgments

The unpublished result included in this chapter from B. Ray's laboratory are from studies funded by the National Livestock Meat Board, Binational Agricultural Research Development program, and Wyoming Economic Development program.

References

Anderson, R. (1986). Inhibition of *Staphylococcus aureus* and spheroplasts of gram-negative bacteria by an antagonistic compound produced by a strain of *Lactobacillus plantarum*. *Int. J. Food Microbiol.* **3,** 149–160.

Bacus, J. N., and Brown, W. L. (1981). Use of microbial cultures: Meat products. *Food Technol.* **35,** 74–78, 83.

Barnes, E. M., Impey, C. S., and Cooper, D. M. (1980). Manipulation of the crop and intestinal flora of the newly hatched chick. *Am. J. Clin. Nutr.* **33,** 2426–2433.

Baumann, H. E., and Foster, E. M. (1956). Characteristics of organisms isolated from the rumen of cows fed high and low roughage rations. *J. Bacteriol.* **71,** 333–338.

Berry, E. D., Liewen, M. B., Mandigo, R. W., and Hutkins, R. W. (1990). Inhibition of *Listeria monocytogenes* by bacteriocin-producing *Pediococcus* during the manufacture of fermented semidry sausage. *J. Food Protect.* **53,** 194–197.

Bhunia, A. K. (1989). Characteristics and mode of action of an antimicrobial peptide produced by *Pediococcus acidilactici* H. Ph.D. thesis, University of Wyoming, Laramie.

Bhunia, A. K., Kim, W. J., Johnson, M. C. and Ray, B. (1987a). Partial purification and characterization of antimicrobial substances from *Pediococcus acidilactici* and *Lactobacillus plantarum*. IFT 87 Program and Abstracts. Annual Meeting, Institute of Food Technologists, Las Vegas, June 16–19. Abstract No. 143.

Bhunia, A. K., Johnson, M. C., and Ray, B. (1987b). Direct detection of an antimicrobial peptide of *Pediococcus acidilactici* in sodium dodecyl sulfate-polyacrylamide gel electrophoresis. *J. Ind. Microbiol.* **2,** 319–322.

Bhunia, A. K., Johnson, M. C., and Ray, B. (1988). Purification, characterization and antimicrobial spectrum of a bacteriocin produced by *Pediococcus acidilactici*. *J. Appl. Bacteriol.* **65,** 261–268.

Bhunia, A. K., Johnson, M. C., and Ray, B. (1990). Antigenic properties of pediocin AcH. *J. Appl. Bacteriol.* **69,** 211–215.

Bhunia, A. K., Johnson, M. C., Ray, B., and Kalchayanand, N. (1991). Mode of action of pediocin AcH from *Pediococcus acidilactici* H on sensitive bacterial strains. *J. Appl. Bacteriol.* **70,** 25–33.

Biswas, S. R., Ray, P., Johnson, M. C., and Ray, B. (1991). Influence of growth conditions on the production of a bacteriocin, pediocin AcH, by *Pediococcus acidilactici* H. *Appl. Environ. Microbiol.* **57,** 1265–1267.

Blood, R. M. (1975). Lactic acid bacteria in marinated herring. In "Lactic Acid Bacteria in Beverages and Food" (J. G. Carr, C. V. Cutting, and G. C. Whiting, eds.), pp. 195–208. Academic Press, Inc., New York.

Chassy, B. M., Mercenier, A., and Flickinger, J. (1988). Transformation of bacteria by electroporation. *Trends Biotechnol.* **6,** 303–309.

Cleale, IV, R. M., Firkins, J. L., Van Der Beek, F., Clark, J. H., Jaster, E. H., McCoy, G. C., and Klusmeyer, T. H. (1990). Effect of inoculation of whole plant corn forage with *Pediococcus acidilactici* and *Lactobacillus xylosus* on preservation of silage and heifer growth. *J. Dairy Sci.* **73,** 711–718.

Daeschel, M. A., and Klaenhammer, T. R. (1985). Association of a 13.6-megadalton plasmid in *Pediococcus pentosaceus* with bacteriocin activity. *Appl. Environ. Microbiol.* **50,** 1538–1541.

Degnan, A. J., Yousef, A. E., and Luchansky, J. B. (1992). Use of *Pediococcus acidilactici* to control *Listeria monoctyogenes* in temperature-abused vacuum-packaged wieners. *J. Food Prot.* **55,** 98–103.

Etchells, J. L., Costilow, R. N., Anderson, T. E., and Bell, T. A. (1964). Pure culture fermentation of brined cucumbers. *Appl. Microbiol.* **12,** 523–535.

Facklam, R., Hollis, D., and Collins, M. D. (1989). Identification of gram-positive coccal and coccobacillary vancomycin-resistant bacteria. *J. Clin. Microbiol.* **27,** 724–730.

Fleming, H. P., and McFeeters, R. F. (1981). Use of microbial cultures: Vegetable products. *Food Technol.* **35,** 84–88.

Fleming, H. P., Etchells, J. L., and Costilow, R. N. (1975). Microbial inhibition by an isolate of *Pediococcus* from cucumber brines. *Appl. Microbiol.* **30,** 1040–1042.

Foegeding, P. M., Thomas, A. B., Pilkington, D. H., and Klaenhammer, T. R. (1992). Enhanced control of *Listeria monocytogenes* by in situ-produced pediocin during dry fermented sausage production. *Appl. Environ. Microbiol.* **58,** 884–890.

Garvie, E. I. (1986). Genus *Pediococcus* Claussen 1903. *In* "Bergey's Manual of Systematic

Bacteriology" (P. H. A. Sneath, ed.), Vol. 2, pp. 1075-1079. Williams and Wilkins, Baltimore.
Gonzalez, C. F., and Kunka, B. S. (1983). Plasmid transfer in *Pediococcus* spp.: Intrageneric and intrageneric transfer of pIP501. *Appl. Environ. Microbiol.* **46,** 81-89.
Gonzalez, C. F., and Kunka, B. S. (1986). Evidence for plasmid linkage of raffinose utilization and associated α-galactosidase and sucrose hydrolase activity in *Pediococcus pentosaceus*. *Appl. Environ. Microbiol.* **51,** 105-109.
Gonzalez, C. F., and Kunka, B. S. (1987). Plasmid-associated bacteriocin production and sucrose fermentation in *Pediococcus acidilactici*. *Appl. Environ. Microbiol.* **53,** 2534-2538.
Graham, D. C., and McKay, L. L. (1985). Plasmid DNA in strains of *Pediococcus cerevisiae* and *Pediococcus pentosaceus*. *Appl. Environ. Microbiol.* **50,** 532-534.
Haines, W. C., and Harmon, L. G. (1973). Effect of selected lactic acid bacteria on growth of *Staphylococcus aureus* and production of enterotoxin. *Appl. Microbiol.* **25,** 436-441.
Harris, L. J., Daeschel, M. A., Stile, M. E., and Klaenhammer, T. R. (1989). Antimicrobial activity of lactic acid bacteria against *Listeria monocytogenes*. *J. Food Protect.* **52,** 384-387.
Harrison, A. P., and Hansen, P. A. (1950). The bacterial flora of the cecal feces of healthy turkeys. *J. Bacteriol.* **54,** 197-210.
Henderson, J. T., Chopko, A. L., and van Wassenaar, P. D. (1992). Purification and primary structure of pediocin PA-1 produced by *Pediococcus acidilactici* PAC-1.0. *Arch. Biochem. Biophys.* **295,** 5-12.
Holla, S. (1990). Efficiency of pediocin AcH on viability loss of pathogenic and spoilage bacteria in food. M.S. thesis, University of Wyoming, Laramie.
Hoover, D. G., Walsh, P. M., Kolaetis, K. M., and Daly, M. M. (1988). A bacteriocin produced by *Pediococcus* species associated with a 5.5-megadalton plasmid. *J. Food Protect.* **51,** 29-31.
Hoover, D. G., Dishart, K. J., and Hermes, M. A. (1989). Antagonistic effect of *Pediococcus* spp. against *Listeria monocytogenes*. *Food Biotechnol.* **3,** 183-196.
Juven, B. J., Meinersmann, R. J., and Stern, N. J. (1991). Antagonistic effects of lactobacilli and pediococci to control intestinal colonization by human enteropathogens in live poultry. *J. Appl. Bacteriol.* **70,** 95-103.
Kalchayanand, N. (1990). Extension of shelf-life of vacuum-packaged refrigerated fresh beef by bacteriocins of lactic acid bacteria. Ph.D. thesis, University of Wyoming, Laramie.
Kayahara, H., Yasuhira, H., and Sekiguchi, J. (1989). Isolation and classification of *Pediococcus halophilus* plasmids. *Agric. Biol. Chem.* **53,** 3039-3041.
Kekessy, D. A., and Piguet, J. D. (1970). New method for detecting bacteriocin production. *Appl. Microbiol.* **20,** 282-283.
Kim, W. J., Johnson, M. C., and Ray, B. (1992). Plasmid transfer by conjugation and electroporation in *Pediococcus acidilactici*. *J. Appl. Bacteriol.* **72,** 201-207.
Klaenhammer, T. R. (1988). Bacteriocins of lactic acid bacteria. *Biochimie* **70,** 337-349.
Lafon-Lafourcade, S. (1983). Wine and brandy. *In* "Biotechnology, Volume 5" (G. Reed, ed.), pp. 81-163. Verlag Chemie, Weinheim Deerfield Beach, Florida.
Lewus, C. B., Kaiser, A., and Montville, T. J. (1991). Inhibition of foodborne bacterial pathogens by bacteriocins from lactic acid bacteria isolated from meat. *Appl. Environ. Microbiol.* **57,** 1683-1688.
Lozano, J. C. N., Meyer, J. N., Sletten, K., Pelaz, C., and Nes, I. F. (1992). Purification and amino acid sequence of a bacteriocin produced by *Pediococcus acidilactici*. *J. Gen. Microbiol.* **138,** 1985-1990.
Marugg, J. D., Gonzalez, C. F., Kunka, B. S., Ledeboer, A. M., Pucci, M. J., Toonen, M. Y., Walker, S. A., Zoetmudler, L. C. M., and Vandenbergh, P. A. (1992). Cloning, expression, and nucleotide sequence of genes involved in production of pediocin PA-1, a bacteriocin from *Pediococcus acidilactici* PAC 1.0. *Appl. Environ. Microbiol.* **58,** 2360-2367.
Motlagh, A. (1991). Antimicrobial efficiency of pediocin AcH and genetic studies of the plasmid encoding for pediocin AcH. Ph.D. thesis, University of Wyoming, Laramie.
Motlagh, A., Johnson, M. C., and Ray, B. (1991). Viability loss of foodborne pathogens by starter culture metabolites. *J. Food Prot.* **54,** 873-878, 884.
Motlagh, A. M., Holla, S., Johnson, M. C., Ray, B. and Field, R. A. (1992a). Inhibition of *Listeria*

spp. in sterile food systems by pediocin AcH, a bacteriocin of *Pediococcus acidilactici* H. *J. Food Prot.* **55,** 337–343.

Motlagh, A. M., Bhunia, A. K., Szostek, F., Hansen, T. R., Johnson, M. G. and Ray, B. (1992b). Nucleotide and amino acid sequence of *pap*-gene (pediocin AcH production) in Pediococcus acidilactici H. *Lett. Appl. Microbiol.* **15,** 45–48.

Mundt, J. O. (1986). Lactic acid streptococci. *In* "Bergey's Manual of Systematic Bacteriology" (P. H. A. Sneath, ed.), Vol. 2, pp. 1065–1066. Williams and Wilkins, Baltimore.

Mundt, J. O., Beattie, W. G., and Weiland, F. R. (1969). Pediococci residing on plants. *J. Bacteriol.* **98,** 938–942.

Nielsen, J. W., Dickson, J. S., and Crouse, J. D. (1990). Use of bacteriocin produced by *Pediococcus acidilactici* to inhibit *Listeria monocytogenes* associated with fresh meat. *Appl. Environ. Microbiol.* **56,** 2142–2145.

Pucci, M. J., Vedamuthu, E. R., Kunka, B. S., and Vandenburgh, P. A. (1988). Inhibition of *Listeria monocytogenes* by using bacteriocin PA-1 produced by *Pediococcus acidilactici* PAC 1.0. *Appl. Environ. Microbiol.* **54,** 2349–2353.

Raccach, M. (1987). Pediococci and biotechnology. *CRC Crit. Rev. Microbiol.* **14,** 291–309.

Raccach, M., McGrath, R., and Daftarian, H. (1989). Antibiosis of some lactic acid bacteria including *Lactobacillus acidophilus* toward *Listeria monocytogenes. Int. J. Food Microbiol.* **9,** 25–32.

Rainbow, C. (1975). Beer spoilage lactic acid bacteria. *In* "Lactic Acid Bacteria in Beverages and Food" (J. G. Carr, C. V. Cutting, and G. C. Whiting, eds.), pp. 149–158. Academic Press, Inc., New York.

Ray, B. (1992a). Bacteriocins of starter culture bacteria as food biopreservatives: an overview. *In* "Food Biopreservatives of Microbial Origin" (B. Ray and M. A. Daeschel, eds.) pp. 177–205. CRC Press Inc., Boca Raton, Florida.

Ray, B. (1992b). Pediocin(s) of *Pediococcus acidilactici* as a food biopreservative. *In* "Food Biopreservatives of Microbial Origin." (B. Ray and M. A. Daeschel, eds.) pp. 265–322. CRC Press Inc., Boca Raton, Florida.

Ray, B. (1993). Sublethal injury, bacteriocins and food microbiology. *ASM News.* (Accepted).

Ray, S. K., Johnson, M. C., and Ray, B. (1989a). Bacteriocin plasmids of *Pediococcus acidilactici. J. Ind. Microbiol.* **4,** 163–171.

Ray, B., Kalchayanand, N., and Field, R. A. (1989b). Isolation of a *Clostridium* sp. from spoiled vacuum-packaged beef and its susceptibility to bacteriocin from *Pediococcus acidilactici. In* "Proceedings of the 35th International Congress of Meat Science and Technology," pp. 285–290. Danish Meat Research Institute, Roskilde, Denmark.

Ray, S. K., Kim, W. J., Johnson, M. C., and Ray, B. (1989c). Conjugal transfer of a plasmid encoding bacteriocin production and immunity in *Pediococcus acidilactici* H. *J. Appl. Bacteriol.* **66,** 393–399.

Ray, B., Nie, H., and Menke, S. (1991). Mapping and cloning of pSMB 74, a plasmid encoding pediocin AcH production trait in *Pediococcus acidilactici* H. Abstract No. 031, ASM Annual Meeting, Dallas, Texas, May 5–9.

Ray, B., Motlagh, A., Johnson, M. C., and Bozoglu, F. (1992). Mapping of pSMB74, a plasmid-encoding bacteriocin, pediocin AcH, production (Pap+) by *Pediococcus acidilactici* H. *Lett. Appl. Microbiol.* **15,** 35–37.

Reuter, G. (1970). Laktobazillen und eng verwandte Mikroorganismen in Fleisch und Fleischerzengnissen. 4. Mitterlung: Die Okologie von Laktobazillen, Leuconostoc-Spezies und Pediococcen. *Fleischwirtschaft* **50,** 1397–1399.

Rozbeh, M., Kalchayanand, N., Ray, B., and Field, R. A. (1991). Shelf-life extension of vacuum-packaged refrigerated beef using starter culture metabolites. *In* "Proceedings Western Section American Society of Animal Science, Annual Meeting," No. 42, pp. 50–53. Laramie, Wyoming.

Rueckert, P. W. (1979). Studies on a bacteriocin-like activity produced by *Pediococcus acidilactici* effective against gram-positive organisms. M.S. thesis, Michigan State University, East Lansing.

Solberg, O., Hegna, I. K., and Clausen, O. G. (1975). *Pediococcus acidilactici* NCIB 6990, a new test organism for microbiological assaying of pantothenic acid. *J. Appl. Bacteriol.* **39,** 119–120.

Spelhaug, S. R., and Harlander, S. K. (1989). Inhibition of foodborne bacterial pathogens by bacteriocins from *Lactococcus lactis* and *Pediococcus pentosaceous. J. Food Protect.* **52,** 856–862.

Steinkraus, K. H. (1983). Indigenous fermented amino acid/peptide sauces and pastes with meatlike flavors. *In* "Handbook of Indigenous Fermented Foods" (K. H. Steinkraus, ed.), pp. 433–571. Marcel Dekker, New York.

Sugihara, T. F. (1985). Microbiology of breadmaking. *In.* "Microbiology of Fermented Foods" (B. J. B. Wood, ed.), Vol. 1, pp. 249–261. Elsevier Applied Science Publishers, London and New York.

Tagg, J. R., and McGiven, A. R. (1971). Assay system for bacteriocins. *Appl. Microbiol.* **21,** 943.

Tagg, J. R., Dajani, A. S., and Wannamaker, L. W. (1976). Bacteriocins of gram-positive bacteria. *Bacteriol. Rev.* **40,** 722–756.

Tanaka, N., Gordon, N. M., Lindsay, R. C., Meske, L. M., Doyle, M. P., and Traisman, E. (1985a). Sensory characteristics of reduced nitrite bacon manufactured by the Wisconsin Process. *J. Food Protect.* **48,** 687–692.

Tanaka, N., Meske, L. M., Doyle, M. P., and Traisman, E. (1985b). Plant trials of bacon prepared with lactic acid bacteria, sucrose and reduced amounts of sodium nitrite. *J. Food Protect.* **48,** 679–686.

Teuber, M., and Geis, A. (1981). The family Streptococcaceae (nonmedical aspects). *In* "The Prokaryotes" (M. P. Starr, H. Stolp, H. G. Truper, A. Balows, and H. G. Schlegel, eds.), Vol. II, pp. 1614–1630. Springer-Verlag, Berlin, Heidelberg, New York.

Thomas, T. D. (1987). Acetate production from lactate and citrate by non-starter bacteria in cheddar cheese. *N.Z.J. Dairy Sci. Technol.* **22,** 25–38.

Thomas, T. D., McKay, L. L., and Morris, H. A. (1985). Lactate metabolism by pediococci isolated from cheese. *Appl. Environ. Microbiol.* **49,** 908–913.

Toda, A., Kayahara, H., Yasuhira, H., and Sekiguchi, J. (1989). Conjugal transfer of pIP501 from *Enterococcus faecalis* to *Pediococcus halophilus. Agric. Biol. Chem.* **53,** 3317–3318.

Vandenbergh, P. A. (1991). Pediocin PA-1, produced by *Pediococcus acidilactici.* Annual Meeting of the Society for Industrial Microbiology, Philadelphia, Pennsylvania. August 4–9. Paper S101.

Wallinder, I. B., and Neujahr, H. Y. (1971). Cell wall and peptidoglycan from *Lactobacillus fermenti. J. Bacteriol.* **105,** 918–926.

Wallinder, I. B. and Neujahr, H. Y. (1971). Cell wall and peptidoglycan from *Lactobacillus fermenti.* J. Bacteriol. **105,** 918–926.

Wu, W. H., Rule, D. C., Busboom, J. R., Field, R. A., and Ray, B. (1991). Starter culture and time/temperature of storage influences on quality of fermented mutton sausage. *J. Food Sci.* **56,** 916–919, 925.

Yang, R., Johnson, M. C., and Ray, B. (1992). Novel method to extract large amounts of bacteriocin from lactic acid bacteria. *Appl. Environ. Microbiol.* **58,** 3355–3359.

Yousef, A. E., Luchansky, J. B., Degnan, A. J., and Doyle, M. P. (1991). Behavior of *Listeria monocytogenes* in wiener exudates in the presence of *Pediococcus acidilactici* H or pediocin AcH during storage at 4 or 25°C. *Appl. Environ. Microbiol.* **57,** 1461–1467.

CHAPTER **9**

Bacteriocins from Carnobacterium *and* Leuconostoc

MICHAEL E. STILES

I. Description of the Genera *Carnobacterium* and *Leuconostoc*

The genus *Carnobacterium* was described by Collins et al. in 1987 to include the atypically nonaciduric lactobacilli from chilled chicken meat (Thornley, 1957) and vacuum packaged meats (Hitchener et al., 1982; Holzapfel and Gerber, 1983; Shaw and Harding, 1984). Studies on these nonaciduric lactobacilli from meats resulted in the proposal of two new species: *Lactobacillus divergens* (Holzapfel and Gerber, 1983) and *Lactobacillus carnis* (Shaw and Harding, 1985). Because these species were not included in earlier taxonomic studies, Collins et al. (1987) studied their biochemical and chemical characteristics and those of the salmonid fish pathogen, *Lactobacillus piscicola* (Hiu et al., 1984). They decided that the proposal of a new genus *Carnobacterium* was justified, based on differences from other lactobacilli and the low DNA relatedness to *Brochothrix thermosphacta*. Four species were proposed, of which *Carnobacterium divergens* is the type species; *Carnobacterium piscicola* includes both *L. carnis* and *L. piscicola* because they are in the same DNA homology group; and *Carnobacterium gallinarum* and *Carnobacterium mobile* are also described. Members of the genus *Carnobacterium* are heterofermentative, Gram-positive, rod-shaped organisms. Gas production is variable and frequently negative depending on the growth substrate. Carnobacteria produce L(+)-lactate from glucose, and they do not grow on acetate agar adjusted to pH 5.6 (Rogosa et al., 1951) or in 8% NaCl. They grow at 15°C and at temperatures as low as 0°C, but not at 45°C (Holzapfel and Gerber, 1983). The carnobacteria are nonaciduric. When grown in modified lactobacilli MRS broth without acetate or citrate they produce a final pH of 5.0–5.3 after 4 days incubation at 25°C (De Man et al., 1960).

The genus *Leuconostoc* has long been recognized and included among the lactic acid bacteria. These bacteria are found on plants and in habitats notably similar to lactobacilli and lactococci. The genus was described for *Bergey's Manual of Systematic Bacteriology* by Garvie (1986). Members of the genus *Leuconostoc* are heterofermentative, Gram-positive, coccus-shaped organisms. Gas production from glucose separates the leuconostocs from the lactococci. Differentiation of leuconostocs from heterofermentative lactobacilli can be tenuous because it is based on morphology. Leuconostocs grown in broth are elongated and can be mistaken as rods (Garvie, 1986). Leuconostocs produce D(−)-lactate from glucose in the absence of malate. L(+)-lactate is produced from malate. Milk is noted as a poor medium for growth; however, *Leuconostoc mesenteroides* subsp. *mesenteroides* can acidify and clot milk with gas formation. *L. mesenteroides* subsp. *cremoris* and *Leuconostoc lactis* have been isolated only from dairy sources. They are used as starter cultures for dairy products because of their dissimilation of citrate to acetoin and diacetyl. *Leuconostoc oenos* is important for the malolactic fermentation in wine. However, the nonacidophilic leuconostocs, *L. mesenteroides* subspp. *mesenteroides, dextranicum,* and *cremoris, L. paramesenteroides,* and *L. lactis* have more in common with *Lactobacillus confusus* and *Lactobacillus viridescens* than with *Leuconostoc oenos*.

Leuconostocs also have an important niche among the lactic acid bacteria of chilled fresh and processed meats packaged under vacuum or in modified atmosphere with elevated CO_2 content (Shaw and Harding, 1984). Nondextran-producing *L. paramesenteroides* has been reported on bacon (Cavett, 1963) and dextran-producing *L. mesenteroides* on vacuum-packaged beef (Hitchener et al., 1982). In processed meats with sucrose added as cure or filler, growth of dextran-producing leuconostocs could cause spoilage because of slime formation. Shaw and Harding (1989) described two new species of leuconostocs isolated from chill-stored meat, *L. carnosum* and *L. gelidum;* both are characterized by their inability to grow at 37°C. The two species are differentiated on the basis of acid production from carbohydrates and *L. gelidum* is β-galactosidase positive, while *L. carnosum* is β-galactosidase negative.

II. Habitats and Sources of *Carnobacterium* and *Leuconostoc* Species

The widespread use of vacuum packaging of meats led to the study of their lactic microflora that enhance the preservation of the product. This revealed the "nonaciduric lactobacilli" (carnobacteria) that are among the dominant lactic acid bacteria on chill-stored raw meats packaged with elevated levels of CO_2 (including vacuum packaging). Little is known of their source in meats, but along with the lactobacilli they are probably inhabitants of the animal intestine and contaminants from the meat handling environment (Kitchell and Shaw, 1975). *C. piscicola* is also a pathogen of salmonid fish and it is a common isolate from processed fish products (J. Leisner, personal communication).

Leuconostocs are widely associated with foods. They are used as starter cultures in many soft (unripened cottage cheese), semisoft (caerphilly), semihard (Gouda), and blue-vein cheeses (Law, 1984); *Leuconostoc mesenteroides* occurs as part of the natural progression of lactic acid bacteria in fermented vegetables (Fleming et al., 1985); and dextran-producing strains of *Leuconostoc* can cause serious spoilage problems in sugar processing (Tilbury, 1975). Leuconostocs, like carnobacteria, have been isolated among the dominant lactic acid bacteria on chill-stored meats packaged with elevated levels of CO_2 (Shaw and Harding, 1984). Leuconostocs are categorically claimed to be nonpathogenic to plants and animals (Garvie, 1986); however, since 1985, there have been several reports of leuconostoc bacteremia (Handwerger et al., 1990). The cases were generally associated with underlying diseases, often involving immunosuppressed patients who had received vancomycin therapy. Clinical and most agricultural isolates of *Leuconostoc* are highly resistant to vancomycin. Approximately 50% of the bacteremias were polymicrobial, often involving staphylococcal species; however, the authors concluded that *Leuconostoc* should be viewed as a potential pathogen.

III. Bacteriocins Produced by *Carnobacterium* and *Leuconostoc* Species

The knowledge of bacteriocins produced by these bacteria is limited. In both genera, bacteriocinogenic (Bac$^+$) strains have been described. Several strains of these bacteria isolated by Shaw and Harding (1984) from vacuum-packaged beef, pork, lamb, and bacon were subsequently shown to produce bacteriocins (Ahn and Stiles, 1990a; Schillinger and Holzapfel, 1990). *C. piscicola* LV17 (Shaw) is of interest because its bacteriocins are detected early in the growth cycle (Ahn and Stiles, 1990a,b). Factors that cause specific strains of lactic acid bacteria to dominate the microflora of a food are not known, but bacteriocin production has been suggested as an important factor in the competitiveness of a bacterium in a mixed bacterial fermentation (Klaenhammer, 1988). The bacteriocin produced by *L. gelidum* UAL 187 is of interest because its activity is also detected early in the growth cycle (Hastings and Stiles, 1991).

A. Bacteriocins Produced by Carnobacterium Species

Production of bacteriocins by *Carnobacterium* was reported by Schillinger and Holzapfel (1990). They found that 18 out of 37 strains tested produced bacteriocins or bacteriocinlike substances, primarily active against other carnobacteria, but some strains were also active against strains of *Enterococcus* and *Listeria* spp. The bacteriocin produced by *C. piscicola* LV61 (Shaw) is heat stable (100°C for 10 minutes) and has a bactericidal mode of action.

In a separate study (Ahn and Stiles, 1990b), it was shown that *Carnobacterium piscicola* LV17 (Shaw) produces bacteriocins with a bactericidal mode of action against strains of other lactic acid bacteria, *Enterococcus* spp., and *Listeria* spp. The bacteriocins are heat resistant (100°C for 30 minutes). Bacteriocin production is detected early in the growth cycle in APT broth, when approximately 5% of growth has occurred from a 1% inoculum. Bacteriocin is not produced if the initial pH of APT broth is adjusted to 5.5. Bacteriocin production and resistance to the bacteriocin are mediated by two large plasmids of 40 and 49 MDa. The plasmids were separated into two mutant strains, LV17A and LV17B, containing the 49- and 40-MDa plasmids, respectively. The mutant strains have different antimicrobial spectra (Ahn and Stiles, 1990b) and only the bacteriocins produced by mutant LV17A are detected early in the growth cycle.

Study on the bacteriocins produced by the two mutant strains is in progress (M. E. Stiles, unpublished data). The chemical data indicate that at least three bacteriocins are produced (two by mutant strain LV17B). Final resolution of the bacteriocins produced by *C. piscicola* LV17 awaits the clarification of the nucleotide sequencing of the bacteriocin genes from these plasmids.

B. Bacteriocins Produced by Leuconostoc Species

Production of bacteriocinlike substances has been reported for leuconostocs of dairy and wine origins (Orberg and Sandine, 1984) and meat origin (Harris et al., 1989; Ahn and Stiles, 1990a). Bacteriocin production by a strain of *L. gelidum* (UAL 187) isolated from meat has been extensively studied (Hastings and Stiles, 1991) and the bacteriocin, leucocin A, has been purified and characterized (Hastings et al., 1991). Production of a bacteriocin, mesenterocin 5, by a strain of *L. mesenteroides* (UL5) isolated from Cheddar cheese has been described (Daba et al., 1991). Mesenterocin 5 is bacteriostatic against *Listeria monocytogenes* strains, but not against several useful lactic acid bacteria. The substance is a relatively heat-stable (100°C for 30 minutes) peptide of approximately 4.5 kDa. Detection of the bacteriocin in the growth medium occurs after 6 hours of incubation at 30°C when approximately 90% of cell growth has occurred. Activity reaches a maximum after 10 hours of incubation, coinciding with achievement of maximum cell population. Thereafter, activity declines dramatically. The relationship between pH and bacteriocin production was not reported; however, the bacteriocin was more stable when pH 5.0 was maintained throughout the stationary phase (Daba et al., 1991).

Leucocin A was also originally described as bacteriostatic (Hastings and Stiles, 1991), but this may depend on the indicator strain tested. The antibacterial spectrum includes a wide range of lactic acid bacteria and *Enterococcus faecalis* and *Listeria monocytogenes*. The substance is a heat-resistant (62°C for 30 minutes) peptide of 3930.3 ± 0.4 Da as determined by mass spectrometry (Hastings et al., 1991). The organism initiated growth and produces bacteriocin over a pH range of 4.0 to 6.5. Bacteriocin production is detected early in the growth cycle when less than 5% of growth of a 1% inoculum has occurred, and activity is still detected after 7 days of incubation at 25°C. *Leuconostoc gelidum* is a psychrotrophic bacterium, and produces bacteriocin during growth at 1°C. Genetic information determining production of, and immunity to, leucocin A is mediated by a 7.6-MDa plasmid.

IV. Chemical Characterization of *Leuconostoc* and *Carnobacterium* Bacteriocins

The N-terminal amino acid sequence of purified leucocin A from *L. gelidum* UAL187 allowed a 23-mer oligonucleotide probe to be synthesized. This was used to probe for the structural gene for leucocin A on the 7.6-MDa plasmid. The nucleotide sequence of the leucocin (*lcn*) gene revealed a 37-amino-acid structural peptide (see Figure 1), preceded by a 23-amino-acid N-terminal extension (Hastings et al., 1991). The *lcn* probe also hybridized with the *C. piscicola* LV17B, but not the LV17A plasmid, indicating homology between leucocin A and carnobacteriocin B (M. E. Stiles, unpublished data). This has been confirmed by N-terminal amino acid sequences for the carnobacteriocin (*cbn*) genes, but complete elucidation of the similarity between these bacteriocins awaits the nucleotide sequencing data and interpretation of the complete amino acid structures.

The study of the amino acid composition of pediocin PA-1 (Marugg et al., 1992) revealed approximately 70% homology with the first 21 amino

```
                        ┌─────── S - S ───────┐
H₂N - Lys - Tyr - Tyr - Gly - Asn - Gly -Val - His - Cys - Thr - Lys - Ser - Gly - Cys -
    Ser - Val - Asn - Trp - Gly - Glu - Ala - Phe - Ser - Ala - Gly - Val - His - Arg -
         Leu - Ala - Asn - Gly - Gly - Asn - Gly - Phe - Trp - COOH
```

Figure 1 Amino acid sequence for leucocin A, a bacteriocin produced by *Leuconostoc gelidum* UAL187. (Adapted from Hastings et al., 1991.)

acids of leucocin A. The prospects for study of the structure and function of these bacteriocins increases dramatically as more bacteriocins are characterized, such as lactococcins A and B (van Belkum et al., 1991).

V. Potential for Application of Carnobacteriocins or Leucocin A in Meat Preservation

The study of bacteriocins produced by meat lactics in our laboratory (Ahn and Stiles, 1990a,b; Hastings and Stiles, 1991; Hastings et al., 1991) is motivated by the possible use of Bac$^+$ strains of *Carnobacterium* and *Leuconostoc* spp. or their bacteriocins to extend the storage life and enhance the safety of chill-stored meats. The potential for this was discussed in a review (Stiles and Hastings, 1991) in which the limitations of nisin or nisin-producing lactococci for use in meat preservation were considered against the advantages and limitations of bacteriocins produced by lactic acid bacteria associated with meats. The antibacterial spectra of *C. piscicola* LV17 and *L. gelidum* UAL187 do not compare favorably with that of nisin, but these bacteria and their bacteriocins possess characteristics that make them suitable to test as prototypes for meat systems.

The early detection of bacteriocin production by both of these strains means that if bacteriocin production is a factor that promotes domination of a strain in a mixed fermentation, then early producers are likely to have a competitive advantage. Although carnobacteria grow at pH 5.0, they do not produce bacteriocin below pH 5.5, which is the pH of normal, raw meat in a vacuum package. The initial pH of processed meats is higher (pH 6.0 or greater), hence carnobacteria would be expected to grow and produce their bacteriocins in these meat products. Leuconostocs are active gas producers and excessive production of CO_2 during meat storage could be a disadvantage; however, CO_2 is more efficiently absorbed by meat stored at low temperatures, so this may not be a limitation in raw meat with limited amounts of fermentable carbohydrate.

Leuconostoc gelidum UAL187 produces leucocin A over a wide range of pH and the bacteriocin is stable. However, *L. gelidum* produces dextran from sucrose (Shaw and Harding, 1989). This would restrict the use of this organism in a processed meat in which sucrose is used as part of the cure or as a filler carbohydrate.

It is probable that neither *Carnobacterium piscicola* nor *Leuconostoc gelidum* nor their bacteriocins will be appropriate for use as meat preservatives. A bacteriocin such as pediocin AcH with its broader antibacterial spectrum (Bhunia et al., 1988) may be more appropriate. Other bacteriocinogenic lactic acid bacteria may be discovered that more appropriately meet the needs for meat preservation. However, the potential to produce "gene cassettes" for production of a range of bacteriocins with enhanced antibacterial spectra or controlled mutation to design appropriate bacteriocins raises possibilities and challenges for the molecular biologist and for the meat industry and its regulators.

References

Ahn, C., and Stiles, M. E. (1990a). Antibacterial activity of lactic acid bacteria isolated from vacuum-packaged meats. *J. Appl. Bacteriol.* **69,** 302–310.

Ahn, C., and Stiles, M. E. (1990b). Plasmid-associated bacteriocin production by a strain of *Carnobacterium piscicola* from meat. *Appl. Environ. Microbiol.* **56,** 2503–2510.

Bhunia, A. K., Johnson, M. C., and Ray, B. (1988). Purification, characterization and antimicrobial spectrum of a bacteriocin produced by *Pediococcus acidilactici*. *J. Appl. Bacteriol.* **65,** 261–268.

Cavett, J. J. (1963). A diagnostic key for identifying the lactic acid bacteria of vacuum packed bacon. *J. Appl. Bacteriol.* **26,** 453–470.

Collins, M. D., Farrow, J. A. E., Phillips, B. A., Ferusu, S., and Jones, D. (1987). Classification of *Lactobacillus divergens*, *Lactobacillus piscicola* and some catalase-negative, asporogenous, rod-shaped bacteria from poultry in a new genus, *Carnobacterium*. *Int. J. Syst. Bacteriol.* **37,** 310–316.

Daba, H., Pandian, S., Gosselin, J. F., Simard, R. E., Huang, J., and Lacroix, C. (1991). Detection and activity of bacteriocin produced by *Leuconostoc mesenteroides*. *Appl. Environ. Microbiol.* **57,** 3450–3455.

De Man, J. C., Rogosa, M., and Sharpe, M. E. (1960). A medium for the cultivation of lactobacilli. *J. Appl. Bacteriol.* **23,** 130–135.

Fleming, H. P., McFeeters, R. F., and Daeschel, M. A. (1985). The lactobacilli, pediococci and leuconostocs: vegetable products. In "Bacterial Starter Cultures for Foods" (S. E. Gilliland, ed.), pp. 97–118. CRC Press, Florida.

Garvie, E. I. (1986). Genus *Leuconostoc*. In "Bergey's Manual of Systematic Bacteriology," vol. 2, (P. H. A. Sneath, N. S. Mair, M. E. Sharpe, and J. G. Holt, eds.), pp. 1071–1075. Williams and Wilkins, Baltimore.

Handwerger, S., Horowitz, H., Coburn, K., Kolokathis, A., and Wormser, G. P. (1990). Infection due to *Leuconostoc* species: six cases and review. *Rev. Infect. Dis* **12,** 602–610.

Harris, L. J., Daeschel, M. A., Stiles, M. E., and Klaenhammer, T. R. (1989). Antimicrobial activity of lactic acid bacteria against *Listeria monocytogenes*. *J. Food Prot.* **52,** 384–387.

Hastings, J. W., and Stiles, M. E. (1991). Antibiosis of *Leuconostoc gelidum* isolated from meat. *J. Appl. Bacteriol.* **70,** 127–134.

Hastings, J. W., Sailer, M., Johnson, K., Roy, K. L., Vederas, J. C., and Stiles, M. E. (1991). Characterization of leucocin A-UAL187 and cloning of the bacteriocin gene from *Leuconostoc gelidum*. *J. Bacteriol.* **173,** 7491–7500.

Hitchener, B. J., Egan, A. F., and Rogers, P. J. (1982). Characteristics of lactic acid bacteria isolated from vacuum-packaged beef. *J. Appl. Bacteriol.* **52,** 31–37.

Hiu, S. F., Holt, R. A., Sriranganathan, N., Seidler, R. J., and Fryer, J. L. (1984). *Lactobacillus piscicola*, a new species from salmonid fish. *Int. J. Syst. Bacteriol.* **34,** 393–400.

Holzapfel, W. H., and Gerber, E. S. (1983). *Lactobacillus divergens* sp. nov., a new heterofermentative *Lactobacillus* species producing L(+)-lactate. *Syst. Appl. Microbiol.* **4,** 522–534.

Kitchell, A. G., and Shaw, B. G. (1975). Lactic acid bacteria in fresh and cured meat. In "Lactic Acid Bacteria in Beverages and Food" (J. G. Carr, C. V. Cutting, and G. C. Whiting, eds.), pp. 209–220. Academic Press, Inc., New York.

Klaenhammer, T. R. (1988). Bacteriocins of lactic acid bacteria. *Biochimie* **70,** 337–349.

Law, B. A. (1984). Flavour development in cheeses. In "Advances in the Microbiology and

Biochemistry of Cheese and Fermented Milk" (F. L. Davies and B. A. Law, eds.), pp. 187–208. Elsevier Applied Science Publishers, London.

Marugg, J. D., Gonzalez, C. F., Kunka, B. S., Ledeboer, A. T., Pucci, M. J., Toonen, M. Y., Walker, S. A., Zoetmulder, L. C. M., and Vandenbergh, P. A. (1992). Cloning, expression, and nucleotide sequence of genes involved in production of pediocin PA-1, a bacteriocin from *Pediococcus acidilactici* PAC1.0. *Appl. Environ. Microbiol.* **58**, 2360–2367.

Orberg, P. K., and Sandine, W. E. (1984). Common occurrence of plasmid DNA and vancomycin resistance in *Leuconostoc* spp. *Appl. Environ. Microbiol.* **48**, 1129–1133.

Rogosa, M., Mitchell, J. A., and Wiseman, R. F. (1951). A selective medium for the isolation and enumeration of oral and fecal lactobacilli. *J. Bacteriol.* **62**, 132–133.

Schillinger, U., and Holzapfel, W. H. (1990). Antibacterial activity of carnobacteria. *Food Microbiol.* **7**, 305–310.

Shaw, B. G., and Harding, C. D. (1984). A numerical taxonomic study of lactic acid bacteria from vacuum-packed beef, pork, lamb and bacon. *J. Appl. Bacteriol.* **56**, 25–40.

Shaw, B. G., and Harding, C. D. (1985). Atypical lactobacilli from vacuum-packaged meats: Comparison by DNA hybridization, cell composition and biochemical tests with a description of *Lactobacillus carnis* sp. nov. *Syst. Appl. Microbiol.* **6**, 291–297.

Shaw, B. G., and Harding, C. D. (1989). *Leuconostoc gelidum* sp. nov. and *Leuconostoc carnosum* sp. nov. from chill-stored meats. *Int. J. Syst. Bacteriol.* **39**, 217–223.

Stiles, M. E., and Hastings, J. W. (1991). Bacteriocin production by lactic acid bacteria: Potential for use in meat preservation. *Trends Food Sci. Technol.* **2**, 247–251.

Thornley, M. J. (1957). Observations on the microflora of minced chicken meat irradiated with 4 MeV cathode rays. *J. Appl. Bacteriol.* **20**, 286–298.

Tilbury, R. H. (1975). Occurrence and effects of lactic acid bacteria in the sugar industry. In "Lactic Acid Bacteria in Beverages and Food" (J. G. Carr, C. V. Cutting, and G. C. Whiting, eds.), pp. 177–191. Academic Press, Inc., New York.

van Belkum, M. J., Hayema, B. J., Jeeninga, R. E., Kok, J., and Venema, G. (1991). Organization and nucleotide sequences of two lactococcal bacteriocin operons. *Appl. Environ. Microbiol.* **57**, 492–498.

CHAPTER 10

Bacteriocins from Dairy Propionibacteria and Inducible Bacteriocins of Lactic Acid Bacteria

SUSAN F. BAREFOOT
DALE A. GRINSTEAD

I. Bacteriocins from Dairy Propionibacteria

A. Introduction to the Propionibacteria

The dairy or classical propionibacteria are taxonomically distant from the lactic acid bacteria discussed in the remainder of this volume. However, the propionibacteria are perhaps best known for the conversion of lactate produced by lactic cultures to the carbon dioxide, propionate, and acetate that contribute "eyes" and flavor to Swiss cheese (Hettinga and Reinbold, 1972; Langsrud and Reinbold, 1973). Although the genus *Propionibacterium* includes cutaneous species of medical importance, the industrially significant classical or dairy species, *P. freudenreichii*, *P. jensenii*, *P. thoenii*, and *P. acidipropionici* (Cummins and Johnson, 1986; Garvie, 1984) and their inhibitors will serve as the focus of this chapter. In addition to the production of Swiss cheese, dairy propionibacteria help to maintain an environment unfavorable for pathogens and spoilage microflora in natural olive and silage fermentations (Woolford, 1975), may be utilized as probiotics (Mantere-Alhonen, 1983; 1987), or may be added as inoculants (Flores-Galarza et al., 1985) to silage.

B. Nonbacteriocin Inhibitors Produced by the Propionibacteria

Dairy propionibacteria produce a number of broad-spectrum inhibitors. Unless care is exercised to eliminate effects of these inhibitors, they may be mistakenly identified as bacteriocins. During growth, the propionibacteria metabolize lactate to propionate, acetate, and carbon dioxide (Hettinga and Reinbold, 1972; Langsrud and Reinbold, 1973); the inhibitory actions of these metabolites are well established. Propionic acid (often in the form of salts) is added to control fungi and ropy spores in bakery products; at accepted maximum concentrations (0.3%) of use, the propionates are effective inhibitors of numerous Gram-negative bacteria (ICMSF, 1980). Antimicrobial activities of both propionic and acetic acids are potentiated by the low pH encountered in fermented foods.

The incompletely defined, broad-spectrum inhibitor Microgard™ is the pasteurized product of the fermentation of grade A skim milk by *P. freudenreichii* subsp. *shermanii*. A recent report indicates that Microgard™ is added to approximately 30% of the cottage cheese in the United States (Daeschel, 1989). Its inhibitory actions have been attributed to diacetyl, propionic, acetic, and lactic acids, and a heat-stable, 700-Da peptide (Daeschel, 1989); however, no reports characterizing the latter have been published. Like propionic acid, Microgard inhibits most Gram-negative bacteria and some fungi (Al-Zoreky et al., 1991). The relative contributions of organic acids and peptide to its inhibitory action have not been identified.

C. Propionibacteria Bacteriocins

Bacteriocins are classically defined as bactericidal proteins with activity against species closely related to the producer culture (Tagg et al., 1976). Some bacteriocins produced by lactic acid bacteria exhibit broad activity extending to unrelated Gram-positive genera; examples include nisin (Hurst, 1981), pediocin AcH (Bhunia et al., 1988), and sakacin A (Schillinger et al., 1991). Both limited- and broad-spectrum protein antagonists have been defined in cutaneous and classical propionibacteria.

Two cell-associated protein antagonists are produced by cutaneous species. One, a protease- and lysozyme-sensitive inhibitor called acnecin is produced by *Propionibacterium acnes* strain CN-8 (Fujimura and Nakamura, 1978). Acnecin has a molecular weight of 60,000 Da, consists of five identical 12,000-Da protein subunits, and contains 3.3% carbohydrate. Predominant amino acids (43%) include aspartic acid, glutamic acid, glycine, and alanine. Acnecin lacks cysteine, and methionine comprises only 15 of its total 10,000 amino acid residues. The limited number of amino acids capable of forming disulfide bonds provides an explanation for its heat sensitivity; a treatment of 60°C for 10 minutes eliminates detectable activity.

Acnecin has not been recovered from concentrated (tenfold) or unconcentrated broth cultures, nor is it released from the surface of producer cells by treatment with 1M NaCl or 8M urea. Only after producer cells are disrupted by sonication is acnecin activity detected. Acnecin demonstrates bacteriostatic activity against strains of *P. acnes* and *Corynebacterium parvum* (Fujimura and Nakamura, 1978). Per the specifications of Tagg et al. (1976), these characteristics are consistent with terming acnecin a bacteriocinlike agent.

A broader spectrum protein antagonist is produced by an oral strain of *Propionibacterium acnes* (RTT 108), exhibits a bacteriostatic mode of action, has a molecular weight of 78,000 Da, and displays activity against Gram-negative and Gram-positive bacteria (Paul and Booth, 1988). Like acnecin (Fujimura and Nakamura, 1978), the agent produced by strain RTT 108 is primarily cell associated; washing cells with 1M NaCl or 8M urea fails to yield detectable activity (Paul and Booth, 1988). In contrast to acnecin (Fujimura and Nakamura, 1978), activity of the inhibitor is detected in fivefold concentrates of spent RTT 108 cultures. The inhibitor produced by RTT 108 is sensitive to treatment with lysozyme or to holding at 55° for 1 hour. It is bacteriostatic, and rather than being designated a bacteriocin (Paul and Booth, 1988), should be termed bacteriocinlike. More precise assessments of the target sites or the specific cellular effects of either acnecin or antagonist RTT 108 on sensitive strains have not been reported to date.

In a more recent study, Razafindrajaona (1989) examined strains of dairy propionibacteria for production of inducible inhibitors. Six of 22 strains of *P. freudenreichii* subsp. *shermanii* produced inducible agents active against other dairy propionibacteria. Although complete bacteriophage particles were observed in some lysates, the inducible inhibitors did not propagate on their hosts and phage indicator strains were not identified. The inhibitory actions of these agents were attributed to both phages and bacteriocins; however, bacteriocin lysates contained defective phage particles and exhibited bacteriolytic activity (Razafindrajaona, 1989) Therefore, they do not fit the definition of bacteriocins (Tagg et al., 1976). Two strains spontaneously released inhibitors active against other propionibacteria during growth in sodium lactate broth; however, the responsible agents were not characterized or identified (Razafindrajaona, 1989).

Specific inhibitory activity among the dairy propionibacteria first was reported by Grinstead (1989) and Grinstead and Glatz (1989). A survey of 150 strains yielded several cultures including *Propionibacterium jensenii* P126 (ATCC 4827) and *P. thoenii* P127 (ATCC 4874) that inhibit related dairy propionibacteria. The inhibitor produced by *P. thoenii* P127 recently was purified and characterized by Lyon and Glatz (1991a,b, 1992). Because the agent responsible for inhibition was sensitive to the action of proteolytic enzymes and displayed bactericidal activity against sensitive propionibacteria, it was classified as a bacteriocin and designated propionicin PLG-1. Propionicin PLG-1 was purified from soft sodium lactate agar producer cultures by ammonium sulfate precipitation; subsequent gel filtration on

Sephadex G-200 yielded two active fractions corresponding to molecular weights of more than 150,000 Da and 10,000 Da. SDS-PAGE resolved the larger active component to one protein with a molecular weight of 10,000 Da and 12 contaminating proteins. SDS-PAGE analysis of the smaller molecular weight fraction yielded a single diffuse protein band with a molecular weight of approximately 10,000 Da (Lyon and Glatz, 1991a). Several lactic acid bacteria produce aggregate bacteriocins including lactacin B (Barefoot and Klaenhammer, 1984) and helveticin J (Joerger and Klaenhammer, 1986). As they did for other bacteriocins, the two fractions containing bacteriocin activity may represent aggregate and monomeric forms of propionicin PLG-1 (Lyon and Glatz, 1991a). Propionicin PLG-1 was produced in sodium lactate broth, beet molasses, and corn steep liquor media by *P. thoenii* P127. It was purified to homogeneity by ammonium sulfate precipitation, ion exchange chromatography, and isoelectric focusing. Analysis of purified propionicin PLG-1 by SDS-PAGE yielded a single band of 10,000 Da (Lyon and Glatz, 1992). Propionicin PLG-1 activity was stable from pH 3 to 9 but sensitive to temperatures greater than 80°C. Partially purified propionicin PLG-1 inhibited strains of *P. thoenii, P. jensenii,* and *P. acidipropionici*, lactic acid bacteria, fungi, and single strains of Gram-negative species including *Campylobacter jejuni, Vibrio parahaemolyticus, Pseudomonas aeruginosa,* and *P. fluorescens* (Lyon and Glatz, 1991a). The spectrum of activity for this bacteriocin produced by a food-related species of propionibacteria is unusually broad. Although numerous bacteriocins of food-related lactic acid bacteria are active against unrelated Gram-positive species (Klaenhammer, 1988), none are active against Gram-negative bacteria.

The inhibitor produced by *Propionibacterium jensenii* P126 has been examined by Grinstead and Barefoot (1991; 1992a,b) and Weinbrenner et al. (1993). The inhibitor, jenseniin G, was produced in both sodium lactate agar and broth cultures. Because of its size (as indicated by its retention in dialysis tubing with a molecular weight cutoff of 12,000 Da) and its sensitivity to proteolytic enzymes, jenseniin G was termed a protein. It acted bacteriostatically against *P. acidipropionici* P5 and displayed bactericidal rather than bacteriolytic activity against the sensitive indicator, *Lactobacillus delbrueckii* subsp. *lactis* ATCC 4797 (Grinstead and Barefoot, 1991; 1992a). Thus, it was classified as a bacteriocin (Tagg et al., 1976). In contrast to the sensitivity of PLG-1 to treatments at 85°C (Lyon and Glatz, 1991a), jenseniin G was stable to 100°C for 20 minutes (Grinstead and Barefoot, 1991). Unlike the broad spectrum activity of propionicin PLG-1 (Lyon and Glatz, 1991a), activity of jenseniin G was confined to dairy propionibacteria, lactococci, lactobacilli (Grinstead and Barefoot, 1991; 1992a) and thermophilic streptococci (Weinbrenner et al., 1993). Jenseniin G displayed greater activity against sensitive lactobacilli than against sensitive propionibacteria; a crude preparation of jenseniin G was 1000 times more active against *L. delbrueckii* subsp. *lactis* than against *P. acidipropionici* P5 (Grinstead and Barefoot, 1992a). The environment in which many propionibacteria are found may provide a partial explanation; propionibacteria typically compete with lactobacilli in dairy and silage fermentations. Production of a compound inhibitory toward lactobacilli would give slowly growing propionibacteria a competitive advantage. An alternative possibility is that crude jenseniin G may contain at least two antagonists, one inhibitory to propionibacteria and the other to lactobacilli. Resolution of this question awaits assessment of the spectrum of purified jenseniin G. To date, other reports of bacteriocins produced by propionibacteria have been restricted to one from our laboratory (Prince et al., 1993).

Clearly, more attention to identification and characterization of bacteriocins produced by dairy propionibacteria is warranted. Several potential applications may be envisioned. To date, only one medium selective for propionibacteria has been developed (Fischer et al., 1991); incorporating propionibacteria bacteriocins as selective agents in that medium may facilitate identification of additional bacteriocinogenic strains. Of major concern is the potential use of broad-spectrum bacteriocins such as propionicin PLG-1 for control of food-borne pathogens and food spoilage species.

D. Genetics of Propionibacteria and Their Bacteriocins

Despite the industrial significance of the genus, genetic systems in the propionibacteria remain poorly defined; related reports to date may be summarized as follows. Manipulations of propionibacteria have been restricted to selection of spontaneous mutants and those generated by mutagenesis with N-methyl-N'-nitro-N-nitroso-guanidine (Hofherr et al., 1983; Glatz and Anderson, 1988). No reports to date have addressed the effects of these treatments on bacteriocinogenesis in dairy propionibacteria. Although several plasmids have been identified in dairy propionibacteria, most are cryptic (Rehberger and Glatz, 1990), and none have been linked conclusively to bacteriocin production. No homologous or heterologous vectors have been developed for the propionibacteria. Protoplast formation and regeneration have been optimized for some strains (Baehman and Glatz, 1989; Woskow and Glatz, 1990), however, electroporation has proven to be the most useful technique for introduction of DNA into propionibacteria (Luchansky et al., 1988; Zirnstein and Rehberger, 1991). Cloning and characterization of propionibacteria genes has been limited to those encoding the adenosylcobalamin-dependent methylmalonyl-CoA mutase from *P. shermanii* (Marsh et al., 1989). Bacteriocin-related genes in propionibacteria have not yet been located or characterized.

Dairy propionibacteria harbor plasmid DNA. Rehberger and Glatz (1990) isolated plasmid DNA molecules from 20 strains of dairy propionibacteria; the plasmids ranged in size from 4.4 MDa to more than 119 MDa. Restriction endonuclease analysis of plasmid DNA from 15 strains identified seven different plasmids that were designated pRG01 through pRG07. Eleven strains contained the same 4.4-MDa plasmid (pRG01); two strains contained an identical 6.3-MDa plasmid (PRG02). Restriction endonuclease maps were constructed for plasmids pRG01 through pRG04. DNA–DNA hybridization studies revealed extensive homology among plasmids pRG01, PRG02, pRG05, and pRG07, pointing to possible plasmid incompatibilities that might interfere with future gene transfer systems (Rehberger and Glatz, 1990).

Rehberger and Glatz (1990) investigated possible linkages of propionibacteria characteristics to these plasmid DNA species. Acriflavin treatment of *Propionibacterium* strains P5 (pRG01), P38 (pRG01, pRG05), P54 (pRG02), and P63 (pRG02, multimer PRG02) generated plasmid-cured derivatives. No detectable differences in antibiotic sensitivity or carbohydrate utilization between cured and parental strains were identified; these results eliminated plasmid linkage of the two traits. Neither parental nor cured strains produced bacteriocinlike antagonists. Phenotypic evidence was presented for linkage of a clumping characteristic with plasmid pRG05. Parental cultures, but not cured derivatives, of strain P38 clumped upon inoculation into fresh broth; however, derivatives lacking pRG05 also lacked pRG01. That other strains carrying pRG01 did not display the clumping phenotype provided support for linkage of the clumping characteristic to plasmid pRG05. The possibility that "cured" plasmids had integrated into the chromosome during the curing process was not eliminated (Rehberger and Glatz, 1990); therefore, conclusive evidence for linkage of the clumping phenotype to pRG05 awaits the development of genetic transfer systems applicable to the dairy propionibacteria.

Like bacteriocins produced by other genera, it appears that determinants of propionibacteria bacteriocins may be chromosomal in origin. Lyon and Glatz (1992) identified a single large (250kb) plasmid in the propionicin PLG-1 producer, *P. thoenii* P127. Treatment of the producer with *N*-methyl-*N'*-nitro-*N*-nitrosoguanidine or acriflavin yielded isolates cured of the plasmid or free of bacteriocin production. However, plasmid loss did not correlate with loss of bacteriocinogeny. The data suggest that the determinants for propionicin PLG-1 are chromosomal (Lyon and Glatz, 1992). The jenseniin G producer, *P. jensenii* P126, does not harbor detectable plasmid DNA (Grinstead and Barefoot, 1992a).

To date, only one report has described the introduction of foreign DNA into propionibacteria by protoplast transformation. Baehman and Glatz (1989) developed a procedure for the production and regeneration of protoplasts for *P. freudenreichii* P104 but did not report transformation of the protoplasts. Protoplasts were produced most efficiently (>99.9%) by subjecting mid-exponential cells to treatment with lysozyme (20 mg/ml); stationary phase cells were recalcitrant to protoplasting. Regeneration of protoplasts occurred in soft agar supplemented with sucrose and gelatin (2.5%) at efficiencies of 10 to 30%. These efficiencies compare favorably with those reported by Lin and Savage (1986) for *Lactobacillus acidophilus* 100-33 (5 to 67%) or by Lee-Wickner and Chassy (1984) for *L. casei* (10 to 40%). However, for propionibacteria, the regeneration process required 21–25 days

(Woskow and Glatz, 1990) and illustrates one of the difficulties in addressing these slow-growing organisms. More recently, the protoplast formation procedure was modified by first growing strain P104 for 24 hours in media containing 1% threonine to destabilize the cell wall and by then treating the cells with lysozyme followed by treatment with chymotrypsin (Woskow and Glatz, 1990). Protoplasts were transformed with pE194 in the presence of PEG and 60 transformed regenerants were detected per µg of plasmid DNA (Woskow and Glatz, 1990).

Electroporation or electrically induced transformation of DNA into propionibacteria has been examined in two studies. Luchansky et al. (1988) used the Bio-Rad Gene Pulser™ to electroporate the 4.4-kb erythromycin (Em^r) and chloramphenicol (Cm^r) resistance, lactococcal-*Escherichia coli-Bacillus subtilis* shuttle vector, pGK12 (Kok et al., 1984), into *P. jensenii* NCK139 (also designated B77). Application of electroporation conditions of 6250 V/cm and 25 µF optimal for *L. acidophilus* resulted in transformation frequencies of 32 CFU/µg plasmid pGK12 DNA for *P. jensenii* NCK139; in comparison, electroporation frequencies for lactobacilli in the same study ranged from 1.7×10^2 to 9.7×10^5/µg. Putative transformants were identified by the pGK12 phenotype (Cm^r and Em^r) and confirmed by the presence of pGK12 in the recipients. Zirnstein and Rehberger (1991) used a BTX electroporation system to introduce the *Staphylococcus aureus* vector pC194 into several strains of propionibacteria. Definition of optimal electroporation conditions for *P. freudenreichii* P7 identified two field-strength and pulse-duration combinations. Application of either 5.4 kV/cm for 5 milliseconds or 37.8 kV/cm for 40 microseconds yielded 1.1×10^2 transformants per µg of plasmid DNA. Plasmid pC194 was not recovered from transformants; however, Southern hybridization and restriction enzyme analysis provided evidence for integration of the plasmid into the chromosome of recipient propionibacteria.

The only genes of propionibacteria characterized to date are those encoding the methylmalonyl-CoA mutase of *P. freudenreichii* subsp. *shermanii* NCIB 9885. Methylmalonyl-CoA mutase catalyzes the rearrangement of succinyl-CoA into methylmalonyl-CoA during the propionic acid fermentation (Marsh et al., 1989). It consists of two nonidentical polypeptides, an α-subunit and a β-subunit, with a combined molecular weight of 146,000 Da. To clone the enzyme, Marsh et al. (1989) generated a minilibrary of *Eco*RI fragments (1.0–1.8 kb) of genomic DNA. Genomic DNA from *P. freudenreichii* subsp. *shermanii* NCIB 9885 was digested to completion with *Eco*RI; appropriately sized fragments were isolated, ligated into pUC13, and transformed into *Escherichia coli*. The use of synthetic oligonucleotides deduced from the N-terminal sequences of the α- and β-subunits permitted identification of clones. Analyses of DNA sequences indicated that the genes for both subunits are in close proximity and probably are transcribed as an operon. No consensus promoter regions upstream of either gene were identified. Sequences homologous to ribosomal binding sites in *E. coli* were present (Marsh et al., 1989). In a subsequent study, placing the linked genes under the control of an inducible promoter resulted in their efficient expression in *E. coli* (McKie et al., 1990).

Thorough characterization of genes encoding production of and immunity to bacteriocins may provide useful tools to increase our understanding of propionibacteria and assist in their genetic manipulation. Selectable markers appropriate for genetic manipulation of food-grade organisms have not yet been mapped in propionibacteria. Vectors capable of being maintained as autonomous plasmids currently have not been developed. Bacteriocin production would provide an easily detectable marker; bacteriocin immunity could be used in recipient selection. As "food-grade" markers, propionibacteria bacteriocins may be used in genetic manipulation of dairy propionibacteria and other food fermenting organisms. Future characterization of genes encoding bacteriocin production, immunity, and related functions, as well as construction of vectors from indigenous plasmids, will provide a better understanding of genetic systems in these industrially significant organisms.

II. Inducible Bacteriocins of Lactic Acid Bacteria

In striking contrast to the paucity of information available concerning propionibacteria bacteriocins, considerable information exists concerning bacteriocins of lactic acid bacteria. As indicated in other chapters of this volume, bacteriocin producers have been identified in all genera and many species included in the lactic acid bacteria (Klaenhammer, 1988; Schillinger, 1990). However, despite extensive biochemical and genetic characterization of many bacteriocins of lactic acid bacteria, few authors report assessments of their inducibility. Only for propionicin PLG-1 (Lyon and Glatz, 1991a), lactocin LP27 (Upreti and Hinsdill, 1973), and the bacteriocin produced by *Lactobacillus fermentum* (DeKlerk and Smit, 1967) has inducibility been investigated; treatment of the producer cultures with mitomycin C or ultraviolet light did not affect production of these bacteriocins. Most bacteriocinogenic lactic acid bacteria including *Carnobacterium piscicola* (Ahn and Stiles, 1990), *L. acidophilus* (Barefoot and Klaenhammer, 1984; Muriana and Klaenhammer, 1991), *L. helveticus* (Joerger and Klaenhammer, 1986), *L. sake* (Schillinger and Lücke, 1989; Mørtvedt and Nes, 1990), *Lactococcus lactis* subsp. *cremoris* (Davey and Richardson, 1981), *L. lactis* subsp. *lactis* (Mattick and Hirsch, 1944; Hurst, 1981), *L. lactis* subsp. *diacetylactis* (Geis et al., 1983; Scherwitz et al., 1983), *Leuconostoc gelidum* (Hastings and Stiles, 1991; Harding and Shaw, 1990), *Pediococcus acidilactici* (Hoover et al., 1988; Bhunia et al., 1988), and *P. pentosaceus* (Daeschel and Klaenhammer, 1985; Gonzalez and Kunka, 1987) produce bacteriocins during growth in broth and/or agar media in the absence of inducing agents. As a consequence, their produc-

10. Bacteriocins from Dairy Propionibacteria

tion has been assumed to be constitutive. To date, only one of the many bacteriocins produced by lactic acid bacteria demonstrates phenotypic induction by treatment with mitomycin C. The bacteriocin, caseicin 80, is bactericidal toward *Lactobacillus brevis* B109 and is produced constitutively at 3.2 arbitrary units (AU) per ml by *L. casei* B 80 in a complex medium (Rammelsberg and Radler, 1990). Treatment of the producer culture with mitomycin C (0.5–1.0 μg/ml) results in the classic phenotypic response that characterizes inducible systems: activity of caseicin 80 in the culture medium increases five- to sevenfold (Rammelsberg et al., 1990). The phenotypic induction of caseicin 80 might be assumed to occur similarly to induction of colicin synthesis. Synthesis of most colicins is controlled by the SOS regulatory system (Pugsley, 1984; Lazdunski et al., 1988). That is, the Lex A repressor protein normally binds the colicin promoter and prevents colicin synthesis. Exposure of colicin producer cultures to DNA-damaging agents such as mitomycin C or ultraviolet light generates an inducing signal that activates the Rec A protein to a proteolytic form. The Rec A protease then cleaves the Lex A repressor resulting in derepression of colicin synthesis (Walker, 1984). Whether the molecular events that increase caseicin 80 production after mitomycin C treatment are similar to those for colicins has not yet been determined.

Numerous authors have speculated that bacteriocin production provides the producer culture with a competitive advantage against closely related organisms seeking the same ecological niche. If so, the presence of bacteriocin-sensitive cells might be expected to elicit a bacteriocin production response from the producer culture. This premise was examined in a recent preliminary, but interesting, study (Hughes and Barefoot, 1990) that addressed the effect of associative cultures on bacteriocin production by the lactacin B producer, *Lactobacillus acidophilus* N2. The heat-stable peptide (6500-Da) bacteriocin, lactacin B, is produced when the producer is cultivated on MRS agar or in MRS broth maintained at pH 6.0 (Barefoot and Klaenhammer, 1983, 1984). The bacteriocin exhibits a limited spectrum of activity against *L. helveticus* and the three species *L. bulgaricus*, *L. lactis*, and *L. leichmannii* that recently were reclassified as *L. delbrueckii* subsp. *bulgaricus* and *lactis*, respectively (Muriana and Klaenhammer, 1991). Hughes and Barefoot (1990) cocultured the bacteriocin producer with a streptomycin-resistant variant of the bacteriocin-sensitive indicator, *L. delbrueckii* subsp. *lactis* ATCC 4797. The producer was propagated alone or with the sensitive indicator in MRS broth (initial pH 6.5) without pH control; antagonistic activity was detected only in associative cultures. Effects of pH were eliminated by cultivating the producer alone or with the sensitive indicator in MRS broth maintained at pH 6.0. Inhibitory activity appeared 4–6 hours earlier in associative cultures but was detected at equal concentrations in both. Like lactacin B (Barefoot and Klaenhammer, 1983), the antagonist from associative cultures was active against *L. helveticus* and *L. delbrueckii* subsp. *bulgaricus* and *lactis*, and stable to heat (3 minutes, 100°C, pH 5) and chaotrophic agents. As a result, Hughes and Barefoot (1990) concluded that the associatively produced antagonist was lactacin B. Other experiments indicated that bacteriocin-enhancing activity was associated with the indicator cells but not with the spent indicator culture medium and that at least 10^6 indicator cells were required to evoke a producer response (Hughes and Barefoot, 1990). A report by Chen et al. (1992) has confirmed the earlier results. Recent data demonstrate that indicator cells that had been subjected to a lethal, but mild, heat treatment (55°C, 7 days) retained the enhancing activity. Addition of cells of insensitive Gram-positive, but not Gram-negative, bacteria elicited bacteriocin production. These results suggest that the moiety responsible for enhancement of bacteriocin production by *L. acidophilus* N2 may be common to Gram-positive cells. The mechanism by which enhancement of bacteriocin production occurs is unknown. Although these results are preliminary, it is tempting to speculate that similar stimulation of bacteriocin production by insensitive cells might occur in the heterogeneous microbial flora of foods.

Acknowledgment

This chapter is Technical Contribution No. 3528 of the South Carolina Agricultural Experiment Station, Clemson University, Clemson, South Carolina.

References

Ahn, C., and Stiles, M. E. (1990). Plasmid-associated bacteriocin production by a strain of *Carnobacterium pisciola* from meat. *Appl. Environ. Microbiol.* **56**, 2503–2510.

Al-Zoreky, N., Ayres, J. W., and Sandine, W. E. (1991). Antimicrobial activity of Microgard™. *J. Dairy Sci.* **74**, 758–763.

Baehman, L. R., and Glatz, B. A. (1989). Protoplast formation and regeneration in *Propionibacterium*. *J. Dairy Sci.* **72**, 2877–2884.

Barefoot, S. F., and Klaenhammer, T. R. (1983). Detection and activity of lactacin B: a bacteriocin produced by *Lactobacillus acidophilus*. *Appl. Environ. Microbiol.* **45**, 1808–1815.

Barefoot, S. F., and Klaenhammer, T. R. (1984). Purification and characterization of the *Lactobacillus acidophilus* bacteriocin lactacin B. *Antimicrob. Agents Chemother.* **26**, 328–334.

Bhunia, A. K., Johnson, M. C., and Ray, B. (1988). Purification, characterization, and antimicrobial spectrum of a bacteriocin produced by *Pediococcus acidilactici*. *J. Appl. Bacteriol.* **65**, 261–268.

Chen, Y. R., Barefoot, S. F., Hughes, M. D., and Hughes, T. A. (1992). The agent from Gram-positive cells that enhances lactacin B production by *Lactobacillus acidophilus* N2 cultures. Abstr. 668. In "Abstracts of the Annual Meeting of the Institute of Food Technologists, New Orleans, Louisiana, June 20–24.

Cummins, C. S., and Johnson, J. L. (1986). *Propionibacterium*. In "Bergey's Manual of Systematic Bacteriology" (P. H. A. Sneath, N. S. Mair, M. E. Sharpe, and J. G. Holt, eds.), Vol. 2, pp. 1346–1353. Williams and Wilkins, Baltimore.

Daeschel, M. A. (1989). Antimicrobial substances from lactic acid bacteria for use as preservatives. *Food Technol.* **43(1)**, 164–167.

Daeschel, M. A., and Klaenhammer, T. R. (1985). Association of a 13.6-megadalton plasmid in *Pediococcus pentosaceus* with bacteriocin activity. *Appl. Environ. Microbiol.* **50**, 1538–1541.

Davey, G. P., and Richardson, B. C. (1981). Purification and some properties of diplococcin from *Streptococcus cremoris*. *Appl. Environ. Microbiol.* **41,** 84–89.
DeKlerk, H. C., and Smit, J. A. (1967). Properties of a *Lactobacillus fermenti* bacteriocin. *J. Gen. Microbiol.* **48,** 309–316.
Fischer, G., Glatz, B., and Tomes, N. (1991). Selective medium for the isolation and enumeration of propionibacteria. P-42. *In* "Abstracts of the Annual Meeting of the American Society for Microbiology," Dallas, Texas, May 5–9.
Flores-Galarza, R. A., Glatz, B. A., Bern, C. J., and Van Fossen, L. D. (1985). Preservation of high-moisture corn by microbial fermentation. *J. Food Prot.* **48,** 407–411.
Fujimura, S., and Nakamura, T. (1978). Purification and properties of a bacteriocin-like substance (acnecin) of oral *Propionibacterium acnes*. *Antimicrob. Ag. Chemother.* **14,** 893–898.
Garvie, E. I. (1984). Taxonomy and identification of bacteria important in cheese and fermented dairy foods. *In* "Advances in the Chemistry and Biochemistry of Cheese and Fermented Milk" (F. L. Davies and B. A. Law, eds.), pp. 35–66. Elsevier Applied Science, New York.
Geis, A., Singh, J., and Teuber, M. (1983). Potential of lactic streptococci to produce bacteriocin. *Appl. Environ. Microbiol.* **45,** 205–211.
Glatz, B. A., and Anderson, K. I. (1988). Isolation and characterization of mutants of *Propionibacterium* strains. *J. Dairy Sci.* **71,** 1769–1776.
Gonzalez, C. F., and Kunka, B. S. (1987). Plasmid-associated bacteriocin production and sucrose fermentation in *Pediococcus acidilactici*. *Appl. Environ. Microbiol.* **53,** 2534–2538.
Grinstead, D. A. (1989). The detection of bacteriophage and bacteriocins among the classical propionibacteria. MS Thesis, Iowa State University, Ames.
Grinstead, D. A., and Barefoot, S. F. (1991). A bacteriocin-like inhibitor produced by *Propionibacterium jensenii*. O-29. *In* "Abstracts of the Annual Meeting of the American Society for Microbiology," Dallas, Texas, May 5–13.
Grinstead, D. A., and Barefoot, S. F. (1992a). Jenseniin G, a heat-stable bacteriocin produced by *Propionibacterium jensenii* P126. *Appl. Environ. Microbiol.* **58,** 215–220.
Grinstead, D. A., and Barefoot, S. F. (1992b). Partial purification and characterization of jenseniin G, a bacteriocin produced by *Propionibacterium jensenii* P126. Abstr. 666. In "Abstracts of the Annual Meeting of the Institute of Food Technologists, New Orleans, Louisiana, June 20–24.
Grinstead, D. A., and Glatz, B. A. (1989). Detection of bacteriophage and bacteriocins among the classical propionibacteria. P-41. *In* "Abstracts of the Annual Meeting of the American Society for Microbiology," New Orleans, Louisiana, May 14–18.
Harding, C. D., and Shaw, B. G. (1990). Antimicrobial activity of *Leuconostoc gelidum* against closely related species and *Listeria monocytogenes*. *J. Appl. Bacteriol.* **69,** 648–654.
Hastings, J. W., and Stiles, M. E. (1991). Antibiosis of *Leuconostoc gelidum* isolated from meat. *J. Appl. Bacteriol.* **70,** 127–134.
Hettinga, D. H., and Reinbold, G. W. (1972). The propionic acid bacteria-a review. I. Growth. *J. Milk Food Technol.* **35,** 295–372.
Hofherr, L. A., Glatz, B. A., and Hammond, E. G. (1983). Mutagenesis of strains of *Propionibacterium* to produce cold-sensitive mutants. *J. Dairy Sci.* **66,** 2482–2487.
Hoover, D. G., Walsh, P. M., Kolaetis, K. M., and Daly, M. M. (1988). A bacteriocin produced by *Pediococcus* species associated with a 5.5-megadalton plasmid. *J. Food Protect.* **51,** 29–31.
Hughes, M. D., and Barefoot, S. F. (1990). Activity of the bacteriocin lactacin B is enhanced during associative growth of *Lactobacillus acidophilus* N2 and *Lactobacillus leichmanii* 4797. O-12. *In* "Abstracts of the Annual Meeting of the American Society for Microbiology," Anaheim, California, May 13–17.
Hurst, A. (1981). Nisin. *In* "Advances in Applied Microbiology" (D. Perlman and A. I. Laskin, eds.), Vol. 27, pp. 85–123. Academic Press, Inc., New York.
ICMSF. (1980). "Microbial Ecology of Foods" Vol. 1. International Commission on Microbiological Specifications for Foods, Academic Press, Inc., New York.
Joerger, M. K., and Klaenhammer, T. R. (1986). Characterization and purification of helveticin J and evidence for a chromosomally determined bacteriocin produced by *Lactobacillus helveticus* 481. *J. Bacteriol.* **167,** 439–446.

Klaenhammer, T. R. (1988). Bacteriocins in lactic acid bacteria. *Biochimie* **70,** 337–349.
Kok, J., Van der Vossen, J. M. B. M., and Venema, G. (1984). Construction of plasmid cloning vectors for lactic streptocci which also replicate in *Bacillus subtilis* and *Escherichia coli. Appl. Environ. Microbiol.* **48,** 726–731.
Langsrud, T., and Reinbold, G. W. (1973). Flavor development and ripening of Swiss cheese—a review. III. Ripening and flavor production. *J. Milk Food Technol.* **36,** 593–609.
Lazdunski, C. J., Baty, D., Geli, V., Cavard, D., Morlon, J., Lloubes, R., Howard, S. P., Knibiehler, M., Chartier, M., Varenne, S., Frenette, M., Dasseux, J. L., and Pattus, F. (1988). The membrane channel-forming colicin A: synthesis, secretion, structure, action, and immunity. *Biochim. Biophys. Acta* **947,** 445–464.
Lee-Wickner, L. J., and Chassy, B. M. (1984). Production and regeneration of *Lactobacillus casei* protoplasts. *Appl. Environ. Microbiol.* **48,** 994–1000.
Lin, J. H. C., and Savage, D. C. (1986). Genetic transformation of rifampicin resistance in *Lactobacillus acidophilus. J. Gen. Microbiol.* **132,** 2107–2111.
Luchansky, J. B., Muriana, P. M., and Klaenhammer, T. R. (1988). Application of electroporation for transfer of plasmid DNA to *Lactobacillus, Lactococcus, Leuconostoc, Listeria, Pediococcus, Bacillus, Staphylococcus, Enterococcus,* and *Propionibacterium. Mol. Microbiol.* **2,** 637–646.
Lyon, W. J., and Glatz, B. A. (1991a). Partial purification and characterization of a bacteriocin produced by *Propionibacterium thoenii. Appl. Environ. Microbiol.* **57,** 701–706.
Lyon, W. J., and Glatz, B. A. (1991b). Isolation and characterization of propionicin PLG-1, a bacteriocin produced by a strain of *Propionibacterium thoenii. In* "Abstracts of the Annual Meeting of the American Society for Microbiology," Dallas, Texas, May 5–13.
Lyon, W. J., and Glatz, B. A. (1992). Production and purification of a bacteriocin from a strain of *Propionibacterium* thoenii. Abstr. 215. In "Abstracts of the Annual Meeting of the Institute of Food Technologists, New Orleans, Louisiana, June 20–24.
Mantere-Alhonen, S. (1983). On the survival of a *Propionibacterium freudenreichii* culture during in vitro gastric digestion. *Meijeritiet. Aikak.* **41,** 19–23.
Mantere-Alhonen, S. (1987). A new type of sour milk with propionibacteria. *Meijeritiet. Aikak.* **45,** 49–61.
Marsh, E. N., McKie, N., Davis, N. K., and Leadlay, P. F. (1989). Cloning and structural characterization of the genes coding for adenosylcobalamin-dependent methylmalonyl-CoA mutase from *Propionibacterium shermanii. Biochem. J.* **260,** 345–352.
Mattick, A. T. R., and Hirsch, A. (1944). A powerful inhibitory substance produced by group N streptocci. *Nature (London)* **154,** 551.
McKie, N., Keep, N. H., Patchett, M. L., and Leadlay, P. F. (1990). Adenosylcobalamin-dependent methylmalonyl-CoA mutase from *Propionibacterium shermanii. Biochem. J.* **269,** 293–298.
Mørtvedt, C. I., and Nes, I. F. (1990). Plasmid-associated bacteriocin production by a *Lactobacillus sake* strain. *J. Gen. Microbiol.* **136,** 1601–1607.
Muriana, P. M., and Klaenhammer, T. R. (1991). Purification and partial characterization of lactacin F, a bacteriocin produced by *Lactobacillus acidophilus* 11088. *Appl. Environ. Microbiol.* **57,** 114–121.
Paul, G. E., and Booth, J. S. (1988). Properties and characteristics of a bacteriocin-like substance produced by *Propionibacterium acnes* isolated from dental plaque. *Can. J. Microbiol.* **34,** 1344–1347.
Prince, L. D., Barefoot, S. F., Bodine, A. B., and Grinstead, D. A. (1993). Characterization of a bacteriocin, jenseniin P, from *Propionibacterium jensenii* P1264. *In* "Abstracts of the Annual Meeting of the American Society for Microbiology," Atlanta, Georgia, May 16–20.
Pugsley, A. P. (1984). The ins and outs of colicins. Part I. production and translocation across membranes. *Microbiol. Sci.* **1,** 168–175.
Rammelsberg, M., and Radler, F. (1990). Antibacterial polypeptides of *Lactobacillus* species. *J. Appl. Bacteriol.* **69,** 177–184.
Rammelsberg, M., Müller, E., and Radler, F. (1990). Caseicin 80: purification and characterization of a new bacteriocin from *Lactobacillus casei. Arch. Microbiol.* **154,** 249–252.
Razafindrajaona, J. M. (1989). Inducible bacteriolytic agents against *Propionibacterium*

freudenreichii subsp. *shermanii*. MS Thesis, Swedish University of Agricultural Sciences, Uppsala.

Rehberger, T. G., and Glatz, B. A. (1990). Characterization of *Propionibacterium* plasmids. *Appl. Environ. Microbiol.* **56,** 867–871.

Scherwitz, K. M., Baldwin, K. A., and McKay, L. L. (1983). Plasmid-linkage of a bacteriocin-like substance in *Streptococcus lactis* subsp. *diacetylactis* strain WM$_4$: transferability to *Streptococcus lactis. Appl. Environ. Microbiol.* **45,** 1506–1512.

Schillinger, U. (1990). Bacteriocins of lactic acid bacteria. *In* "Biotechnology and Food Safety" (D. D. Bills and S. Kung, eds.), pp. 55–74. Butterworth-Heinemann, Boston.

Schillinger, U., and Lücke, F.-K. (1989). Antibacterial activity of *Lactobacillus sake* isolated from meat. *Appl. Environ. Microbiol.* **55,** 1901–1906.

Schillinger, U., Kaya, M., and Lücke, F.-K. (1991). Behaviour of *Listeria monocytogenes* in meat and its control by a bacteriocin-producing strain of *Lactobacillus sake. J. Appl. Bacteriol.* **70,** 473–478.

Tagg, J. R., Dajani, A. S., and Wannamaker, L. W. (1976). Bacteriocins of gram-positive bacteria. *Bacteriol. Rev.* **40,** 722–756.

Upreti, G. C., and Hinsdill, R. D. (1973). Isolation and characterization of a bacteriocin from a homofermentative *Lactobacillus. Antimicrob. Agents Chemother.* **4,** 487–494.

Walker, G. C. (1984). Mutagenesis and inducible responses to deoxyribonucleic acid damage in *Escherichia coli. Microbiol. Rev.* **48,** 60–93.

Weinbrenner, D. R., Grinstead, D. A., and Barefoot, S. F. 1993. Inhibitory activity of jenseniin G against *Lactobacillus delbrueckii* ssp. *bulgaricus* and *Streptococcus salivarius* ssp. *thermophilus.* In "Abstracts of the Annual Meeting of the American Society for Microbiology," Atlanta, Georgia, May 16–20.

Woolford, M. K. (1975). The significance of *Propionibacterium* species and *Micrococcus lactilyticus* to the ensiling process. *J. Appl. Bact.* **39,** 301–306.

Woskow, S. A., and Glatz, B. A. (1990). Improved method for production of protoplasts and polyethylene glycol-induced transformation of *Propionibacterium freudenreichii. In* "Abstracts of the Annual Meeting of the American Society for Microbiology," Anaheim, California, May 13–17.

Zirnstein, G. W., and Rehberger, T. G. (1991). Electroporation-mediated gene transfer of pC194 into the *Propionibacterium* genome. *In* "Abstracts of the Annual Meeting of the American Society for Microbiology," Dallas, Texas, May 5–9.

CHAPTER 11

Regulatory Aspects of Bacteriocin Use

SUSAN K. HARLANDER

I. Introduction

Recent approval by the U.S. Food and Drug Administration (FDA) of the bacteriocin nisin for use in processed cheese spreads has stimulated interest in the potential application of other antimicrobial compounds produced by food-grade microorganisms. Bacteriocins are particularly attractive preservatives, as they are naturally produced by many strains of lactic acid bacteria used for the production of fermented foods, and thus have been consumed safely by humans for hundreds of years. In addition, bacteriocins are protein in nature and therefore should be readily digested in the human gastrointestinal tract. They can function as natural food preservatives through the inhibition of spoilage or pathogenic bacteria and ultimately contribute to food safety. Heightened consumer concern over "chemical" food additives has led to the search for alternative methods for control of food-borne pathogens. The fact that bacteriocins are produced as normal by-products of microbial metabolism make them attractive as "natural" preservatives. Using genetic engineering, the gene(s) encoding bacteriocin production could be transferred into starter cultures used for the production of fermented foods to inhibit the growth of pathogenic and spoilage organisms *in situ* and extend the shelf-life of the products. Alternatively, bacteriocins could be produced via fermentation by native or genetically engineered organisms, purified, and added to foods as pure chemicals. This chapter will explore the regulatory climate and some of the criteria FDA might consider important for documenting the safety of and granting approval for the use of genetically engineered bacteriocin-producing starter cultures or fermentation-derived bacteriocins as natural preservatives in foods.

II. Characteristics of Bacteriocins

Many species of lactic acid bacteria (*Lactococcus, Lactobacillus, Streptococcus, Pediococcus, Leuconostoc*) used for the production of fermented foods have been shown to be strongly antagonistic toward other bacteria, including representatives from the same genera, as well as toward unrelated spoilage organisms and pathogens. In many cases, inhibition is caused by organic acids or hydrogen peroxide, normal by-products of metabolism. In some cases, however, bactericidal proteins have been demonstrated to be responsible for inhibition. According to the classical definition outlined by Tagg et al. (1976), bacteriocins are proteins produced by a wide variety of microorganisms, which are bactericidal to a limited number of other closely related strains. At least some bacteriocins demonstrate a broad inhibitory spectrum, although most, if not all bacteriocins produced by the lactic acid bacteria are inhibitory against only Gram-positive organisms. It has been proposed that the definition of bacteriocin be expanded, as many bacteriocins are capable of inhibiting a broad spectrum of related and unrelated species.

Bacteriocins can be small proteins, with molecular sizes of a few thousand daltons, or complex structures containing subunits in excess of 10^6 Da, with associated carbohydrate or lipid moieties. The heterogeneity of bacteriocins is reflected in differences in optimal conditions for activity, mode of action, optimal conditions for production, and genetic basis (plasmid or chromosomal). Bacteriocin-producing strains have evolved mechanisms of immunity to the inhibitory action of their own bacteriocins, and immunity gene(s) are usually genetically linked to production gene(s).

Nisin is produced by some strains of *Lactococcus lactis* subsp. *lactis* and has been used as a food preservative for over 30 years in 52 different countries throughout the world. It has been used to inhibit spore-forming organisms in processed cheese spreads, canned foods, and hot-plate bakery products, to extend shelf-life of pasteurized milk, to control lactic acid bacteria in beer production, and to control *Clostridium botulinum* type E in modified-atmosphere packaged fresh fish. Nisin has been approved for use in the United States as an antibotulinal agent in processed cheese spreads (Federal Register, 1988). Nisin is the only bacteriocin produced by lactic acid bacteria that has been thoroughly characterized. Nisin is a pentacyclic peptide containing three unusual amino acids, dehydroalanine, lanthionine and β-methyl-lanthionine, and has a molecular weight of 3510 Da. It is inactivated by α-chymotrypsin, but is resistant to treatments with pronase, trypsin, and heat (100°C, 10 minutes) under acidic conditions. Nisin is effective against Gram-positive pathogens and prevents outgrowth of *Clostridium* and *Bacillus* spores.

III. Potential Uses of Bacteriocins or Bacteriocin-Producing Organisms in Foods and Pharmaceuticals

A. Food Preservatives

Although nisin is the only approved bacteriocin for use in the United States, there is a great deal of interest in other bacteriocins that have similar properties and exhibit broad-spectrum inhibitory activity. Bacteriocins produced by fermentation could be purified and added to foods as pure chemicals to inhibit food-borne pathogens and spoilage organisms. Bacteriocins have several characteristics that make them ideal food preservatives. Many bacteriocins are capable of resisting inactivation at the relatively high temperatures used in food processing and can remain functional over a broad pH range. Bacteriocins are usually inactivated by one or more of the proteolytic enzymes present in the digestive tract of humans and would be digested just like any other protein in the diet. Bacteriocins are nontoxic, odorless, colorless, and tasteless. Finally, bacteriocins may be perceived by consumers to be more natural than chemical preservatives.

The efficacy of using bacteriocins as food preservatives will need to be determined for each food system. Solubility, stability, sensory impact, heat and pH tolerance, and types and numbers of organisms inhibited will need to be evaluated for each bacteriocin in each food product category under a variety of storage conditions.

B. Health Care Products

Because nisin inhibits a broad spectrum of Gram-positive organisms, it has been used in teat dips for prevention of mastitis in cows; in oral health care products, such as toothpaste and mouthwash, for inhibition of dental caries and periodontal disease; and in soap, skin care products, and cosmetics for treatment of acne. The worldwide market for mastitis treatment is approximately $100 million and is expected to grow over 30% in the next five years. In the oral health care market, the mouthwash market alone is $500 million annually in the United States; toothpaste is an even larger market. Skin care is also a potentially large market worldwide.

C. Starter Cultures

Lactic acid bacteria are used extensively for the production of fermented dairy, meat, and vegetable products. Bacteriocin-producing strains could be used to enhance the safety of these products, since many have been shown to inhibit Gram-positive pathogens such as *Listeria monocytogenes*, *Staphylococcus aureus*, and *Clostridium botulinum*. Naturally occurring bacteriocin-producing strains could also be used in nonfermented products. Since bacteriocin production and immunity are frequently plasmid-mediated traits in the lactic acid bacteria, once identified and characterized, natural gene transfer systems such as conjugation could be used to transfer these plasmids to other starter cultures (Klaenhammer, 1988).

D. Genetically Engineered Starter Cultures

Alternatively, bacteriocin production and immunity genes could be genetically engineered into dairy and meat starter cultures to inhibit pathogens, into yeast to inhibit lactic spoilage organisms, or into silage inocula to inhibit competing organisms during fermentation. Bacteriocin production and/or immunity genes localized on specific DNA fragments could be inserted into cloning vectors using recombinant DNA techniques. Recombinant plasmids can be transferred to bacterial hosts using transformation or electroporation techniques.

E. Probiotic Organisms

Bacteriocin-producing organisms, particularly lactobacilli that are naturally present in the gut of humans or animals, could be used as probiotics to influence the ecology of the gut. It has been postulated that certain gut microorganisms provide health benefits that include stimulation of the immune system, inactivation of potentially carcinogenic compounds, and reduction of serum cholesterol. Bacteriocins might enhance the ability of these organisms to colonize and compete with indigenous as well as potentially pathogenic gut microflora.

F. Markers for Food-Grade Cloning Vector Construction

Genes encoding resistance to therapeutic antibiotics (e.g., erythromycin, tetracycline) are frequently used as selectable markers on cloning vectors. These markers are unacceptable for engineering of starter cultures because of the concern over possible transfer of antibiotic resistance to gut microflora. Bacteriocin immunity gene(s) could be used as alternative selectable markers for the construction of food-grade cloning vectors. Since bacteriocins are not used therapeutically, transfer of resistance to gut microorganisms would not be an issue.

IV. Factors Affecting Regulatory Approval of Bacteriocins As Food Ingredients

The U.S. Food and Drug Administration's (FDA) primary responsibility is to ensure a safe and wholesome food supply for the nation. The burden of proof of safety of new foods or food ingredients, however, rests with the supplier. The Federal Food, Drug, and Cosmetic Act (FDC Act) of 1938 provides broad authority to the FDA to accomplish its mission, but at the same time provides discretion to apply the law as needed to protect public health. The provisions that apply primarily to the safety of foods are encompassed in the adulteration provisions (21CFR, Section 402). FDA may take action against a food if it establishes that a substance added to a food may render the food injurious to health. In 1958 the FDC Act was amended to require premarket approval for new chemicals (i.e., food additives) that would be added to food for technical effects such as sweetness, texture, and preservation.

A food additive is defined as a substance, natural or synthetic, whose intended use results or may reasonably be expected to result in its becoming a component of food, directly or indirectly. Substances subject to prior sanction, in use prior to 1958, or covered by other provisions or statutes were exempt from this provision. Substances in use prior to 1958 were generally recognized as safe (GRAS) and were subjected to either of two safety standards: (1) scientific information must establish that safety is predicated on information published by scientists, who as qualified experts recognize the safety of the proposed use of the substance; or (2) history of the substance's use is dependent on documented safe use in food before 1958 (Maryanski, 1990).

A. Are Bacteriocins Food Additives or GRAS Ingredients?

Purified bacteriocins to be added to foods as preservatives would be regulated by FDA as either food additives or GRAS substances. If the bacteriocin is a substance for which history of use prior to 1958 could not be established, the manufacturer would need to file a food additive petition containing extensive information on the physical and chemical properties of the bacteriocin, its intended use, and its safety. If, on evaluation, FDA agrees that the proposed use is safe, the agency would issue a food additive regulation describing the additive and the conditions under which it may be used. If it can be documented that the bacteriocin was present in the food supply prior to 1958 and has been safely consumed, two options are available. The manufacturer could self-affirm that the bacteriocin is GRAS and proceed with using it, or they could file a GRAS petition and request that FDA affirm the ingredient as GRAS.

Food additive petitions and petitions for GRAS affirmation contain similar information; however, access to the information by the general public differs. GRAS petitions are placed on file and the information is readily available for public inspection. For food additive petitions, on the other hand, some of the information of a proprietary nature need not be disclosed to the public. Obtaining approval for a new food additive can be costly and time consuming, since the burden of proof of safety rests with the petitioner. If a reasonable case can be made that the substance is a GRAS ingredient, safety documentation is based on history of safe use.

For the first products of biotechnology, it may not always be clear which regulatory route might be required by FDA. In the case of recombinant rennet, one of the first products of biotechnology to gain regulatory approval, the company chose to file both a GRAS affirmation petition as well as a food additive petition. FDA affirmed GRAS status for this ingredient in March 1990. In the case of bacteriocins that have probably been safely consumed as an integral part of fermented foods for decades, a strong case could be made that these are GRAS ingredients rather than new food additives.

B. Data Required by FDA

When either a food additive petition or GRAS affirmation is filed, FDA will require information about the bacteriocin.

1. Identity and Proposed Use

The identity and the technical effect of the bacteriocin must be defined. Information that will be of interest to FDA includes the formal name or synonyms of the bacteriocin, structure and chemical formula, molecular weight, purity and limits of impurities of the bacteriocin preparation to be added to foods and potential by-products or breakdown products, and sensitivity of the bacteriocin to pH, temperature, and proteolytic digestion.

2. Description of the Manufacturing Process

The safety of bacteriocins produced via fermentation will depend on all the steps of the process, including (1) the source organism, its characteristics, and how it has been developed to perform its intended function; (2) the fermentation process, including the substrates and other growth materials and the conditions of growth; (3) the isolation procedures and all the steps in the purification process, when appropriate; and (4) the final standardization of the product. Current good manufacturing practices must be fundamental to any production process (WHO, 1991). Impurities must be identified and specifications established for each product.

3. Analytical Methods

The manufacturer must supply detailed descriptions of the analytical procedures used to determine the presence, potency, and stability of the bacteriocin and its breakdown products in foods. Analytical methods should be specific, precise, accurate and reliable, and capable of being performed in standard laboratories by personnel trained in routine laboratory procedures. All data accumulated with purified bacteriocins, as well as data from actual food systems, should be provided.

4. Efficacy

Evidence that the bacteriocin performs as an antimicrobial agent against specific food-borne pathogens or spoilage organisms at expected levels of addition must be provided. The bacteriocin should be tested at several levels of addition in those food products for which approval will be requested.

5. Estimated Intake

Estimated daily intake of the bacteriocins in the food products to be supplemented should be determined. If the bacteriocin will be used in a number of different food products, an estimate of how much the typical consumer might ingest from all sources must be provided. Petitioners should be aware if a particular segment of the population consumes relatively high levels of a specific product to which bacteriocins will be added.

6. Safety Evaluations and Toxicity Testing

Although bacteriocins have been safely consumed in foods for decades, FDA might have some concern about the toxicological and the allergenic properties that might be associated with elevated levels of bacteriocins in food products. The amount of bacteriocin to be added to foods to achieve preservation should be compared to the amount produced naturally in foods by bacteriocin-producing organisms. If the level to be added is significantly higher than that naturally produced, additional safety evaluation may be necessary.

V. Factors Affecting Regulatory Approval of Naturally Occurring Bacteriocin-Producing Strains

Microorganisms added to foods, such as yeast in bread, wine and beer making, and bacteria used for the production of fermented dairy, meat, and vegetable products, have been safely consumed by humans for literally hundreds of years. The microorganisms are generally recognized as safe by the FDA and have been sanctioned for specific applications. The organisms themselves, however, have not been evaluated as GRAS substances. The agency considers them to be an integral constituent of fermented foods and even though they have been safely used for one purpose, safety must be documented if they are to be used for a different purpose. For example, *Lactococcus lactis* subsp. *cremoris* is considered GRAS for the production of certain fermented dairy products; however, it may not necessarily be considered GRAS if used as a preservative in nonfermented refrigerated salads or if used for the production of an ingredient to be added to foods. If naturally occurring bacteriocin-producing microorganisms are to be used for a new purpose, many of the conditions outlined above for pure ingredients should be considered.

Theoretically, the bacteriocin-producing ability of an organism could be increased by chemical mutagenesis. In addition, plasmid-mediated bacteriocin production could be transferred to other food-approved organisms via conjugation. In either case, genetic improvement by traditional means does not require regulatory approval if the organism will be used for a sanctioned purpose, even though the result would be increased concentrations of the bacteriocin in the final food product and increased exposure to consumers.

VI. Factors Affecting Regulatory Approval of Genetically Engineered Bacteriocin-Producing Strains

Genetic engineering is an alternative strategy for construction of bacteriocin-producing microbial starter cultures. Genes encoding bacteriocin production and immunity have been identified and characterized for nisin, as well as other bacteriocins. To date, FDA has not promulgated new regulatory guidelines for genetically engineered organisms. Applications have been handled on a case-by-case basis applying regulations already in use; however, specific safety assessment considerations were identified in a report of the Joint FAO/WHO Consultation (WHO, 1991) held in Geneva in 1990. These considerations are discussed briefly below.

1. Characterization of the Genetic Modification

It is essential that the DNA to be inserted into organisms be characterized to ensure that it does not code for a harmful substance. The use of DNA isolated from organisms previously used in food will be of less concern than DNA isolated from an organism not previously consumed by humans.

The overall construction strategy must be outlined in detail. In the case of bacteriocins, the DNA sequence for open reading frames and regulatory signals can be documented and confirmed by sequence analysis, restriction mapping, and hybridization techniques to ensure that additional DNA is not present. Data must be provided on the stability of the construct. All organisms that are used to donate genetic material or used as intermediate hosts must also be well characterized.

2. Origin and Identity of the Host Organism

The use of host organisms that have been shown to be safe for use in food is preferred. Taxonomic identity must be confirmed by comparison with reference cultures, particularly if the host organism is isolated from the natural environment. This information is useful in predicting the potential presence of toxins, virulence factors, or impurities that need to be evaluated.

3. Vectors

Vectors must be well characterized to ensure that no harmful substances are produced, since the final construct will become an integral part of the food and will be consumed by humans. Food-grade vectors derived from organisms recognized to have a history of safe use in food are most desirable. These vectors do not contain selectable marker genes that encode resistance to antibiotics used therapeutically. It is also desirable for the vectors to be stably maintained and for transfer of traits to other microorganisms to be minimized.

4. Selectable Markers

Although antibiotic resistance markers are useful selection systems, they are not acceptable for genetic engineering of microbial starter cultures. Since viable engineered organisms will be consumed, there is the possibility that DNA encoding antibiotic resistance could be transferred to pathogenic microbes in the gut, rendering them refractory to clinical therapy. For this reason, functional marker genes should not be present in the final production organisms when the viable organism is to be used in food. Bacteriocins to be produced by genetically engineered organisms and used as pure ingredients must be examined for the presence of biologically active DNA that could encode resistance to antibiotics.

5. Pleiotropic Effects

Insertion of genetic material into a bacterial host might cause unintentional pleiotropic effects. This will depend in part upon the vector system used; vectors that insert into the bacterial chromosome would be more likely to cause secondary effects, although these may not be detectable. Performance of engineered strains would need to be evaluated under actual manufacturing conditions. Final product characteristics, including chemical, structural, and organoleptic properties, must be carefully evaluated and compared to nonengineered counterparts to identify unanticipated differences between parental and engineered strains.

6. Pathogenicity and Toxigenicity

Microorganisms intended for use in food processing should be derived from organisms that are known, or that have been shown by appropriate tests in animals to be free of traits that confer pathogenicity. If the host organism has a long history of safe use, and the source of the DNA is from a similar nonpathogenic, food-grade microorganism, there is little possibility that the introduced DNA might code for a microbial enterotoxin, endotoxin, mycotoxin, or antimicrobial substance. However, introduced DNA could be examined for the absence of DNA sequences of the size and pattern of known toxin sequences, and in cases of doubt, animal and *in vitro* tests could be used to confirm that toxins are absent.

7. Allergenicity

Certain proteins and glycoproteins elicit allergic responses in sensitive individuals. Most proteins, however, are largely digested in the gastrointestinal tract. Allergenicity of elevated levels of bacteriocins produced by genetically engineered organisms might be of concern to FDA; therefore, it would be important to document the susceptibility of the specific bacteriocin to proteolytic cleavage by digestive enzymes. Comparison of gene sequences with data on known allergens may become increasingly useful as the information on such proteins increases.

8. Level of Intake

Assessment of changes in dietary use or exposure patterns resulting from the use of a genetically engineered bacteriocin-producing starter culture should be determined. Population groups that might be expected to consume relatively high levels of the product should be identified.

9. Overall Assessment

A general principle for genetic engineering of microbial starter cultures is that the design should be directed toward minimizing genetic transfer to other microorganisms. In addition, the possibility of any significant adverse change in the nutritional composition of a food as a consequence of microbial action should be evaluated by comparison with the nutritional composition of the conventional counterpart.

Since bacteriocins are produced by organisms frequently used in food, and bacteriocin-producing capability would be transferred into other food-grade microorganisms, adverse effects are not anticipated. Providing that antibiotic resistance markers are not used in the construction of strains, it is unlikely that FDA would have serious problems with this application of biotechnology. One concern might be elevated production of bacteriocins, particularly if significantly higher concentrations are required.

FDA has repeatedly indicated that no new laws or regulations should be required for regulation of genetically engineered organisms to be used in foods. They have focused on the product, not the process by which a food ingredient or food is produced. The review process for genetically engineered organisms will likely be case-by-case. Each organism will be analyzed independently and approval will depend highly on the product characteristic and the intended use of the particular organism or product. On May 29, 1992, the FDA published their statement of policy regarding foods derived from new plant varieties (Federal Register, 1992). As expected, the policy indicates that no new regulations are required for oversight of biotechnology-derived food crops and that regulation will be identical in principle to that applied to foods developed by traditional plant breeding. Regulatory status is dependent upon objective characteristics of the food and its intended use rather than the process by which it is produced. As in the past with traditionally derived crops, producers have an obligation to ensure that the foods are safe and in compliance with applicable legal requirements. Producers are encouraged to informally consult with FDA prior to marketing new foods and a "decision tree" approach provides guidance as to when contact with FDA would be required. For example, insertion of a gene which codes for a product not currently consumed in the human diet would be a trigger for FDA consultation. Such a crop might be considered a new food additive and subject to oversight under Section 409 of the FDC. Unfortunately, the policy does *not* include reference to genetically engineered starter cultures; this application will require additional FDA policy guidelines.

VII. Players in the Regulatory Arena

A. U.S. Food and Drug Administration

As outlined above, the FDA has the most obvious role in the regulation of bacteriocins as pure food ingredients, as well as naturally occurring and genetically engineered bacteriocin-producing strains. Ultimately, it is their decision which criteria will be used to ensure the safety of these products. However, several other players have become involved in the regulatory arena.

B. The International Food Biotechnology Council

In December 1990, the International Food Biotechnology Council (IFBC), a consortium of food processing and food biotechnology companies, published a report that identified issues and assembled a set of scientific criteria to evaluate and ensure the safety of biotechnology-derived food and food ingredients from plants and microorganisms (IFBC, 1990). The IFBC recognized the need to build a consensus on appropriate safety evaluation criteria before the widespread development of new products that might require such evaluation prior to their commercialization. The report was published in a peer-reviewed journal with the hope that it would be useful to government regulatory agencies, the food industry, and the public. IFBC advocated a decision tree approach and developed such decision trees for (1) food ingredients derived from genetically modified microorganisms; (2) single chemical entities and simple chemically defined mixtures; and (3) whole food and other complex mixtures. One of the IFBC decision trees outlined above would apply to each of the different bacteriocin applications described in this chapter.

C. Individual States

In the absence of a coordinated federal oversight system for agricultural and food biotechnology, several states have considered developing their own regulations. Introduction of genetically engineered plants, animals, and microorganisms into the environment has raised public awareness and concern over possible negative environmental effects. The state of Minnesota, for example, recently adopted regulations to be administered by the Minnesota Environmental Quality Board on the release of genetically engineered organisms into the environment (Minnesota State Register, 1991). Other states have considered similar legislation; states could assume an ever increasing role in biotechnology regulation in the future. Adoption of state-by-state regulation of biotechnology could severely inhibit growth and development of this industry and commercialization of the technology.

D. The President's Council on Competitiveness

In February 1992, the President's Council on Competitiveness released a landmark document that sets forth the proper basis for federal agency oversight authority of biotechnology (Federal Register, 1992). It describes a risk-based, scientifically sound approach to the oversight of planned introductions of biotechnology products into the environment that focuses on the characteristics of the biotechnology products and the environment into which it is being introduced, not the process by which the product is created.

The 1986 Coordinated Framework (Federal Register, 1986) established *who* should have oversight authority for specific biotechnology applications, but did not address *how* that authority should be exercised in the frequent situations in which a statute leaves the implementing agency latitude for discretion. The Council on Competitiveness report clearly states that "... to ensure that limited federal oversight resources are applied where they will accomplish the greatest net beneficial protection of public health and the environment, oversight will be exercised only where the risk posed by the introduction is unreasonable, that is, when the value of the reduction in risk obtained by additional oversight is greater than the cost thereby imposed." This report affirms FDA's stated philosophy about engineered foods and food ingredients. A strong case could be made for the fact that based on their long history of safe use, fermentation-derived bacteriocins and naturally occurring bacteriocin-producing strains pose a small risk to public health and the environment. Similarly, if food-grade hosts and vectors are utilized in genetic improvement of microorganisms, a strong case could be made for genetically engineered microbial starter cultures.

E. Others

In addition to other federal regulatory agencies with biotechnology oversight, such as the U.S. Department of Agriculture and the Environmental Protection Agency, other interested parties will have an impact on regulation of biotechnology. Countries throughout the world are independently developing biotechnology regulations that could have a dramatic impact on international trade.

VIII. Future Challenges

A. Consumer Acceptance

Genetically engineered drugs such as insulin, human growth hormone, interferon, and tissue plasminogen activator have been well accepted by the public. Genetically engineered foods and food ingredients may not be as readily accepted. How the public feels about food is fundamentally different than how they feel about drugs: risks associated with food are unacceptable, whereas risks associated with drugs are a fact of life. This dichotomy is due, in part, to the fact that the average consumer has little understanding of agriculture and food processing. The food supply is taken for granted—high quality, safe, nutritious, and inexpensive food is expected. To make educated decisions about biotechnology, the public must understand what biotechnology is and how it compares with standard agricultural and food processing practices.

With all of the concern about microbiological hazards in the food supply, the development of natural preservatives to control pathogenic and spoilage organisms might be an application of biotechnology that would be perceived as beneficial by the public. Further, many consumers have a rudimentary understanding of fermented foods and understand that they have been consuming viable microorganisms in yogurt and other cultured products.

B. Competition in the Marketplace

Biotechnology provides a means for producing food ingredients more cost effectively than traditional methods, and companies that adopt the technology early will be more competitive in the marketplace. Companies will need to suppress the temptation to undermine consumer confidence in biotechnology-derived products manufactured by their competitors.

C. Labeling of Biotechnology-Derived Foods

It has been suggested that foods be labeled to inform consumers about addition of biotechnology-derived food ingredients. In a recent report, the Environmental Defense Fund (1991) proposed that FDA issue three new regulations to assure the safety of and to provide consumers information about certain genetically engineered foods. These regulations would require (1) premarket approval by FDA of new substances added to foods via genetic engineering; (2) labeling of foods to disclose substances added via genetic engineering; and (3) premarket notification of FDA of changes in the composition of all genetically engineered whole foods.

IX. Conclusion

The primary objective of this chapter was to provide an overview of the kind of issues regulatory agencies will consider for ensuring the safety of bacteriocins or bacteriocin-producing organisms in foods. An understanding of FDA's informational needs should allow more efficient development of bacteriocin applications in the future.

References

Environmental Defense Fund. (1991). A Mutable Feast: Assuring Food Safety in the Era of Genetic Engineering. Washington, D.C.
Federal Register. (1986). Coordinated framework for the regulation of biotechnology. Vol. 51, No. 123, June 26.

11. Regulatory Aspects of Bacteriocin Use

Federal Register. (1988). Nisin preparation: Affirmation of GRAS status as a direct human food ingredient. Vol. 53, No. 66, April 6.

Federal Register. (1992). Exercise of federal oversight within scope of statutory authority: Planned introductions of biotechnology products into the environment. Vol. 57, No. 39, February 27.

Federal Register. (1992). Statement of policy regarding foods derived from new plant varieties. Vol. 57, No. 104, May 29.

International Food Biotechnology Council (IFBC). (1990). Biotechnologies and Food: Assuring the Safety of Food Produced by Genetic Modification. International Food Biotechnology Council Report. *Regulatory Toxicology and Pharmacology* 12(3).

Klaenhammer, T. R. (1988). Bacteriocins of lactic acid bacteria. *Biochimie* 70, 337–349.

Maryanski, J. H. (1989). Special challenges of novel foods (biotechnology). *Food Drug Cosmetic Law J.* 45(5), 545–550.

Maryanski, J. H. (1990). Regulation of new food substances made by biotechnology. Presentation at the 34th Annual Educational Conference sponsored by the Food and Drug Law Institute. Washington, D.C. December 11–12, 1990.

Minnesota State Register. (1991). Proposed permanent rules relating to the release of genetically engineered organisms. CITE 16S.R.422. August 26.

Tagg, J. R., Dajani, A. S., and Wannamaker, L. W. (1976). Bacteriocins of gram-positive bacteria. *Bacteriol. Rev.* 40, 722–756.

WHO. (1991). Report of a Joint FAO/WHO Consultation: Strategies for assessing the safety of foods produced by biotechnology. Office of Publications, World Health Organization, Geneva, Switzerland.

CHAPTER 12

Future Prospects for Research and Applications of Nisin and Other Bacteriocins

WILLEM M. DE VOS

I. From Past to Present

Although it has been suggested that the food industry started about two million years ago, it is generally assumed that food fermentations developed since the Neolithic times when humans adopted a lifestyle that allowed agriculture to develop (Toussaint-Samat et al., 1991). Ever since, it is likely that lactic acid bacteria have played an important role in the production of fermented foods, although based on recorded history this can be traced back only a few millennia. This time frame is important in considering the extent to which lactic acid bacteria have adapted to their new ecological niche, that is, the food environment. In view of the fact that traditional food fermentations, and even modern, large-scale production processes, are operated under nonsterile conditions, it is no surprise that many lactic acid bacteria produce antagonistic compounds that increase their competitive value.

This century has seen a major effort in describing, cataloging, and characterizing the wide variety of antagonistic compounds produced by lactic acid bacteria. One important class considered in this book are the bacteriocins, being operationally defined here as ribosomally synthesized proteins with antimicrobial activity to which the producer strains show immunity (see Montville and Kaiser, Chapter 1, this volume). It is often cited that the first report on a bacteriocin from lactic acid bacteria dates from 1928 (Rogers, 1928; Rogers and Whittier, 1928). However, closer reading of these studies, which describe an inhibitory effect of a lactic streptococcal strain on *Lactobacillus bulgaricus,* raises the question as to whether the dialyzable and heat-stable substance, which reportedly is a common property (Rogers, 1928), meets the definition of a bacteriocin, since it also appeared to inhibit the producer culture itself (Rogers and Whittier, 1928). Even if it is a bacteriocin, there is no evidence to support the suggestion made by various authors (see Hansen, Chapter 5, this volume; Kolter and Moreno, 1992) that this compound is nisin and does not represent one of the various other bacteriocins produced by lactic streptococci. Better documented early examples of true bacteriocins include the substances produced by two lactic streptococci, strains 9S and D1 (Whitehead, 1933). These substances are sensitive to proteolysis by trypsin but not by pepsin, they are unstable at pH 8, and they appear relatively heat stable. Since nisin meets all these criteria, it is possible that this report represents a first description of nisin, especially since one of the producer strains, strain D1, is capable of fermenting sucrose, a property that is phenotypically and genetically coupled to nisin production (LeBlanc et al., 1980; Gasson, 1984; Rauch and de Vos, 1992). The fact that the producer strain D1 resembled an unusual *Streptococcus* sp. does not invalidate this speculation since it is now well known that nisin-producing strains are usually atypical *Lactococcus lactis* strains (Hirsch, 1953; de Vos et al., 1992a). Other early studies on bacteriocins from lactic streptococci include the description and denomination of diplococcin (Oxford, 1944) and nisin (Mattick and Hirsch, 1944). Since then a large number of bacteriocins have been detected and characterized in many species of lactic acid bacteria, amounting to more than 50 different compounds that for the sake of simplicity can be divided into three groups (see Klaenhammer et al., Chapter 7, this volume): lantibiotics, small hydrophobic peptides, and large heat-labile proteins.

In view of its early discovery, its wide application, and unusual properties, it may be explained in retrospect that research on the lantibiotic nisin (Hurst, 1981; Hansen, Chapter 5, this volume) has dominated the field of bacteriocins by providing the first structure to be determined (Gross and Morell, 1971), the first gene to be isolated and sequenced (Buchman et al., 1988), the first three-dimensional structure to be proposed (Van de Ven et al., 1991b), and the first bacteriocin to be improved by protein engineering (Kuipers et al., 1991). However, major progress has also been realized, mainly in the last decade, with other bacteriocins, resulting in various significant milestones, such as the first characterization on the protein and DNA level of representatives of small hydrophobic peptide bacteriocins (lactococcin A from *Lactococcus lactis;* Holo et al., 1991) and the large heat-labile protein bacteriocins (helveticin J from *Lactobacillus helveticus;* Joerger and Klaenhammer, 1990).

In view of the growing scientific and industrial interest in the bacteriocins of lactic acid bacteria, which is evident from the increasing number of publications and patent applications, the near future will see considerable progress beyond the state of the art described in this volume. What are the incentives for this future research; in which direction will this research develop and at what rate? Below, a number of important areas will be described in research and application that already are actively pursued at present. In addition, new developments will be indicated, some of which have already been initiated, whereas others are likely to be started soon. As stated above, research on the lantibiotic nisin is the most advanced of all bacteriocins. Moreover, it is produced by a lactic acid bacterium, *Lactococcus lactis*, for which well-characterized genetic systems have been developed (de Vos, 1987). Therefore, in selected cases, the progress obtained with nisin will be used to illustrate the present and future developments in the field of bacteriocins.

II. Research Approaches and Incentives

All bacteriocins studied so far are produced as a primary translation product, are transported across the cellular membrane, and are subject to one or more processing events before showing antimicrobial activity. An illustrative scheme that summarizes several stages in the research on bacteriocins is presented in Figure 1. It introduces the levels at which bacteriocins are usually characterized, that is, the activity, protein and gene level, and the major areas of interest including biosynthesis, evolution, three-dimensional (3D) structure, and mode of action of bacteriocins. In addition, it shows the sequence and direction of the events that lead to the expression of a bacteriocin gene and the processing of the primary translation product to yield an active bacteriocin. Finally, the scheme depicts the well-known reversed genetics approach, aimed at characterizing the structural bacteriocin genes.

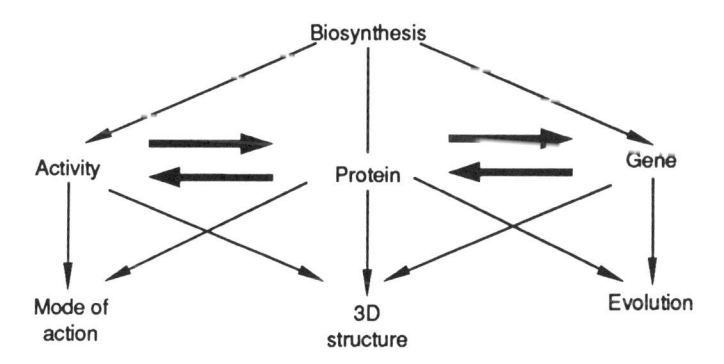

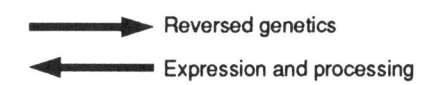

Figure 1 General scheme summarizing research approaches with bacteriocins.

Although bacteriocins can be produced on a large scale and applied in the food industry without detailed knowledge of their properties, there are at least three major reasons to have some, and in other cases detailed, information on the biosynthesis, evolution, 3D structure, and mode of action of the present bacteriocins and those that are to be developed in the future.

The first reason concerns the registration and legal approval of bacteriocins. This is a vital stage in all cases in which bacteriocins are applied as additives. Important issues to be addressed include the biochemical characterization of the compound, its potential toxicity and sensitivity to gastric enzymes, its spectrum and mode of action, and its induction of cross-resistance against therapeutically used drugs. Experience with nisin, that was already well-defined in the early 1970s (Gross and Morell, 1971), showed that approval by the appropriate bodies is time consuming, limited to specific applications, or still denied in some countries (Hurst, 1981; Delves-Broughton, 1990; Food and Drug Administration, 1988). Although recently sensible regulations or deregulations have been proposed to allow streamlining of federal regulations of biotechnology products (Bush, 1992), it cannot be expected that the current approval procedures will change drastically in the near future.

Second, knowledge of the above-mentioned properties of bacteriocins is useful, and in various cases essential, for further improving their application. Thus, substantial information of the biochemical properties of a bacteriocin is important if procedures are to be developed for industrial-scale purification. Furthermore, insight into the mode of action may be advantageous in improving the antimicrobial spectrum of a bacteriocin. This can be illustrated by the use of nisin to inhibit Gram-negative bacteria, which are usually insensitive to this lantibiotic (Hurst, 1981). Since it was known that nisin could dissipate the membrane potential in vesicles of Gram-negative bacteria, similar to its action on vesicles of Gram-positive bacteria (Kordel and Sahl, 1986), it was to be expected that, in combination with treatments that expose their cellular membrane, Gram-negative bacteria would also be sensitive to nisin. It was demonstrated that *Salmonella* and other undesired Gram-negative bacteria can be antagonized with nisin in combination with additions of EDTA or citrate that are known to disintegrate their outer membrane (Blackburn et al., 1989).

Finally, detailed information on the molecular properties of bacteriocins is essential to allow for developing existing and novel bacteriocins based on genetic engineering, molecular screening, or protein engineering strategies, as is exemplified below.

III. Genetic Engineering

In general, there are two approaches aimed at characterizing bacteriocins at the gene level; an entirely genetic one, and one starting with the purification and subsequent biochemical characterization of the antimicrobial activity. Both have their specific advantages and weaknesses and the approach taken determines to a large extent the direction of present and future studies of bacteriocins. It should be noted, however, that both approaches are complementary in providing complete evidence for the cloning of a structural bacteriocin gene, that is, the gene encoding the primary translation product that is eventually converted into the component with antimicrobial activity. This evidence should be based on the comparison of the properties of the expressed gene product with those of the purified bacteriocin. This criterium has been met only by a small number of the described bacteriocins as is illustrated in Table I. Moreover, since many bacteriocins are subject to processing, a detailed biochemical approach is imperative to determine the exact processing site(s).

In many cases a genetic approach based on curing and transfer studies has allowed the localization of genes involved in bacteriocin production. This appears to be a rewarding strategy, since many bacteriocin production and immunity genes appear to reside on mobile DNA elements (Table I). Although abundantly present in many lactic acid bacteria, plasmids are not the only elements carrying bacteriocin genes, as studies with the nisin biosynthesis genes have demonstrated. While it was assumed for about a decade that the tightly linked genes for nisin biosynthesis, immunity, and sucrose utilization were located on (unstable and conjugative) plasmids (LeBlanc et al., 1980; Gasson, 1984; Gonzales and Kunka, 1984), it required molecular tools in combination with chromosome analysis before the first convincing data were reported on their location on a 70-kb conjugative, chromosomally located transposon (Rauch and de Vos, 1990). This transposon, Tn5276, can be considered as a prototype for other similar sucrose-nisin transposons (Horn et al., 1991; Rauch et al., 1991), and now has been characterized in detail (Rauch and de Vos, 1992). Sequence analysis in con-

TABLE I
Genetically and Biochemically Characterized Bacteriocins from Lactic Acid Bacteria

Bacteriocin	Strain	Location[a]	Reference
Large heat-stable proteins			
Helveticin J	*Lactobacillus helveticus*	X	Joerger and Klaenhammer, 1990
Small hydrophobic proteins			
Lactacin F	*Lactobacillus acidophilus*	P	Muriana and Klaenhammer, 1991
Pediocin PA-1	*Pediococcus acidilactici*	P	Marugg et al., 1991
Lactococcin A	*Lactococcus lactis*	P	Holo et al., 1991; Van Belkum et al., 1991a
Leucocin A	*Leuconostoc gelidum*	P	Hastings et al., 1991
Lantibiotics			
Nisin A	*Lactococcus lactis*	T	Horn et al., 1991; Rauch and de Vos, 1990; 1992
Nisin Z	*Lactococcus lactis*	T	Rauch et al., 1992
Lacticin 481	*Lactococcus lactis*	P	Piard et al., 1992

[a]Location of the structural gene on a plasmid (P), transposon (T), or the chromosome (X).

junction with transcription, complementation, and expression studies allowed the identification of the Tn5276-located operons that are involved in (1) nisin biosynthesis, secretion, and immunity as well as its regulation, (2) the utilization of sucrose and its regulation, and (3) the excision and integration during conjugative transfer (Rauch et al., 1992).

An important consequence of the location of bacteriocin genes on mobile elements is their possible transfer to other strains using natural conjugation systems. If the transferred elements carry the required bacteriocin production and immunity genes, this property provides the opportunity to construct novel bacteriocin-producing strains. This possibility is extremely relevant when it comes to application, since the naturally found bacteriocin-producing strains are not necessarily the best production strains or those with the appropriate industrial properties. An example is the capacity to produce nisin, usually associated with nondairy *Lactococcus lactis* strains (Hirsch, 1953; de Vos et al., 1993), which are not suitable as starter strains in industrial dairy fermentations. Therefore, the location of the nisin biosynthesis and immunity genes on a single conjugative transposon that also encodes sucrose utilization, an easily selectable property, allows the construction of the required industrial strains (Broadbent and Kondo, 1991; Hugenholtz and de Veer, 1991).

Subsequent genetic analysis of bacteriocin genes with known location is usually hampered because of (1) difficulties in detecting, isolating, and manipulating large plasmids or transposons, in case they are linked to a mobile genetic element; (2) the absence of genetic tools for the producer organism that prevent site-specific mutation and complementation studies; and (3) failure to express active bacteriocin in heterologous and even homologous systems, since specific host functions are often required. An interesting exception is the elegant work on the genes for lactococcin A production and immunity (later designated *lcn*A and *lci*A, respectively), which have been located on a 60-kb plasmid and further characterized by subcloning, sequencing, and expression studies (van Belkum et al., 1991a; van Belkum, 1991). This study showed that the deduced amino acid sequence of the putative *lcn*A expression product included the sequence of the purified lactococcin A (Holo et al., 1991). Unexpectedly, a cloned DNA fragment, containing the *lcn*A and *lci*A genes, yielded low expression of active bacteriocin in only one *L. lactis* strain (strain IL 1403) and not in other tested *L. lactis* strains (Holo et al., 1991). Similar inability to produce antimicrobial activity has also been found when an expression vector carrying a structural gene for nisin production was introduced in nonnisin-producing *L. lactis* strains (Kuipers et al., 1991). Another exception, illustrating an approach for analyzing chromosomally located bacteriocin operons, is the recent work on the generation and detailed analysis of mutants in Tn5276 that are deficient in the biosynthesis of nisin. Mutants belonging to different complementation groups, some of which were deficient in *nis*A transcription and showed reduced nisin immunity, were obtained by both classical mutation of the *nis*-genes (Rauch et al., 1991) and site-directed mutagenesis using a replacement recombination approach (O. Kuipers, personal communication). The *nis*A structural gene could be expressed using a T7 expression system in *Escherichia coli* but the produced prenisin showed no detectable antimicrobial activity (de Vuyst et al., 1990). This contrasts with the bacteriocins lactococcin M and pediocin PA-1 that were found to be produced in an active but apparently unprocessed form by *E. coli* (van Belkum et al., 1992; Marugg et al., 1991).

The second approach starts at the level of the component with antimicrobial activity that has to be purified, preferably in sufficient amounts to allow its further characterization using a variety of biochemical methods, including N-terminal or complete amino acid sequencing, amino acid composition determination, molecular mass analysis, and in some cases NMR analysis to determine the bacteriocin structure. The information on the primary structure of a bacteriocin allows the isolation and further characterization of the bacteriocin structural gene by reversed genetics. This has been done with all bacteriocins shown in Table I, with the exception of helveticin J, whose gene has been detected in an *E. coli* expression library of genomic *Lactobacillus helveticus* DNA that was screened with antibodies obtained against the purified bacteriocin (Joerger and Klaenhammer, 1990).

Immunological studies using antibodies against purified bacteriocins or their synthetic peptides are also helpful in studying the bacteriocin processing events, as has been shown for nisin, using antibodies against the purified lantibiotic and its leader peptide (J. Van der Meer, personal communication). It should be noted, however, that it is sometimes difficult to obtain an immune response to small hydrophobic peptides also when coupled to large carriers. In addition, antibodies against bacteriocins can be used to develop sensitive immunological detection methods as was shown for competition ELISA for nisin (de Vos et al., unpublished results).

The availability of the bacteriocin genes allows for increasing the bacteriocin production and/or immunity by using well-established methods such as raising copy numbers, increasing the expression, and alleviating regulation of relevant genes. A successful overproduction (approximately threefold) was obtained by introducing the nisin structural gene under control of an efficient promoter on a high copy number vector in a *L. lactis* transconjugant containing a single copy of Tn5276 (Kuipers et al., 1991). However, at present there are more reports on the inability to produce a bacteriocin than published examples of overproduction. This is due to various reasons, the main one being that for the production of bacteriocins the required genetic information is complex and the limiting steps have not yet been determined. It is, however, fair to assume that for various already well-developed bacteriocins (Table I) such engineering will be possible in the near future, especially for those that have an established market position (nisin, pediocin PA-1), have market potential (leucocin, lacticin 481, lactacin F), or can be regarded as models (lactococcin A, helveticin J).

IV. Traditional and Molecular Screening

Screening of lactic acid bacteria for the production of antimicrobials is an important activity in detecting novel bacteriocins with useful properties. Those may include completely new antimicrobial proteins or natural variants of existing bacteriocins. The simplicity, efficiency, and speed of the screening procedures determine their eventual success. Established methods focus on the antimicrobial activity of bacteriocins and include screening for deferred antagonism (Kekessy and Piguet, 1970), followed by a direct test of antimicrobial activity of cell supernatant (Tagg and McGiven, 1971). These have the great advantage that specific indicator strains can be applied and that rough quantification of antimicrobial activity is possible. These and other methods have been used in many studies and have yielded the large number of present bacteriocins. Because new products containing lactic acid bacteria are still being discovered and included in those screening activities, and because hand-screening strategies are being improved (for instance, by automation), it is expected that many more bacteriocins will be found.

An important new development is the use of molecular methods in the screening of lactic acid bacteria, designated here as molecular screening. This approach is based on a molecular characterization of the property that is screened for. In case of bacteriocins, traditional screening is on the activity level, while molecular screening could then take place on the protein or gene level (see Figure 1). Although established biochemical or genetic procedures are suitable, there are methods of choice for each approach. For screening at the protein level an immunological approach is likely to meet the requirements for efficient screening. To this end antibodies can be applied that are raised against purified bacteriocins, proteolytic fragments thereof, or synthetic peptides. At the genetic level methods can be based on the polymerase chain reaction (PCR; Saiki et al., 1988) that has revolutionarized diagnostic strategies. This method has considerable potential since the amount of specific PCR targets is great, varying from structural bacteriocin genes to genes involved in bacteriocin processing, secretion, or immunity. The limitation of screening at the genetic level is that there is no selection for a property but instead for a gene, which always could turn out to be silent. This, however, could also be advantageous since it is, for instance, the only way to detect cryptic bacteriocin genes that could be activated in other hosts.

An important aspect of any screening approach, using either traditional or molecular methods, is the fact that it does not involve the use of genetic modification techniques and identifies bacteriocins in natural strains. It is to be expected that for industrial applications of bacteriocins in the near future those and other natural bacteriocin-producing strains are the most suitable.

Since molecular screening is based on established properties it cannot be expected that the strains to be found produce completely new bacteriocins. It is possible however to select for specific properties such as

12. Future Prospects for Research and Applications of Nisin and Other Bacteriocins

bacteriocin overproduction or natural bacteriocin variants or hybrids. Examples of molecular screening are limited to natural variants of nisin. In a first traditional screening approach aimed at finding natural nisin-overproducing strains, culture supernatants of a limited number of sucrose-fermenting lactic acid bacteria were analyzed in a standard agar diffusion assay (Tramer and Fowler, 1964) with various nisin-sensitive indicators. The rationale for specifically including sucrose-fermenting strains was based on the finding that the nisin biosynthesis and sucrose fermentation genes indeed were physically linked (Rauch and de Vos, 1992). It appeared that the supernatant of *L. lactis* NIZO 22186 produced much larger inhibition zones than that of other strains (see Figure 2). Further studies showed that strain NIZO 22186 produced a lanthionine-containing 3.5-kDa polypeptide, which showed high similarities but no complete identity to commercial nisin with respect to retention on reversed-phase (RP) HPLC and amino acid composition (Mulders et al., 1991). NMR studies of the purified product and sequence analysis of its structural gene showed that this lantibiotic is a natural nisin variant that has a similar structure as nisin but contains the substitution His27Asn (Mulders et al., 1991). To avoid confusion between the natural nisin variants and since no parent–progeny relation can be established, the nisin variant described previously (Gross and Morrell, 1971) was designated nisin A, encoded by the *nis*A gene (Kaletta and Entian, 1989), and the nisin variant produced by *L. lactis* strain NIZO 22186 was described as nisin Z, encoded by the *nis*Z gene, which follows the proposed general nomenclature of lantibiotics (de Vos et al., 1991). Comparison of the *nis*A and *nis*Z gene sequences showed that the His27Asn substitution is based on a single base-pair mutation (C to A transversion) and that there is no sequence difference in the part of the gene encoding the leader peptide (Mulders et al., 1991).

Since the high activity of the culture supernatant strain NIZO 22186 could be attributed to either a high production level or inherent properties of nisin Z or both, the antimicrobial activities of purified preparations of nisin Z and nisin A were studied using six relevant species of Gram positive bacteria as indicators (de Vos et al., 1993). This analysis showed that (1) the inhibition zones with nisin Z at practically used lantibiotic concentrations are invariably larger than those with nisin A; (2) no difference in inhibition zones is observed between both natural variants at low lantibiotic concentration; (3) the minimal inhibitor concentrations are similar; and (4) halos obtained with nisin Z are more diffuse than those obtained with nisin A (Figure 2). These results were interpreted as implying that because of the His27Asn substitution, nisin Z has better diffusion properties than nisin A, which may be of considerable practical significance, since nisin is applied in many systems that are diffusion limited (Hurst, 1981). Several aspects of nisin Z applications in the dairy technology have been discussed (Hugenholtz and De Veer, 1991).

Based on the sequence differences between the *nis*A and *nis*Z genes, molecular tools were developed to allow differentiation between the two

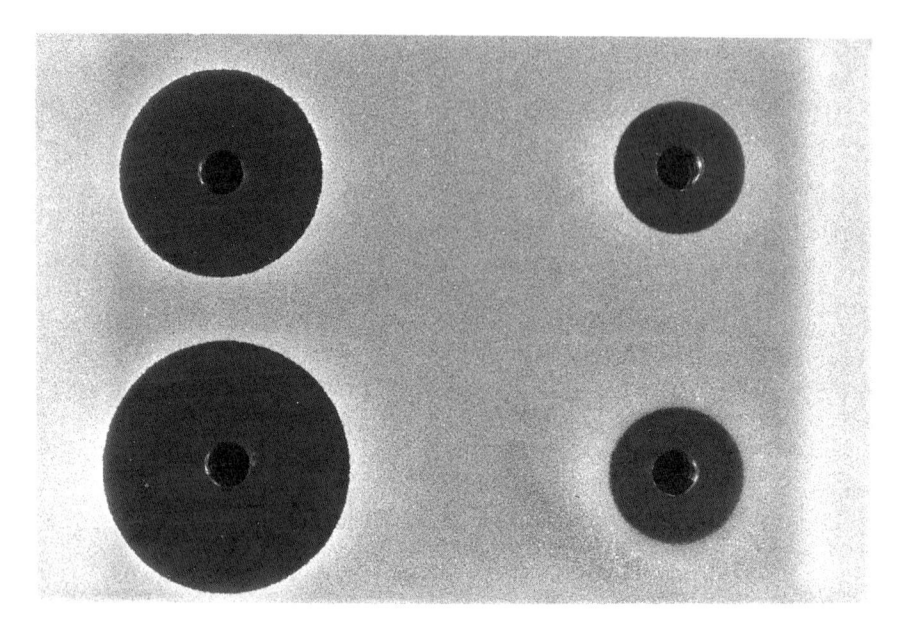

Figure 2 Antimicrobial activity of purified nisin A and nisin Z (de Vos et al., 1992). Identical amounts of nisin A (*top*) and nisin Z (*bottom*) were subjected to an agar diffusion assay using *Micrococcus flavus* as indicator (Tramer and Fowler, 1964). Samples at the left contained 1 μg/well and samples at the right, 0.05 μg/well.

natural variants at the gene level. To this purpose total DNA was isolated from 27 sucrose-fermenting strains from a variety of sources, including new natural isolates and strains that were not known to produce nisin, and a putative *nis* gene was amplified using flanking PCR primers (de Vos et al., 1992a). In all cases an amplified product with the expected size of 0.4 kb was obtained that was analyzed by direct sequencing. Using this molecular approach (Kuipers et al., 1991; de Vos et al., 1993) and by determining the capacity of conjugative transfer of nisin and sucrose genes (Rauch et al., 1991), it was shown that (1) 16 of the 27 strains contained the *nis*Z gene; (2) there were no other natural variants in the tested strains; and (3) various nisin Z-producing strains were present that, in contrast to strain NIZO 22186, showed efficient transfer of a *nis*Z gene-containing transposon, designated Tn*5278* (de Vos et al., 1993). Apart from showing that *nis*Z genes are widespread, this example of molecular screening has resulted in appropriate donors that can be used to construct, by conjugation, novel industrial strains that produce nisin Z.

V. Protein Engineering

Protein engineering of bacteriocins is an upcoming challenge now that several bacteriocins have been analyzed at the biochemical and genetic level

12. Future Prospects for Research and Applications of Nisin and Other Bacteriocins

(see Table I). In many cases, protein engineering is critically dependent on knowledge of the three-dimensional structure of the protein, or a related one. Only in this way can a systematic approach be realized that allows the rational design of mutations required to optimize protein function. A major bottleneck in this approach is the determination of the tertiary structure of a protein by x-ray crystallography. Although this applies particularly to the bacteriocins belonging to the class of large, heat-labile proteins, it does not hold to the same extent for the small proteins belonging to the classes of lantibiotics and small hydrophobic proteins. Engineering of these peptides is also possible in the absence of a 3D structure, because there is only a limited number of residues. This allows for site-specific mutagenesis by trial and error and renders it feasible to consider random mutagenesis. In addition, several predictions can be made from the primary and secondary structures and comparison with structures of other small peptides. Those permit specific mutation strategies to be developed based on an educated guess approach and may also result in the creation of hybrid bacteriocins. Moreover, because of their size these small peptides are excellent substrates to be studied by proton NMR spectroscopy in order to determine their secondary and spatial structures. This approach has the additional advantage that the structure of the small bacteriocins can be studied in different solvents, which is extremely relevant since it allows determining their spatial conformation in lipophilic environments mimicking the membrane, which is the likely target of both hydrophobic bacteriocins and lantibiotics.

Because of their unusual thioether bridges, lantibiotics are especially interesting molecules to analyze using spectroscopic techniques. Recently, a series of 2D NMR studies of nisin A, its degradation products, and specifically synthesized fragments, has been published (van de Ven et al., 1991a; Lian et al., 1991; Goodman et al., 1991), and the spatial structure of purified nisin A in aqueous solution has been solved (van de Ven et al., 1991b; Lian et al., 1992). The nisin A molecule appears to be quite flexible and consists of two domains; an N-terminal domain (residues 3–19) comprising the first three lanthionine rings, and a C-terminal domain (residues 23–28) consisting of the two coupled β-methyllanthionine rings, that are connected by a flexible hinge region around methionine 21. This is illustrated in Figure 3, showing a model of nisin Z based on the spatial conformation of nisin A. These and other spectroscopic studies have confirmed the proposed amphiphilic character of nisin A (Jung, 1991), suggesting that its mode of action on the cytoplasmic membrane, in analogy with that of other amphiphilic peptides such as mellitin, could be an insertion model or a wedge model (van de Ven et al., 1991a,b). Protein engineering in conjunction with detailed studies on the mode of action will allow experimental verification of these models and may also be used to explain the recent finding that nisin activity is affected by the phospholipid composition of membranes (Gao et al., 1991).

It is likely that similar spectroscopic studies with other small bacteriocins will follow soon, as is corroborated by the recent report on the NMR analysis of the first representative of the small hydrophobic bacteriocins, leucocin A

NISIN Z

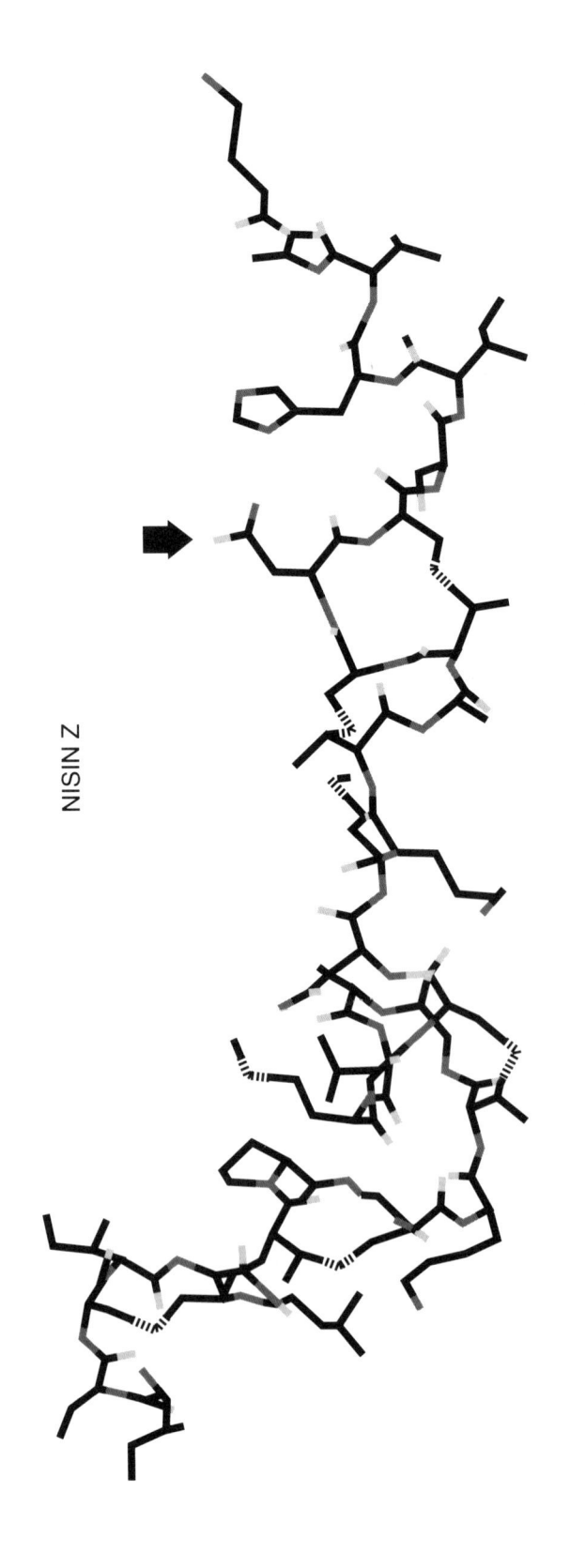

(Henkel et al., 1992). These and other studies will allow a further analysis of the structure-function relation of small hydrophobic bacteriocins and lantibiotics.

An important question is which properties should be engineered. In general, two types of engineering approaches can be distinguished: one directed at increasing the insight in the structure–activity relation of bacteriocins and the creation of novel proteins; and one directed toward improving the properties of a bacteriocin for industrial applications. It is not easy to separate these approaches, especially not at this moment, when the first engineered bacteriocins have just been reported (Kuipers et al., 1991; 1992). Therefore, a few self-explanatory examples that indicate the types of engineering feasible in the near future are summarized in Table II.

Apart from insight into what residues should be changed, it is important to have the tools to (over)express the mutated bacteriocin genes. As described above, only a few expression systems for bacteriocins are presently available and usually the production levels are low. An exception is nisin for which three different expression systems have been developed (Kuipers et al., 1991; O. Kuipers, personal communication). One is based on expressing mutated *nisZ* genes in a nisin A producer, another employs complementation of a classical nisin-deficient mutant, and yet another is based on complementation of a nisin-deficient mutant in which a site-directed deletion has been introduced in the *nisA* gene using a food-grade replacement recombination approach. The resulting strains have been used as hosts for the production of novel nisin species, obtained by protein engineering, that were subsequently purified and characterized using proton NMR. In this way the feasibility of engineering new dehydrated amino acids into nisin Z has been demonstrated (Kuipers et al., 1991). Finally, this engineering approach has generated a mutant of nisin Z with increased antimicrobial activity (Kuipers et al., 1992).

VI. Concluding Remarks

The areas described above illustrate some trends in the present and future research on bacteriocins and are not intended to be complete. Apart from studies on the biosynthesis, including the role of the leader peptide, of bacteriocins and, in particular, lantibiotics (Jung, 1991), at least one other area is presently developing at a fast rate and includes research on essential

---

Figure 3 Proposed model of nisin Z obtained by molecular modeling. The model is based on one of the nisin A conformations that have been determined using 2D-NMR (van de Ven et al., 1991b). Sulphur (⊞) present in lanthionine bridges and methionine; nitrogen (■); oxygen ( ); and carbon (■). Protons are not shown. The amino terminus is at the left and the carboxy terminus is at the right. The arrow indicates residue Asn27 (which is His in nisin A); it is located in the last of the five lanthionine rings.

TABLE II
Some Present and Future Targets
for Engineering of Bacteriocins

Bacteriocin production
Solubility of bacteriocins
Bacteriocin stability
Spectrum of antagonistic activity
Specific antagonistic activity
Diffusion in target systems
Bacteriocin immunity

---

processes involving the membrane of producer and target organisms. Putative membrane proteins have been discovered that confer bacteriocin immunity (van Belkum et al., 1991b; 1992) or resistance (Froseth and McKay, 1991), while recently also proteins belonging to the superfamily of ATP-dependent translocators have been detected that are probably involved in export of small bacteriocins (Stoddard et al., 1992; O. Kuipers, personal communication). In some target organisms specific membrane-located bacteriocin receptors have been identified (van Belkum et al., 1991b) and studies have been initiated aimed at using vesicles, liposomes, or micelles to analyze the structure-function relation of bacteriocins (Gao et al., 1991; van de Ven et al., 1991a,b; Kuipers et al., 1992).

Independent of the accuracy of the predictions described here, it is clear that this decade will see many challenging developments in the research on bacteriocins, which will further contribute to recognizing lactic acid bacteria as well-defined biological models with great potential for providing scientific problems on one hand and solving industrial problems on the other.

Acknowledgments

I am grateful to Roland J. Siezen for critically reading this manuscript and offering useful suggestions, Harry S. Rollema and Hein Boot for assistance in preparing the figures, and Marke M. Beerthuyzen, Jeroen Hugenholtz, Oscar P. Kuipers, Jan Roelof van der Meer, Jean Christophe Piard, and Peter J. G. Rauch for stimulating discussions. Part of this work was financed by contract BIOT-CT91-0265 in the framework of the BRIDGE project Lantibiotics of the Commission of European Communities.

References

Blackburn, P., Polak, J., Gusik, S., and Rubino, S. D. (1989). Nisin compositions for use as enhanced broad host range bacteriocins. Int. Patent Appl. PCT/US89/02625.

12. Future Prospects for Research and Applications of Nisin and Other Bacteriocins

Broadbent, J. R., and Kondo, J. K. (1991). Genetic construction of nisin-producing *Lactococcus lactis* subsp. *cremoris* and analysis of a rapid method for conjugation. *Appl. Environ. Microbiol.* 57, 517–524.

Buchman, W. B., Banerjee, S., and Hansen, J. R. (1988). Structure, expression and evolution of a gene encoding the precursor of nisin, a small protein antibiotic. *J. Biol. Chem.* 263, 16260–16266.

Bush, G. (1992). Streamlining federal regulation on biotechnology products. Fact sheet from the President's Council on Competitiveness. February.

Delves-Broughton, J. (1990). Nisin and its uses as a preservative. *Food Technol.* 44, 100–102.

de Vos, W. M. (1987). Gene cloning and expression in lactic streptococci. *FEMS Microbiol. Rev.* 46, 281–295.

de Vos, W. M., Jung, G., and Sahl, H.-G. (1991). Appendix: definitions and nomenclature of lantibiotics. *In* "Nisin and Novel Lantibiotics" (G. Jung and H.-G. Sahl, eds.), pp. 457–464. ESCOM, Leiden, The Netherlands.

de Vos, W. M., Mulders, J. W. M., Siezen, R. J., Hugenholtz, J., and Kuipers, O. P. (1993). Properties of Nisin Z and the distribution of its gene, *nisZ*, in *Lactococcus lactis Appl. Environ. Microbial* 59, 213–218.

de Vuyst, L., Mulder, J. W. M., Rauch, P. J. G., de Vos, W. M., Jensen, S. E., and Vandamme, E. J. (1990). Cloning, expresion of the *Lactococcus lactis* ssp. *lactis* pronisin gene and isolation of pre- and pronisin. *FEMS Microbiol. Rev.* 87, A26, P41.

Food and Drug Administration. (1988). Nisin preparation: affirmation of GRAS status as a direct human food ingredient. *Fed. Regis.* 53, 12247.

Froseth, B. R., and McKay, L. L. (1991). Molecular characterization of the nisin resistance region of *Lactococcus lactis* subsp. *lactis* biovar. *diacetylactis* DRC3. *Appl. Environ. Microbiol.* 57, 804–811.

Gao, F. H., Abee, T., and Konings, W. N. (1991). Mechanism of action of the peptide antibiotic nisin in liposomes and cytochrome *c* oxidase-containing proteoliposomes. *Appl. Environ. Microbiol.* 57, 2164–2170.

Gasson, M. J. (1984). Transfer of sucrose fermenting ability, nisin resistance and nisin production in *Streptococcus lactis* 712. *FEMS Microbiol. Lett.* 21, 7–10.

Gonzales, C. F., and Kunka, B. S. (1984). Transfer of sucrose-fermenting ability and nisin production among lactic streptococci. *Appl. Environ. Microbiol.* 49, 627–633.

Goodman, M., Plamer, D. E., Mierke, D., Ro, S., Nunami, K., Wakmiya, T., Fukase, K., Horimotio, S., Kitazawa, M., Fujita, H., Kubo, A., and Shiba, T. (1991). Conformation of nisin and its fragments using synthesis, NMR and computer simulations. *In* "Nisin and Novel Lantibiotics" (G. Jung and H.-G. Sahl, eds.), pp 59–76. ESCOM, Leiden, The Netherlands.

Gross, E., and Morell, J. L. (1971). The structure of nisin. *J. Am. Chem. Soc.* 93, 4634–4635.

Hastings, J. W., Sailer, M., Johnson, K., Roy, K. L., Vederas, J. C., and Stiles, M. E. (1991). Characterization of leucocin A-UAL 187 and cloning of the bacteriocin gene from *Leuconostoc gelidum*. *J. Bacteriol.* 173, 7491–7500.

Henkel, T., Sailer, M., Helms, G. L., Stiles, M. E., and Vederas, J.C. (1992). NMR assignment of leucocin A, a bacteriocin from Leuconostoc gelidum, supported by a stable isotope labeling technique for peptides and proteins. *J. Am. Chem. Soc.* 114, 1898–1900.

Hirsch, A. (1953). The evolution of the lactic streptococci. *J. Dairy Res.* 20, 290–293.

Holo, H., Nilssen, O., and Nes, I. F. (1991). Lactococcin A, a new bacteriocin from *Lactococcus lactis* subsp. *cremoris:* isolation and characterization of the protein and its gene. *J. Bacteriol.* 173, 3878–3887.

Horn, N., Dodd, H. M., and Gasson, M. J. (1991). Nisin biosynthesis genes are encoded by a novel conjugative transposon. *Mol. Gen. Genet.* 228, 129–135.

Hugenholtz, J., and de Veer, G. J. C. M. (1991). Application of nisin A and nisin Z in dairy technology. *In* "Nisin and Novel Lantibiotics" (G. Jung and H.-G. Sahl, eds.), pp. 440–448. ESCOM, Leiden, The Netherlands.

Hurst, A. (1981). Nisin. *Adv. Appl. Microbiol.* 27, 85–123.

Joerger, C., and Klaenhammer, T. R. (1990). Cloning, expression, and nucleotide sequence of the *Lactobacillus helveticus* 481 gene encoding the bacteriocin helveticin J. *J. Bacteriol.* 171, 6339–6347.

Jung, G. (1991). Lantibiotics: a survey. In "Nisin and Novel Lantibiotics" (G. Jung and H.-G. Sahl, eds.), pp. 1–34. ESCOM, Leiden, The Netherlands.

Kaletta, C., and Entian, K.-D. (1989). Nisin, a peptide antibiotic: cloning and sequencing of the *nisA* gene and post-translational processing of its peptide product. *J. Bacteriol.* 171, 1597–1601.

Kekessy, D. A., and Piguet, J. D. (1970). New method for detecting bacteriocin production. *Appl. Microbiol.* 20, 282–283.

Kolter, R., and Moreno, F. (1992). Genetics of ribosomally synthesized peptide antibiotics. *Ann. Rev. Microbiol.* 46, 141–165.

Kordel, M., and Sahl, H.-G. (1986). Susceptibility of bacterial, eukaryotic, and artificial membranes to the disruptive action of cationic peptides Pep5 and nisin. *FEMS Microbiol. Lett.* 34, 139–144.

Kuipers, O. P., Yap, W. M. G. J., Rollema, H. S., Beerthuyzen, M. M., Siezen, R. J., and de Vos, W. M. (1991). Expression of wild-type and mutant nisin genes in *Lactococcus lactis*. In "Nisin and Novel Lantibiotics" (G. Jung and H.-G. Sahl, eds.), pp. 250–260. ESCOM, Leiden, The Netherlands.

Kuipers, O. P., Rollema, H. S., Yap, W. M. G. J., Boot, H. J., Siezen, R. J., and de Vos, W. M. (1992). Engineering dehydrated amino acid residues in the antimicrobial peptide nisin. *J. Biol. Chem.* 267, 24340–24346.

LeBlanc, D. J., Crow, V. L., and Lee, L. N. (1980). Plasmid mediated carbohydrate catabolic enzymes among strains of *Streptococcus lactis*. In "Plasmids and Transposons, Environmental Effects and Maintenance Mechanisms" (C. Stuttard and K. R. Rosee, eds.), pp. 31–41. Academic Press, Inc., New York.

Lian, L.-Y., Chang, W. C., Morley, S. D., Roberts, G. C. K., Bycroft, B. W., and Jackson, D. (1991). NMR studies of the solution structure of nisin A and related peptides. In "Nisin and Novel Lantibiotics" (G. Jung and H.-G. Sahl, eds.), pp. 43–59. ESCOM, Leiden, The Netherlands.

Lian, L.-Y., Chang, W. C., Morley, S. D., Roberts, G. C. K., Bycroft, B. W., and Jackson, D. (1992). NMR studies of the solution structure of nisin A. *Biochem. J.* 283, 413–420.

Marugg, J. D., Gonzales, C. F., Kunka, B. S., Ledeboer, A. M., Pucci, M. J., Toonen, M. Y., Walker, S. A., Zoetmulder, L. C. M., and Vandenbergh, P. A. (1991). Cloning, expression, and nucleotide sequence of genes involved in production of pediocin PA-1, a bacteriocin from *Pediococcus acidilacti* PAC1.0. *Appl. Environ. Microbiol.* 58, 2360–2367.

Mattick, A. T. R., and Hirsch, A. (1944). A powerful inhibitory substance produced by group N streptococci. *Nature (London)* 154, 551.

Mulders, J. W. M., Boerrigter, I. J., Rollema, H. S., Siezen, R. J., and de Vos, W. M. (1991). Identification and characterization of the lantibiotic nisin Z, a natural nisin variant. *Eur. J. Biochem.* 201, 581–584.

Muriana, P. M., and Klaenhammer, T. R. (1991). Cloning, phenotypic expression and DNA sequence of the gene for lactacin F, an antimicrobial peptide produced by *Lactobacillus* spp. *J. Bacteriol.* 173, 1779–1788.

Oxford, A. E. (1944). Diplococcin, an anti-bacterial protein elaborated by certain milk streptococci. *Biochemistry* 38, 178–182.

Piard, J.-C., Kuipers, O. P., Rollema, H. S., Desmazeaud, M. J., and de Vos, W. M. (1992). Lacticin 481, a novel lantibiotic produced by *Lactococcus lactis* CNRZ 481 (submitted for publication).

Rauch, P. J. G., and de Vos, W. M. (1990). Molecular analysis of the *Lactococcus lactis* nisin-sucrose conjugative transposon. Abstracts Book 3rd International Conference on Streptococcal Genetics, American Society for Microbiology, Minneapolis, 23, A/46.

Rauch, P. J. G., and de Vos, W. M. (1992). Characterization of the novel nisin-sucrose conjugative transposon Tn5276 and its insertion in *Lactococcus lactis*. *J. Bacteriol.* 174, 1280–1287.

Rauch, P. J. G., Beerthuyzen, M. M., and de Vos, W. M. (1991). Molecular analysis and evolution of conjugative transposons encoding nisin production and sucrose fermentation in *Lactococcus lactis*. *In* "Nisin and Novel Lantibiotics" (G. Jung and H.-G. Sahl, eds.), pp. 243–249. ESCOM, Leiden, The Netherlands.

Rauch, P. J. G., Kuipers, O. P., Siezen, R. J., and de Vos, W. M. (1992). Genetics and protein engineering of nisin. *In* "Bacteriocin of Lactic Acid Bacteria: Microbiology, Genetics and Application" (L. de Vuyst and E. J. Vandamme, eds.). Elsevier Applied Sciences, London (in press).

Rogers, L. A. (1928). The inhibiting effect of *Streptococcus lactis* on *Lactobacillus bulgaricus*. *J. Bacteriol.* 16, 321–325.

Rogers, L. A., and Whittier, E. O. (1928). Limiting factors in the lactic fermentation. *J. Bacteriol.* 16, 211–229.

Saiki, R. K., Gelfland, D. H., Stoffel, E. J., Scharf, S. J., Higuchi, R. G., Horn, G. T., Mullis, K. B., and Ehrlich, H. A. (1988). Primer-directed enzymatic amplification of DNA with a thermostable DNA polymerase. *Science* 239, 487–491.

Stoddard, G. W., Petzel, J. P., van Belkum, M. J., Kok, J., and McKay, K. L. (1992). Molecular analysis of the lactococcin A gene cluster from *Lactococcus lactis* subsp. *lactis* biovar. *diacetylactis* WM4. *Appl. Environ. Microbiol.* 58, 1952–1961.

Tagg, J. R., and McGiven, A. R. (1971). Assay system for bacteriocins. *Appl. Microbiol.* 21, 943.

Toussaint-Samat, M., Alberny, R., and Horman, I. (1991). "Two Million Years of the Food Industry" (R. Montavon, ed.). Nestle S. A., Vevey, Switzerland.

Tramer, J., and Fowler, G. G. (1964). Estimation of nisin in foods. *J. Sci. Food Agric.* 15, 522–528.

van Belkum, M. J. (1991). Lactococcal bacteriocins: genetics and mode of action. Ph.D. Thesis, University of Groningen, The Netherlands.

van Belkum, M., Hayema, B. J., Jeeninga, R. E., Kok, J., and Venema, G. (1991a). Organization and nucleotide sequence of two lactococcal bacteriocin operons. *Appl. Environ. Microbiol.* 57, 492–498.

van Belkum, M. J., Kok, J., Venema, G., Holo, H., Nes, I. F., Konings, W. N., and Abee, T. (1991b). The bacteriocin lactococcin A specifically increases the permeability of lactococcal cytoplasmic membranes in a voltage-dependent, protein-mediated manner. *J. Bacteriol.* 173, 7934–7941.

van Belkum, M. J., Kok, J., and Venema, G. (1992). Cloning, sequencing, and expression in *Escherichia coli* of *lcnB*, a third bacteriocin determinant from the lactococcal bacteriocin plasmid p9B4-6. *Appl. Environ. Microbiol.* 58, 572–577.

van de Ven, F. J. M., van den Hooven, H. W., Konings, R. N H., and Hilbers, C. W. (1991a). The spatial structure of nisin in aqueous solution. *In* "Nisin and Novel Lantibiotics" (G. Jung and H.-G. Sahl, eds.), pp. 35–43. ESCOM, Leiden, The Netherlands.

van de Ven, F. J. M., van den Hooven, H. W., Konings, R. N. H., and Hilbers, C. W. (1991b). NMR studies of lantibiotics: The structure of nisin in aqueous solution. *Eur. J. Biochem.* 202, 1181–1188.

Whitehead, H. R. (1933). A substance inhibiting bacterial growth, produced by certain strains of lactic streptococci. *Biochem. J.* 27, 1793–1800.

Index

Acidophilucin A, characteristics, 153, 156
Acnecin
  antimicrobial activity, 221
  characteristics, 220–221
Aeromonas hydrophila, pediocin activity against, 187–188
Agar diffusion techniques
  adaptations of, 24–27, 31–35
  diffusion of antibiotic compounds in, 26–27
  factors affecting, 24–27
  history of, 24
  inhibition zones in, 24
Agar spot plating method, 32–33
  media composition effect on zone diameter in, 33–34
Agar-well diffusion, bacteriocin production and, 43
Alcoholic beverages, bacteriocins in, 74, 75
Anthocyanin, effect on nisin activity, 79
Antibiotics, bacteriocins and, 4–5
Antimicrobial proteins
  animal, 14–16
  bacteriocins as, *see* Bacteriocin
  thionins as plant, 11–14
  yeast, 10–11
ATP luminescence assay, nisin, 85

Bacillus cereus, nisin resistance, 100
Bacillus lichenformis, nisin activity against, sodium chloride effects, 80
Bacillus subtilis
  leucine uptake, effect of lactococcin A, 143
  nisin analogues, production of natural and engineered, 115–116
  nisin expression, 114
  as producer of lantibiotics, 113–114
  subtilin production, transfer between strains, 105–108
  subtilin restriction map, 107
Bacteriocin
  in alcoholic beverages, 74, 75
  amino acid sequence determination, 56
  antibiotics and, comparison, 4–5
  assays, 84–85
  bacterial protection from, 4
  binding sites, 3
  Carnobacterium, see Carnobacterium, bacteriocins
  characteristics, 3–4, 234
    future research and knowledge of, 252
  definition
    classical, 2–4
    new, 4, 234
  as food preservative, 53–55, 237–238
    factors affecting regulatory approval, 237–239
    factors compromising efficacy, 76
    FDA data requirement for, 238, 239
  gene location, 3–4
  genera producing, 6
  genes, problems in sequencing, 254
  genetically engineered
    consumer acceptance, 245–246
    future research involving, 252–255
    present and future targets for, 262
  in health care products, 235
  immunological studies, 55–56
  inhibitory activity, 3
  interactions with food components, 76–80
  lactic acid bacteria, *see* Lactic acid bacteria, bacteriocins
  Lactobacillus, see Lactobacillus, bacteriocins

267

Bacteriocin (continued)
  lactococcal, nomenclature, 122–123
  Leuconostoc, see Leuconostoc, bacteriocins
  microorganism resistance to, 82–83
  molecular screening, 256–258
  nomenclature, 5–6
  occurrence, 6
  Pediococcus, see Pediocin
  physical properties, 55
  prepeptide, common processing sites in, 173–174
  production
    on agar, 42–43
    in broth, 44–48
    methods of testing for, 42
  protein engineering, 258–261
  purification, biochemical methods for, 48–50
  purified, applications, 53–54
  regulation, agencies involved in, 243–245
  research approaches with, 251–252
  screening procedures for, 6–7
  use in foods and pharmaceuticals, 235–236
Bacteriocin activity
  agar diffusion techniques, see Agar diffusion techniques
  detection and assay, 42
  enhancement, 80–81
  history, 23–24
  incubation conditions affecting, 30
  inducibility, 30–31
  lactic acid bacteria, adaptations of agar diffusion for, 31–35
  liquid media for screening, 35–36
  media composition affecting, 29–30
  plating methods, 27–29
  survivor counts in measuring, 37
  titration, 36–37
Bacteriocin-producing strains
  genetically engineered, 236
    factors affecting approval, 240–243
  naturally-occurring, factors affecting approval, 239–240
    allergenicity as, 242
    genetic modification as, 240
    level of intake as, 242
    origin and identity of host organism as, 241
    overall assessment as, 242–243
    pathogenicity and toxigenicity as, 242
    pleiotropic effects as, 241–242
    selectable markers as, 241
    vectors as, 241
  as probiotics, 236
  regulation, players involved in, 243–245
  as starter cultures, 235–236

Bacteriophages, agar diffusion plating methods and, 28
Beef, effect of pediocin AcH, 202–204
Beer fermentation
  bacteriocin applications in, 75
  lactic acid bacteria contaminants, 72
Bentonite, nisin inactivation by, 79
Beverages
  bacteriocins in, factors affecting, 76
Binding sites, bacteriocin, 3
Brandy production, nisin as bacterial inhibitor in, 73
Brevicin 37, characteristics, 153, 156
Brochothrix thermosphacta, effect of pediocin AcH, 201

Carnobacteriocin, characteristics, 156
Carnobacteriocin B, chemical characterization, 215
Carnobacterium
  bacteriocins, 214
    applications, 216
    chemical characterization, 215
    characteristics, 211–212
    habitats, 212–213
    sources of, 212–213
Carnobacterium piscicola, 152, 213
  bacteriocins, 158, 214
  in meat preservation, 216
Carnocin U149, characteristics, 153, 156
Caseicin 80
  characteristics, 153
  inducibility, 227
  purification, 51
Catechin, effect on nisin activity, 79
Cationic detergent mechanism, nisin action and, 116
Cell receptors, colicin, 8
Cheese
  fermentation
    diplococcin activity in, 72
    nisin-producing lactococci as starters in, 71
  nisin antibotulinal activity in, 76
  nisin retention in, 78
Chicken, nisin activity in, 78
Chloroform, agar diffusion plating methods and, 27
Chromatography, bacteriocin purification with, 49
Clostridia, food spoilage, 71
Clostridium acetobutyricum
  leucine uptake, effect of lactococcin A, 143

Index

Clostridium botulinum
  in foods, nisin effects, 76–77, 234
  nisin activity against
    nitrate effects, 81
    nitrite effects, 80–81
    sodium chloride effects, 80
  sensitivity to lactic acid bacteria bacteriocins, 69–70
Clostridium laramie, effect of pediocin AcH, 201
Clostridium sporogenes
  nisin activity against, effect of nitrate, 81
  sensitivity to nisin, 69
Colicin
  cell membrane interaction, 9
  cell receptors, 8
  conditions for production and purification, 44, 51
  enzyme activity, 9
  general features, 7
  immunity, 10
  mode of action, 8–10
  origin, 5
  pore-forming, 9
  single hit inactivation kinetics, 8
  in wine making, 73, 74
Crambin, structure, 12–13
Critical dilution method, bacteriocin activity and, 36, 42

Defensins, as animal antimicrobial proteins, 11–16
Dehydro residues, lantibiotics and, 116–118
Diacetyl, beer fermentation and, 79
Dialysis, bacteriocin purification with, 49
Diplococcin
  application in cheese production, 72
  effect on bacterial cells, 124
  identification, 123
  plasmids in, 124–125
  production and immunity, 124–125
  purification, 123
Distilled spirit mashes, bacteriocin application in, 75
DNase, colicin, 8, 9

Enterobacteriaceae, colicins of, 7–10
Enzyme immunoassay, nisin, 85
Erwinia chrysanthemi
  protease B, secretion genes for, 135
Escherichia coli
  cloning lactococcin determinants, 129
  colicins, 7–10

hemolysin
  secretion genes, 134–135
  translocation mechanism, 103
leucine uptake, effect of lactococcin A, 143

Fermentation
  beer, *see* Beer fermentation
  causes for disruption of, 71
  nisin-producing lactococci in, 71
  vegetable, *see* Vegetable fermentation
  wine, *see* Wine fermentation
Flip spot plating method, 32
Flip streak plating method, 32, 43
Food
  bacteriocins in, 76–80
    assays for, 84–85
    factors affecting efficacy, 76
  bacteriocins as preservatives in, 235
    history, 249–251
    regulatory approval, 237–239
  biotechnology-derived, labeling, 246
  fermented, consumer image, 65
  genetically engineered, consumer acceptance, 245–246
  nisin retention in, 78
  refrigerated, *Listeria* contamination protection, 66
Food-borne disease, cases of, 63–64
Food and Drug Administration, as regulatory agency for bacteriocin use, 237–240, 243
Food emulsifiers, effect on nisin activity, 81
Food preservatives
  bacteriocins as, 53–55
  pediocin PA-1 as, 185
Food processing, recent trends in, 64

Gallic acid, effect on nisin activity, 79
Gasericin A, characteristics, 154
Gel electrophoresis, bacteriocin purification with, 49
Genetic engineering, future research for bacteriocin in, 252–255

Heat, nisin activity affected by, 79–80
Helveticin J
  amino acid composition, 159
  characteristics, 154, 156, 158
  cloning, 160–161
  conditions for production and purification, 48, 50
  expression in heterologous hosts, 161

Helveticin J (*continued*)
 genes, molecular organization, 161–167
 immunological studies, 56
 operon
  nucleotide sequence, 162–165
  proteins encoded by, 166
Hemolysin
 Escherichia coli, secretion genes for, 134–135

Immunity
 colicin, 10
 lactococcin, 132–134
Immunity protein, lactococcin, 133
International Food Biotechnology Council, 244

Jenseniin G, activity of, 222

K1 toxin, 10–11
Killer toxins, 10–11

Lactacin B
 conditions for production and purification, 48, 50
 characteristics, 153, 156, 158
 inducibility, 31, 227–228
Lactacin F
 amino acid sequence, 171
 amino acid composition, 158–159
 characteristics, 153, 156, 158
 conditions for production and purification, 48, 50
 export, genes associated with, 173
 gene cloning for, 167–168
 heat stability, 167
 hydropathic profile, 172
 N-terminal amino acid presequences, 140
 operon
  molecular organization, 168–173
  nucleotide sequence, 169–171
 processing sites, 173–175
Lactic acid bacteria, *see also specific bacteria*
 agar spot plating method for, 32–33
 bacteriocinogenic
  factors compromising efficacy, 77
  as starter cultures in vegetable fermentations, 75
 bacteriocins
  agar diffusion methods, 31–35
  Clostridium botulinum sensitivity to, 69–70
  common processing sites, 173–175
  conditions for purification, 48–53
  detection and assay, 42
  genetically and biochemically characterized, 253
  inducible, 226–228
  Listeria monocytogenes sensitivity to, 66–68
  nomenclature, 122–123
  N-terminal amino acid presequences, 140
  production, 42–44, 48
  properties shared by, 41
  resistance to, 83
  safety, 65
  study, reasons for, 63–64, 65
  survivor counts, 37
  titration, 36–37
 as contaminants in beer fermentation, 72–73
 flip spot plating method, 32
 flip streak plating method, 32
 nisin sensitivity, 73–74
 in wine, 73–74
Lacticin, 152
 purification, 52
 characteristics, 153, 156
Lactobacillus
 bacteriocins
  amino acid composition, 158–160
  bacteria inhibited by, 160
  biochemical types and classes, 156
  evidence of, 152
  genetic engineering targets, 175–176
  improvements in expression, 176
  characteristics, 151–152
  jenseniin G activity against, 222–223
  lantibiotics, 157
  nisin resistance, 83
  products synthesized by, 152
Lactobacillus acidophilus, 152
 bacteriocins, 153
  conditions for production and purification, 44, 50
  helveticin J expression, 161
Lactobacillus brevis, bacteriocins, 153
Lactobacillus carnis, bacteriocins, 153
Lactobacillus casei, bacteriocins, 153
Lactobacillus delbrueckii, bacteriocins, 152, 153
Lactobacillus fermenti
 bacteriocins, 154, 157
  amino acid composition, 159
Lactobacillus gaseri, bacteriocins, 154
Lactobacillus helveticus
 bacteriocins, 154, 157

Index

conditions for production and purification, 44, 50
Lactobacillus lactis, as indicator in assays for nisin, 84
Lactobacillus plantarum
  bacteriocins, 152, 154
  effect of *Pediococcus cerevisiae* on growth, 183
  pediocin AcH activity against, 193–194, 200
Lactobacillus sake
  bacteriocins, 155, 158
  action on *Listeria monocytogenes,* 67–68
  as indicator in assay for nisin, 84
Lactocin, purification, 50–51
Lactocin 27
  amino acid composition, 159
  characteristics, 154, 156, 157
Lactocin S
  amino acid composition, 159
  characteristics, 155
Lactococci
  mode of action of lactococcin A on, 141
  sensitivity to lactococcin A, 138
Lactococcin, *see also specific lactococcins*
  determinants
    cloning, 128–129
    mutation analysis, 131–132
    nucleotide sequence, 131–132
  genetics, 127–134
  immunity and resistance, 132–134
  plasmid map, 130
  plasmid role in production, 127
  secretion, *Lactococcus lactis* IL1403, 136–137
  structural genes, 131–132
Lactococcin A
  amino acid sequence, 139
  cloning determinants, 129
  cytoplasmic membrane effects, 141–143
  genes, sequence analysis, 254
  hydrophobicity, 141
  mode of action, 141–144
  model for secretion, 135
  molecular weight, 141
  N-terminal amino acid presequences, 140
  nucleotide sequence, 131–132
  physical characteristics, 139–141
  production, 137–139
  properties, 139–141
  purification, 51, 138–139
  secretion machinery, 134–137
Lactococcin B
  N-terminal amino acid presequences, 140
  processing sites, 173–175

Lactococcin M
  determinants, nucleotide sequence, 131–132
  immunity, 133
  N-terminal amino acid presequences, 140
Lactococcin Ma, processing sites, 173–175
Lactococcus
  bacteriocin inhibitors, 72
  distinguishing *Pediococcus* from, 182–183
Lactococcus cremoris, bacteriocin purification, 51
Lactococcus lactis
  chromosomal lactococcin secretion, 136–137
  cytoplasmic membrane, effect on lactococcin A, 141–143
  diplococcin, *see* Diplococcin
  glutamate uptake, effect of lactococcin A, 142
  lactococcin, purification, 52
  lactococcin A effects, 141–142
  lactococcin immunity genes, 131–134
  leucine uptake, effect of lactococcin A, 143
  nisin biosynthesis, 104
  nisin production transfer between strains, 108–109
  nisin resistance, 100
  plasmid transfer, 127–128
  in sauerkraut fermentation, 75
  secretion genes, 134–135
Lactosin S, characteristics, 156
Lactostrepcin
  defined, 125
  identification, 125
  mode of action, 125–126
  purification, 125–126
Lactostrepcin 5
  mode of action, 125–126
  plasmid curing and production, 126
  purification, 125–126
Lantibiotic, 94
  analogues, structural and functional analysis, 116–118
  dehydro residues in, antimicrobial activity and, 116–118
  modification, 97
  multiple purposes, 97–98
  production, conservation of steps in, 112–114
  production, transfer of genes for, 106
  resistance factors, 99
lcnA, sequence, 131
lcnC, lactococcin production and, 134
lcnD, lactococcin production and, 134

lcnM, operon, 131–132
Leucocin, purification, 53
Leucocin A
 amino acid sequence, 215
 characteristics and activity, 215
 in meat preservation, 216
 processing sites, 173–175
Leucocin A-UAL 187, N-terminal amino acid presequences, 140
Leuconostoc
 bacteriocins, 214–215
  chemical characteristics, 215–216
 characteristics, 212
 commercial value, 212, 213
 distinguishing Pediococcus from, 182–183
 habitats, 213
 sources, 213
Leuconostoc gelidum, 214, 215
 bacteriocin
  purification, 53
  action on Listeria monocytogenes, 68
 in meat preservation, 216
Leuconostoc mesenteroides
 bacteriocin
  action on Listeria monocytogenes, 68
  in meat preservation, 216
  effect of pediocin AcH, 194, 197
  in foods, effect of pediocin AcH, 201, 202–203
 in sauerkraut fermentation, 75
Leuconostoc oenos
 nisin resistance, 83
 in wine making, 73, 74
Leuconostoc paramesenteroides, bacteriocin, purification, 53
Lipoteichoic acid, as pediocin binding site, 194
Listeria ivanovii, effect of pediocin AcH, 197–198, 201
Listeria monocytogenes
 effect of
  nisin, 66
  pediocin AcH, 195, 197–198
  in foods, effect of pediocin AcH, 200, 201
  inhibition by pediocins, 54
  in meat systems, effect of pediocins, 204–205
 nisin activity against, 77
  emulsifier effects, 81
  sodium chloride effects, 80
 nisin resistance, 83
 pediocin inhibition, sauerkraut fermentation and, 75
 pediocin PA-1 activity against, 185

Pediococcus bacteriocin activity against, 186
 sensitivity to lactic acid bacteria bacteriocins, 66–68
Lutri-Plate, agar diffusion plating methods and, 28

Malolactic fermentation (MLF)
 nisin in prevention or promotion of, 74
 in wine, factors affecting, 73
Meat preservation
 carnobacteriocins in, 216–217
 leucocin A in, 216–217
 nisin activity in, 77–78
Media
 composition, bacteriocin activity and, 29–30
Mesenterocin 5, 214–215
Methylmalonyl-CoA mutase, characterization of gene for, 225
Micrococcus luteus, as indicator in assays for nisin, 84
Microgard, inhibitory activities, 220
Milk
 nisin activity in, factors affecting, 77
Mitomycin C, bacteriocin inducibility, 31
Molecular screening, 256–258
Monocins, origin, 5
Myricetin, nisin interaction with, 78

Nisin
 action, mechanisms, 116
 agar plate diffusion assay, 31
 analogues, production of natural and engineered, 115–116
 antibiotics, mutation as selection process for, 99
 antibotulinal activity, 76–80
 assays, 84–85
 ATP luminescence assay, 84–85
 bacterial resistance, 82–83
 in beer fermentation, 72–73
 in brandy production, 73
 chimeric precursor peptide, processing of, 114–115
 enzyme immunoassay, 85
 as food preservative, 54, 78, 84, 234
 gene expression systems, 261
 genes
  expression, 103–104
  organization, 110
  sequence, 94
  structural, 101–103, 108

Index

heterologous expression, 114–115
historical perspective, 93–95
immunological studies, 255
inactivation, 117–118
malolactic fermentation and, 74
molecular screening, 257
mRNA
  characterization, 103–104
  expression, 110
myricetin interaction with, 78
NMR spectrum, 117
physical properties, 55
posttranslational signals for synthesis, 102–103
precursor peptides, gene cloning, 100–101
processing, 112–114
purification, 51
regulatory and safety considerations for use, 85–86
resistance, 98–100
sensitivity
  Clostridium botulinum, 69
  Clostridium sporogenes, 69
  Listeria monocytogenes, 66, 67
structure, 94
sulfhydryl group inactivation by, 116–117
tannin interaction, 78
translocation mechanisms, 102–103
in wine processing, 73
  lactic acid bacteria sensitivity to, 73–74
Nisin A
  antimicrobial activity, 257–258
  spatial conformation, 259
Nisin activity
  EDTA effects, 81
  food emulsifier effects, 81
  heat effects, 79–80
  nisinase effects, 82
  pH effects, 79–80
  sodium chloride effects, 80
Nisinase, *Streptococcus thermophilus*, 82
Nisin biosynthesis, 94
  complexity, 100
  genetic transfer, 108–109
  growth stages, 104
  organization of genes associated with, 109–110
Nisin Z
  antimicrobial activity, 257–258
  spatial conformation, 259
Nitrate
  effect on nisin activity, 81
  substitutes, food spoilage control and, 71–72

Nitrite, effect on nisin activity, 80–81

Pediocin
  agar diffusion assay, 31–32
  as food preservative, 54
  in meat systems, 204–205
  production and purification, 44, 48, 52–53
Pediocin A
  antimicrobial activity, 184
  sensitivity to *Clostridium botulinum*, 70
Pediocin AcH
  activity, effect of chemical and physical treatment, 188–189
  adsorption properties, 193–194
  antibacterial effectiveness, 198–199
  applications, 205–206
  in cosmetics, 206
  effects of pH, 196
  factors affecting production, 196–197
  in foods, 199–204
  genetics, 197–198
  mode of action, 192–195
  plasmid biology, 197–198
  properties, 188–191
  purification, 189, 191
  receptors for binding, 194
  in silage production, 206
  toxicity, 191–192
Pediocin PA-1
  activity on *Listeria monocytogenes*, 67
  characteristics, 184–185
  as food preservative, 184–185
  N-terminal amino acid presequences, 140
  processing sites, 173–175
  purification and activity, 185
Pediococcus
  bacteriocin activity, 183–188
  characteristics of genus, 181–183
  commercial uses, 181
  conditions for growth, 182
  distinguishing *Lactococcus* and *Leuconostoc* from, 182–183
  plasmid biology, 184
Pediococcus acidilactici, bacteriocin, *see also* Pediocin AcH
  activity, 185
  conditions for production and purification, 44, 52
  action on *Listeria monocytogenes*, 67, 186–188
  conditions for growth, 182
  effect in meat systems, 204–205

Pediococcus acidilactici (continued)
  in sauerkraut fermentation, 75
  in bacon processing, 185
Pediococcus cerevisiae, bacteriocin activity, 183
Pediococcus damnosus
  bacteriocin, action on *Listeria monocytogenes,* 67
  conditions for growth, 182
  in wine making, 74
Pediococcus dextrinicus, conditions for growth, 182
Pediococcus halophilus, 206
Pediococcus inopinatus, conditions for growth, 182
Pediococcus parvulus, conditions for growth, 182
Pediococcus pentosaceus, bacteriocins
  and *Clostridium botulinum,* 70
  and *Listeria,* 67, 187–188
Pediococcus urinaequi, conditions for growth, 182
Peptides, posttranslational modification, 95, 97
pH, nisin activity affected by, 79–80
Phagocytosis, defensins and, 14–15
Pharmaceuticals, bacteriocins in, 235
Phenol, effect on nisin activity, 78–79
Phoratoxin, structure, 12
Plaf promoter, lactacin F gene transcription and, 171
Plantacin B, characteristics, 154
Plantaricin A, 152, 154
Plasmids, *Lactococcus lactis,* 124–125
Plating methods
  in agar diffusion techniques, 27–29
  agar spot, 32–33
  bacteriocin
    deferred, 27–28
    simultaneous, 28–29
  flip spot, 32
  flip streak, 32
Polymerase chain reaction, bacteriocin screening and, 256
Pork, nisin antibotulinal activity, 76
Precipitation, bacteriocin purification with, 49
Preservatives
  lactic fermentation as, 64–65
  recent trends in use of, 64
President's Council on Competitiveness, biotechnology products and, 244
Probiotics, bacteriocin-producing strains as, 236
Propionibacteria
  bacteriocins, 220
    genetics, 223–226

characteristics, 219
genes characterized, 225
nonbacteriocin inhibitors, 220
plasmids, 223–224
Propionibacterium, commercially significant, 219
Propionibacterium acnes
  acnecin produced from, 220–221
  RTT 108 bacteriocin, 221
Propionibacterium freudenreichii, 220
  inducible inhibitors from, 221
  protoplast formation, 224–225
Propionibacterium jensenii
  bacteriocins, 222
  inhibitory activity, 221
Propionibacterium thoenii, bacteriocins, 221
Propionic acid, inhibitory properties, 220
Propionicin PLG-1
  antimicrobial activity, 222
  determinants, 224
  properties, 221–222
  purification, 222
Protease B
  Erwinia chrysanthemi, secretion genes, 134–135
Proteinases, lactostrepcin 5 activity and, 126
Protein engineering, 258–261
Proton motive force, colicin effects, 8
Purothionins, 13

Quercetin, effect on nisin activity, 79

Respiratory burst, defensins and, 15
RNase, colicin, 8, 9

Saccharomyces cerevisiae, killer toxins, 10–11
Sakacin A, characteristics, 155
Sauerkraut, fermentation, starter cultures, 75
Sausage, fermentation, lactic starters, 75
Sec system, 102–103
Slam-deferred transfer, agar diffusion plating methods and, 27–28
Sodium chloride, bacteriocin activity enhancement by, 80
Spot-on-lawn procedure, agar diffusion plating methods and, 29
Staphylocin, origin, 5
Staphylococcus aureus, pediocin activity against, 187–188
Streptococcus agalactiae, as indicator in assays for nisin, 84

Index

Streptococcus cremoris
  agar diffusion plating methods, 27
  bacteriocin, purification, 51
  as indicator in assays for nisin, 84
Streptococcus thermophilus, nisinase activity, 82
Subtilin
  biosynthesis
    conversion of *Bacillus subtilis* 168 to, 106–108
    genetic transfer, 105–108
    growth stages, 104
    organization of genes associated with, 110–112
  chimeric precursor peptide, processing of, 114–115
  heterologous expression, 114–115
  inactivation, 117–118
  mRNA, characterization, 104–105
  and nisin
    comparison of structural genes, 101–103
    conservation of processing steps for, 112–114
  nisinase inactivation of, 82
  operon, organization of open reading frames within, 112
  origin, 5
  posttranslational signals for synthesis, 102–103
  precursor peptides, gene cloning, 100–101
  production, genes for expression of, 104–105
  translocation mechanisms, 102–103
Sucrose, effect on bacteriocin activity, 99

Tannin, nisin interaction with, 78–79
Thionins, 11
  function, 13–14
  Pyrularia, toxicity, 13–14
  structure, 12

Ultraviolet light, bacteriocin inducibility, 31

Vacuum concentration, bacteriocin purification and, 49
Vegetable fermentation
  bacteriocinogenic lactic acid bacteria in, 75
  controlling and directing, 74–75
Viruslike particle (VLP), M_1, 11
Viscotoxins
  structure, 12
  function, 14

Wine fermentation
  bacteriocin applications in, 75
  lactic acid bacteria in, 73–74
  nisin in
    retention, 78
    sensitivity of lactic acid bacteria to, 73–74

Yeast, antimicrobial proteins, 10–11

FOOD SCIENCE AND TECHNOLOGY

International Series

Maynard A. Amerine, Rose Marie Pangborn, and Edward B. Roessler, *Principles of Sensory Evaluation of Food.* 1965.
Martin Glicksman, *Gum Technology in the Food Industry.* 1970.
C. R. Stumbo, *Thermobacteriology in Food Processing,* second edition. 1973.
Aaron M. Altschul (ed.), *New Protein Foods*: Volume 1, Technology, Part A—1974. Volume 2, Technology, Part B—1976. Volume 3, Animal Protein Supplies,Part A—1978. Volume 4, Animal Protein Supplies, Part B—1981. Volume 5, Seed Storage Proteins—1985.
S. A. Goldblith, L. Rey, and W. W. Rothmayr, *Freeze Drying and Advanced Food Technology.* 1975.
R. B. Duckworth (ed.), *Water Relations of Food.* 1975.
John A. Troller and J. H. B. Christian, *Water Activity and Food.* 1978.
A. E. Bender, *Food Processing and Nutrition.* 1978.
D. R. Osborne and P. Voogt, *The Analysis of Nutrients in Foods.* 1978.
Marcel Loncin and R. L. Merson, *Food Engineering: Principles and Selected Applications.* 1979.
J. G. Vaughan (ed.), *Food Microscopy.* 1979.
J. R. A. Pollock (ed.), *Brewing Science,* Volume 1—1979. Volume 2—1980. Volume 3—1987.
J. Christopher Bauernfeind (ed.), *Carotenoids as Colorants and Vitamin A Precursors: Technological and Nutritional Applications.* 1981.
Pericles Markakis (ed.), *Anthocyanins as Food Colors.* 1982.
George F. Stewart and Maynard A. Amerine (eds.), *Introduction to Food Science and Technology,* second edition. 1982.
Malcolm C. Bourne, *Food Texture and Viscosity: Concept and Measurement.* 1982.
Héctor A. Iglesias and Jorge Chirife, *Handbook of Food Isotherms: Water Sorption Parameters for Food and Food Components.* 1982.
Colin Dennis (ed.), *Post-Harvest Pathology of Fruits and Vegetables.* 1983.
P. J. Barnes (ed.), *Lipids in Cereal Technology.* 1983.
David Pimentel and Carl W. Hall (eds.), *Food and Energy Resources.* 1984.
Joe M. Regenstein and Carrie E. Regenstein, *Food Protein Chemistry: An Introduction for Food Scientists.* 1984.
Maximo C. Gacula, Jr., and Jagbir Singh, *Statistical Methods in Food and Consumer Research.* 1984.
Fergus M. Clydesdale and Kathryn L. Wiemer (eds.), *Iron Fortification of Foods.* 1985.

Robert V. Decareau, *Microwaves in the Food Processing Industry.* 1985.
S. M. Herschdoerfer (ed.), *Quality Control in the Food Industry*, second edition. Volume 1—1985. Volume 2—1985. Volume 3—1986. Volume 4—1987.
Walter M. Urbain, *Food Irradiation.* 1986.
Peter J. Bechtel (ed.), *Muscle as Food.* 1986.
H. W.-S. Chan (ed.), *Autoxidation of Unsaturated Lipids.* 1986.
F. E. Cunningham and N. A. Cox (eds.), *Microbiology of Poultry Meat Products.* 1987.
Chester O. McCorkle, Jr. (ed.), *Economics of Food Processing in the United States.* 1987.
J. Solms, D. A. Booth, R. M. Dangborn, and O. Raunhardt (eds.), *Food Acceptance and Nutrition.* 1987.
Jethro Jagtiani, Harvey T. Chan, Jr., and Williams S. Sakai, *Tropical Fruit Processing,* 1988.
R. Macrae (ed.), *HPLC in Food Analysis*, second edition. 1988.
A. M. Pearson and R. B. Young, *Muscle and Meat Biochemistry.* 1989.
Dean O. Cliver (ed.), *Foodborne Diseases.* 1990.
Majorie P. Penfield and Ada Marie Campbell, *Experimental Food Science*, third edition. 1990.
Leroy C. Blankenship (ed.), *Colonization Control of Human Bacterial Enteropathogens in Poultry.* 1991.
Yeshajahu Pomeranz, *Functional Properties of Food Components*, second edition. 1991.
Reginald H. Walter (ed.), *The Chemistry and Technology of Pectin.* 1991.
Herbert Stone and Joel L. Sidel, *Sensory Evaluation Practices*, second edition. 1993.
Robert L. Shewfelt and Stanley E. Prussia (eds.), *Postharvest Handling: A Systems Approach.* 1993.
R. Paul Singh and Dennis R. Heldman, *Introduction to Food Engineering*, second edition. 1993.
Tilak Nagodawithana and Gerald Reed (eds.), *Enzymes in Food Processing*, third edition. 1993.
Takayaki Shibamoto and Leonard Bjeldanes, *Introduction to Food Toxicology.* 1993.
Dallas G. Hoover and Larry R. Steenson (eds.), *Bacteriocins of Lactic Acid Bacteria.* 1993.
John A. Troller, *Sanitation in Food Processing*, second edition. 1993.
Ronald S. Jackson, *Wine Science: Principles and Applications.* In Preparation.
Robert G. Jensen and Marvin P. Thompson (eds.), *Handbook of Milk Composition.* In Preparation
Tom Brody, *Nutritional Biochemistry.* In Preparation.